Springer Proceedings in Mathematics & Statistics

Volume 210

Springer Proceedings in Mathematics & Statistics

This book series features volumes composed of selected contributions from workshops and conferences in all areas of current research in mathematics and statistics, including operation research and optimization. In addition to an overall evaluation of the interest, scientific quality, and timeliness of each proposal at the hands of the publisher, individual contributions are all refereed to the high quality standards of leading journals in the field. Thus, this series provides the research community with well-edited, authoritative reports on developments in the most exciting areas of mathematical and statistical research today.

More information about this series at http://www.springer.com/series/10533

Paola Cappanera · Jingshan Li
Andrea Matta · Evren Sahin
Nico J. Vandaele · Filippo Visintin
Editors

Health Care Systems Engineering

HCSE, Florence, Italy, May 2017

 Springer

Editors

Paola Cappanera
Dipartimento di Ingegneria
  dell'Informazione
University of Florence
Florence
Italy

Jingshan Li
Department of Industrial and Systems
  Engineering
University of Wisconsin–Madison
Madison, WI
USA

Andrea Matta
Department of Mechanical Engineering
Politecnico di Milano
Milan
Italy

Evren Sahin
Laboratoire Génie Industriel
École Centrale Paris
Châtenay-Malabry
France

Nico J. Vandaele
OG Productie en Logistiek Leuven
Katholieke Universiteit Leuven
Leuven
Belgium

Filippo Visintin
Dipartimento di Ingegneria Industriale
University of Florence
Florence
Italy

ISSN 2194-1009   ISSN 2194-1017   (electronic)
Springer Proceedings in Mathematics & Statistics
ISBN 978-3-319-66145-2   ISBN 978-3-319-66146-9   (eBook)
https://doi.org/10.1007/978-3-319-66146-9

Library of Congress Control Number: 2017955271

Mathematics Subject Classification (2010): S15007, T22008, H27002, 527030, I18030

Foreword Proceedings HCSE 2017

The International Conference on Health Care Systems Engineering provides an opportunity to discuss state-of-the-art operations management issues in healthcare delivery systems. The third edition of this conference took place in Florence at the Meyer Children's Hospital—a unique location in a beautiful city. One facet of the uniqueness of the Meyer Children's Hospital is that it was built not to look like a hospital. This inspiring atmosphere carried over directly to the participants. Scientists and practitioners used the opportunity to discuss new ideas, methods, and technologies and how they can be brought into practice.

Topics included predictive models and information systems in healthcare, healthcare logistics, planning and location, clinical pathways, patient-centered services, emergency department management, and many more.

What makes HCSE a very special conference is the broadness of the topics and at the same time the dialogue between researchers and practitioners to better evaluate the value of the presented ideas. Moreover, for each session, there is not only a chair person assigned, but also a discussant. This improves the quality and intensity of the discussion after each talk enormously.

The hospital tour at the Meyer Children's Hospital as well as the perfect social program completed the positive impression of HCSE 2017 in Florence. I am very much looking forward to future editions of HCSE.

Prof. Stefan Nickel
Karlsruhe Institute of Technology—KIT, Karlsruhe
Germany

Preface

This volume is dedicated to the peer-reviewed contributions accepted for presentation at the Third International Conference on Health Care Systems Engineering (HCSE 2017) that took place in Florence, Italy, from May 29–31, 2017. This conference aims at giving evidence of how quantitative operations management (OM) methods can support complex decisions arising in healthcare processes.

A distinguishing feature of the conference with respect to the other OM conferences on similar topics is to bring together medical staff and OM people to (i) share knowledge, (ii) share perspectives—often different and possibly conflicting —with the aim of finally (iii) come to a problem solution. The conference is purposely organized in plenary and single-stream regular sessions so as to foster the discussions and collaborations between OM scientists and clinicians.

Another peculiar feature of the conference is the location which is quite unusual for an OM conference: Each edition was hosted inside a hospital and for HCSE 2017 the hosting hospital was the Meyer Children's Hospital in Florence. We would like to express our gratitude to the General Director Dr. Alberto Zanobini for letting us visit and use the (outstanding) hospital facilities throughout the conference.

In this edition of the conference, we had the honor of host two keynote speakers to whom we express our truly deep gratitude: Prof. Stefan Nickel from the Karlsruhe Institute of Technology—KIT (Germany) and Prof. Angela Testi from the University of Genoa (Italy). Their contributions were perfectly fitting the objective of the conference and were very much appreciated by the audience.

Each of the contributions submitted to the conference has undergone a peer-review process which involved at least two reviewers selected by the Conference Scientific Committee. A total of 29 contributions were finally accepted and organized in six regular sessions (25 contributions) and a young session (4 contributions). Each session was chaired by an academic and a clinician discussant, both experts in the field. This conference was indeed multidisciplinary and international with 84 authors spread in 18 countries and in 3 continents.

The regular sessions cover a wide spectrum of topics in the management of healthcare systems, and they feature the following topics: (i) predictive models and information system, (ii) planning and location, (iii) protection and analytical approaches, (iv) emergency department management, (v) robustness management, and (vi) scheduling.

Young session is a novelty of this edition: Young scholars, i.e., graduated students and Ph.D. students, have been given the opportunity to submit an extended abstract instead of a full paper and to present their ongoing research on emerging topics in healthcare management in a dedicated session. Presentations have been held in a friendly, inspiring, and bridge-building environment, where young scholars received valuable feedback from the audience to finalize their research. The most promising research was awarded by a committee made of members of the Scientific Committee and experts in the field, and the winner was Fabian Schäfer presenting the paper: F. Schäfer, M. Walther, A. Hübner, Patient-Bed Allocation in Large Hospitals.

We would like to thank Prof. Nickel who accepted to write the foreword of this volume, all the members of the Scientific Committee who gave us the opportunity to host HCSE 2017, all the speakers, authors, attendees, discussants, and reviewers for their valuable help in guaranteeing the quality of the accepted contributions. Each of them has contributed to the success of HCSE 2017.

We are deeply indebted to the young people from the local organizing committee, i.e., to Caterina Caprara and Roberta Rossi because without their help everything would have been more difficult to manage.

Finally, we would like to express our sincere wishes for the success of the next edition of the conference, HCSE 2019!

Florence, Italy   Paola Cappanera
                                                                                       Filippo Visintin

Contents

Contributors

Waleed Abo-Hamad College of Business, Dublin Institute of Technology (DIT), Dublin, Ireland

Roshanghalb Afsaneh Department of Management, Economics & Industrial Engineering, Politecnico di Milano, Milan, Italy

Davide Aloini Department of Energy, Systems, Territory and Construction Engineering, University of Pisa, Pisa, Italy

Ozgur M. Araz College of Business Administration, Supply Chain Management and Analytics, University of Nebraska Lincoln, Lincoln, NE, USA

Roberto Aringhieri Dipartimento di Informatica, Università degli Studi di Torino, Torino, Italy

Cristina Azcarate Public University of Navarre, Pamplona, Navarre, Spain

Hari Balasubramanian Department of Mechanical and Industrial Engineering, University of Massachusetts, Amherst, United States

Nicholas Bambos Stanford University Department of Electrical Engineering, Stanford University Department of Management Science and Engineering, and Lucile Packard Children's Hospital Stanford, Palo Alto, CA, USA

Margaret L. Brandeau Department of Management Science and Engineering, Stanford University, Stanford, CA, USA

Valérie Bélanger Department of Logistics and Operations Management, HEC Montréal, Montréal, Canada

Paola Cappanera Dipartimento di Ingegneria dell'Informazione, University of Florence, Florence, Italy

Caterina Caprara IBIS Lab, University of Florence, Florence, Italy

Giuliana Carello Politecnico di Milano, DEIB, Milan, Italy

Sergio Cavalieri CELS - Research Group on Industrial Engineering, Logistics and Service Operations, Department of Management, Information and Production Engineering, Università degli Studi di Bergamo, Bergamo, Italy

Yan Chen Department of Decision Sciences, Macau University of Science and Technology, Taipa, Macau, China

Marta Cildoz Public University of Navarre, Pamplona, Navarre, Spain

Domenico Conforti DIMEG, University of Calabria, Rende, Italy

Justine Coton Grenoble INP (Institute of Engineering), G-SCOP, University Grenoble Alpes, CNRS, Grenoble, France

Mazzali Cristina Department of Management, Economics & Industrial Engineering, Politecnico di Milano, Milan, Italy

Paul Damien McCombs School of Business, Information Risk and Operations Management, University of Texas at Austin, Austin, TX, USA

Maria Di Mascolo Grenoble INP (Institute of Engineering), G-SCOP, Univ. Grenoble Alpes, CNRS, Grenoble, France

Stefano Dotti CELS - Research Group on Industrial Engineering, Logistics and Service Operations, Department of Management, Information and Production Engineering, Università degli Studi di Bergamo, Bergamo, Italy

Riccardo Dulmin Department of Energy, Systems, Territory and Construction Engineering, University of Pisa, Pisa, Italy

Davide Duma Dipartimento di Informatica, Università degli Studi di Torino, Torino, Italy

Lettieri Emanuele Department of Management, Economics & Industrial Engineering, Politecnico di Milano, Milan, Italy

Marie-Laure Espinouse Grenoble INP (Institute of Engineering), G-SCOP, Univ. Grenoble Alpes, CNRS, Grenoble, France

Paolo Gaiardelli CELS - Research Group on Industrial Engineering, Logistics and Service Operations, Department of Management, Information and Production Engineering, Università degli Studi di Bergamo, Bergamo, Italy

Vittorio Giudici Bolognini Hospital, Seriate, Italy

Pierre Gruau CEA Grenoble, Grenoble, France

Rosita Guido DIMEG, University of Calabria, Rende, Italy

Alain Guinet DISP, Institut National Des Sciences Appliquées de Lyon, (Université de Lyon), Villeurbanne, France

Kwei-Long Huang Institute of Industrial Engineering, National Taiwan University, Taipei, Taiwan

Kyosang Hwang KAIST, Yuseong-gu, Deajeon, Republic of Korea

Alexander Hübner Catholic University Eichstätt-Ingolstadt, Operations Management, Ingolstadt, Germany

Amaia Ibarra Hospital Compound of Navarre, Navarre, Spain

Giuseppe Ielpa DIMEG, University of Calabria, Rende, Italy

Hoon Jang Georgia Institute of Technology, Atlanta, GA, USA

Minji Kim National Medical Center, Jung-gu, Seoul, Republic of Korea

Melik Koyuncu Engineering-Architecture Faculty, Industrial Engineering Department, Cukurova University, Adana, Turkey

Yong-Hong Kuo Stanley Ho Big Data Decision Analytics Research Centre, The Chinese University of Hong Kong, Shatin, New Territories, Hong Kong

Nadia Lahrichi Polytechnique Montreal, Mathematics and Industrial Engineering, Montreal, Canada

Paolo Landa Medical School, University of Exeter, Exeter, UK

Ettore Lanzarone Istituto di Matematica Applicata E Tecnologie Informatiche (IMATI), Consiglio Nazionale Delle Ricerche (CNR), Milan, Italy

Ettore Lanzarone CNR–IMATI, Milan, Italy

Daniele Laricini Politecnico di Milano, DEIB, Milan, Italy

Taeho Lee National Medical Center, Jung-gu, Seoul, Republic of Korea

Taesik Lee KAIST, Yuseong-gu, Deajeon, Republic of Korea

Fermin Mallor Public University of Navarre, Pamplona, Navarre, Spain

Paganoni Anna Maria Department of Mathematics, Politecnico Di Milano, Milan, Italy

Andrea Matta Dipartimento di Meccanica, Politecnico di Milano, Milan, Italy

Daniel Miller Stanford University Department of Electrical Engineering, Stanford University Department of Management Science and Engineering, and Lucile Packard Children's Hospital Stanford, Palo Alto, CA, USA

Valeria Mininno Department of Energy, Systems, Territory and Construction Engineering, University of Pisa, Pisa, Italy

Pınar Miç Engineering-Architecture Faculty, Industrial Engineering Department, Cukurova University, Adana, Turkey

Vittorio Nicoletta Department of Operations and Decision Systems, Université Laval, Québec, Laval, Canada

Maddalena Nonato Dipartimento di Ingegneria, University of Ferrara, Ferrara, Italy

Maxime Painchaud Department of Operations and Decision Systems, Université Laval, Québec, Canada

Francesco Puggelli Meyer Children's Hospital, Florence, Italy

Jérôme Radureau Adomni-Adhap Services, Lyon and Bourgoin Jallieu, France

Barbara Resta CELS - Research Group on Industrial Engineering, Logistics and Service Operations, Department of Management, Information and Production Engineering, Università degli Studi di Bergamo, Bergamo, Italy

Marina Resta Department of Economics and Business Studies, University of Genova, Genoa, Italy

Roberta Rossi Dipartimento di Ingegneria dell'Informazione, University of Florence, Florence, Italy

Angel Ruiz Department of Operations and Decision Systems, Université Laval, Québec, Canada

David Scheinker Stanford University Department of Electrical Engineering, Stanford University Department of Management Science and Engineering, and Lucile Packard Children's Hospital Stanford, Stanford, CA, USA

Fabian Schäfer Catholic University Eichstätt-Ingolstadt, Operations Management, Ingolstadt, Germany

Maria Grazia Scutellà Dipartimento di Informatica, University of Pisa, Pisa, Italy

Mara Servilio CNR–IASI, Rome, Italy

Hansu Shin National Medical Center, Jung-gu, Seoul, Republic of Korea

Wei Deng Solvang Faculty of Engineering Science and Technology, Department of Industrial Engineering, UiT – the Arctic University of Norway, Tromsø, Norway

Michele Sonnessa Department of Economics and Business Studies, University of Genova, Genoa, Italy

Alessandro Stefanini Department of Enterprise Engineering, University of Rome Tor Vergata, Rome, Italy; Department of Energy, Systems, Territory and Construction Engineering, University of Pisa, Pisa, Italy

Angela Testi Department of Economics and Business Studies, University of Genova, Genoa, Italy

Guillaume Thomann Grenoble INP (Institute of Engineering), G-SCOP, University Grenoble Alpes, CNRS, Grenoble, France

Elena Tànfani Department of Economics and Business Studies, University of Genova, Genoa, Italy

François Villeneuve Grenoble INP (Institute of Engineering), G-SCOP, University Grenoble Alpes, CNRS, Grenoble, France

D. Vincent-Genod Escale PMR department, Hospices Civils de Lyon, University of Lyon, Lyon, France

Filippo Visintin IBIS Lab, University of Florence, Florence, Italy

Carole Vuillerot Escale PMR department, Hospices Civils de Lyon, University of Lyon, Lyon, France

Manuel Walther Catholic University Eichstätt-Ingolstadt, Operations Management, Ingolstadt, Germany

Semih Yalçındağ Industrial and Systems Engineering Department, Yeditepe University, Istanbul, Turkey

Jiun-Yu Yu Department of Business Administration, National Taiwan University, Taipei, Taiwan

Wes Zeger Emergency Medicine Department, University of Nebraska Medical Center, Omaha, NE, USA

Part I
Regular Contributions

How to Protect a Hospital Against Cyber Attacks

Alain Guinet

Abstract Hospitals have not been prepared to face cyber attacks. Their core objective is to take care about patients by curing them efficiently. In this paper we propose a vulnerability assessment approach, to highlight the information system weaknesses of a hospital. By defining a map of the information system which considers the most critical assets (i.e., the units which manage the core information), the most likely attack scenarios with the worst consequences are constructed. By studying these scenarios, we suggest mitigation countermeasures, based on a reorganization of the digital information system into isolated sub-systems. Our objective is to be more resilient to cyber attacks, by increasing the required complexity for hacker's crimes and by limiting the damages of attacks.

keywords Cyber attacks · Hospital vulnerability
Information system mapping · Partitioning · Defence

1 Introduction

A study published by the Institute for Critical Infrastructure Technology in 2016, specifies that 72% of the US health care societies have been targeted by cyber attacks during 2012, 47% of the US people which accessed to the health care system have been victims of corrupted medical data, due to hackers. The cyber attacks are a sad reality in health care systems. Hospitals have not been prepared to face such a cyber war. Their core objective is to take care about patients by curing them efficiently. In this paper we propose a vulnerability assessment approach, to highlight the weaknesses of the digital information system of a hospital. By defining a map of the information system which considers the most critical assets (i.e., the units which manage the core information), the most likely attack scenarios

A. Guinet (✉)
DISP, Institut National Des Sciences Appliquées de Lyon, (Université de Lyon),
21 Av. Jean Capelle, 69621 Villeurbanne, France
e-mail: alain.guinet@insa-lyon.fr

© Springer International Publishing AG 2017
P. Cappanera et al. (eds.), *Health Care Systems Engineering*, Springer Proceedings
in Mathematics & Statistics 210, https://doi.org/10.1007/978-3-319-66146-9_1

with the worst consequences are constructed. By studying these scenarios, we suggest mitigation countermeasures, based on a reorganization of the digital information system into isolated sub-systems. Our objective is to be more resilient to cyber attacks, by increasing the required complexity for hacker's crimes and by limiting the damages of hacker's attacks.

In a second section, we present the different types of hackers and we illustrate their motivations with some cyber attack stories, from the health care sector. In a third section, we propose an approach to reduce the hospital vulnerability in cyber attack situations. This approach is based on a consensus mapping of the information system which enables us to find possible breaches which can be used by hackers. After having defined some scenarios, we discuss in the fourth section how to use the map of the hospital information system to define countermeasures, in order to mitigate the cyber attack scenarios.

2 The Cyber Attack Sources

Cyber attacks can be motivated by claiming, illegal gains or violence, as described hereafter.

2.1 Hacktivists

Hacktivists try to make propaganda by obtaining media attention in order to promote their values. These hackers can also act for enjoyable reasons. Cyber attacks from hacktivits are disruptive. The most common hacktivist attack is a denial of service (DDoS) attack, which overloads a hospital's server with undesired traffics in order to halt the regular operations of the server. A watering hole attack is another cyber attack in which the hacktivist seeks to infect websites in order to send its propaganda or malware to end users. The goal is to infect a server and gain access to the end users by the compromised server. The capabilities of hacktivists are: a low level of organisational support, a poor financial backing, a small network of members and limited capacities for computing/networking.

On April 2014, protesting on a child custody case, who was being kept at Boston Children's hospital against the wishes of him/her parents, the hacktivist group Anonymous launched multiple distributed denial-of-service attacks that targeted the hospital's servers and hamstrung its operations for a week (https://www. bostonglobe.com/business/2014/04/24/hacker-group-anonymous-targets-children-hospital-over-justina-pelletier-case/jSd3EE5VVHbSGTJdS5YrfM/story.html).

2.2   Terrorists

The main terrorist objectives are to spread terror and to kill people. Terrorists try to cause violence and damage, in order to destabilize the hospital and the patients. Cyber attacks from terrorists are destructive. The different forms of cyber attacks are: espionage, theft, sabotage and personal abuse. A possible attack can be information system hacking, in order to change patient prescription, medical test results or the formulation of pharmaceutical receipts. The capabilities of terrorists are: a high level of organisational support, a good financial backing, an efficient network of members, numerous possibilities to subcontract to hackers without ethic for computer virus developments.

At the Def Con hacking conference in 2011, Jay Radcliffe has suggested a technique for attacking a insulin pump that it is used to deliver insulin to a patient. He hacked into the pump by remotely accessing the wireless monitoring system. He demonstrated that the approach could have been used to deliver lethal insulin doses to a patient (http://www.reuters.com/article/us-rapid7-radcliffe-idUSKBN0E929K201 40529).

2.3   Criminals or State Sponsored Hackers

Cyber criminals try to generate profit through the exploitation or through the racketeering of hospital data or patients' data. Cyber attacks from criminals can be disruptive or destructive. The most common criminal attack is the ransom-ware attack, which encrypts files in an information system. If the hospital does not pay the ransom, the files are destroyed. Phishing is another form of criminal attack which consists to steal personal information data (e.g. bank identity) without violence. The user opens a malicious email from a false government agency and gives confidential information about the hospital or a patient. A delinquent has the same capacities such as a Hacktivist. The state sponsored hackers or cybercriminal groups such as mafia versus triad can be viewed as terrorists. Criminals always seek easier targets because they often do not have a plethora of resources.

On February 2016, a Southern California hospital was a victim of a cyber attack, which interfered with day-to-day operations. The CEO said that the shutdown has not affected patient care. Several hospital staff members said the computer system was hacked and was being held on a ransom of 3.4 billion in bit coins, and in exchange, the hackers would send back the key codes to restore the system. In reality the California's Hollywood Presbyterian Medical Centre shelled out 40 bit coins, or about $17,000, in order to regain access to its electronic health records. (http://www.nbclosangeles.com/news/local/FBI-LAPD-Investigating-Hollywood-Hospital-Cyber-Attack-368703121.html).

3 The Cyber Attack Approach

We propose the following vulnerability assessment approach to face the above situations.

1. Find the threat sources: Reviewing historical data on cyber attacks, we specify the adversary profile, their potential actions, their capabilities, and their motivation, as done above.
2. Define critical assets: Identifying the care units and the technical units of the hospital which use digital information as input or output or for management. We locate the critical assets regarding to their contribution to the information system, i.e. the units which are the most likely and the easiest to be faced with a computer virus and which are the most damageable regarding to lost information and the hospital added value … An IDEFØ model enables us to map the critical assets.
3. Calculate critical asset attractiveness: Realize an analysis based on pairing of each critical asset and of each threat source by brainstorming, in order to identify potential vulnerabilities per adversary, and to better evaluate the ease of causing damages, according to adversary's motivation.
4. Define Threat Scenarios: Based on the attractiveness of the critical assets per adversary, the most likely (i.e. the easiest target for the most motivated adversary) attack scenarios with the worst consequences are constructed.
5. Assess Threat Scenarios: Scenarios are studied to evaluate their consequences, to propose possible counter-measures in order to reduce the risk to an acceptable level by a cost/benefit analysis. This step is repeated until all relevant scenarios are mitigated and are instrumented with an efficient response plan. The IDEFØ map of the critical assets is studied to mitigate the threats regarding to their propagation, knowing that infection is propagated through digital exchanges. A mixed linear program is proposed to help the decision maker regarding to mitigation.

3.1 The Information System Mapping as a Tool

IDEFØ [1] is a method designed to model the events, data, and activities of an organization or a system. The IDEFØ model helps to organize the analysis of a system and allows promoting good communication between the analysts and the users. It enhances user involvement and allows us to obtain consensus models [2] which are a basic requirement when actors are multidisciplinary such as physicians, nurses, technicians, engineers, administrative staff, managers, etc. The analysis of the system is represented as a collection of hierarchically organized diagrams with a limited number of elements: boxes which represent activities or physical units, or equipment, and arrows to model physical, information, order flows, data stores etc.

IDEFØ will assist us in identifying on one hand units and services which manage digital information, i.e. which could be a source of computer virus by Internet, USB keys, user access privileges, and on the other hand connections which could be a propagation way for contamination by exploiting the data flows. An IDEFØ model will be firstly created for our hospital analysis (Figs. 1 and 2) and it will be used to identify the critical assets regarding to information management, and secondly to generate a network mapping (Fig. 3) in order to evaluate the weaknesses of the network and to study scenarios including countermeasures. Arrows model only information and data stores are represented as mechanism arrows. By default, reading and writing are allowed for data stores.

In Fig. 1, the hospital units/services use different information files which contain the core elements of their missions (bed planning for the central booking, patient medical files for care units, etc.). The patient identification is the connexion data between these different files to enforce the integrity of the information system but could be a contamination way. Internet access is provided to each unit/service and it defines a potential infection source. Regarding this user view, information is decentralized and can be physically isolated, mainly if information is virtually partitioned on the same server.

In Fig. 2, all of the units use the patient medical files, and can corrupt them. Internet is reserved to hospital wards. The patient monitoring system is dedicated to ICU but it is linked to other care units by the patient medical files. Information is centralized and cannot be easily fragmented except maybe for ICU.

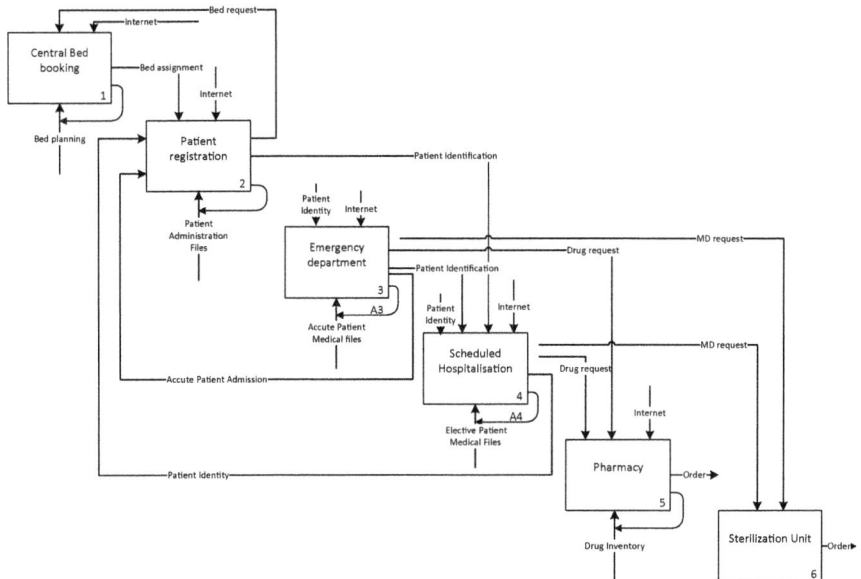

Fig. 1 The global model of the hospital

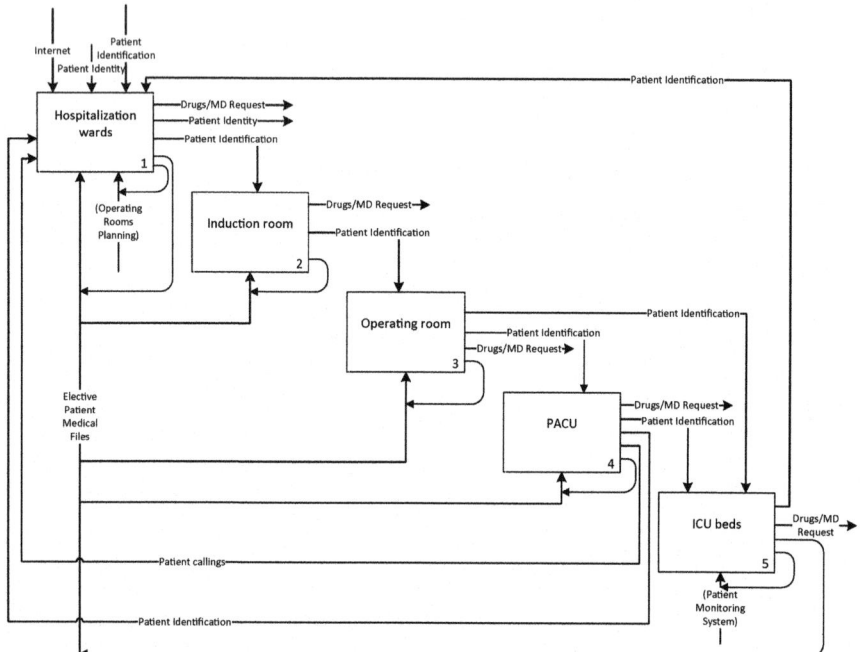

Fig. 2 The detailed model of the scheduled hospitalization

To calculate the information system mapping, we export from the IDEFØ model the different connexions between the leaves (services/units) of the decomposition tree of IDEFØ diagrams.

The result is modelled by a binary matrix (see Fig. 3) where the lines/columns represent the leaves, and the intersections between lines and columns identify the presence (1) or absence (0) of a digital connexion which could be a contamination way between two critical assets. Such binary matrix will be used to define the most resilient information system architecture. Data stores are supposed to be attached to users. If two data stores of two different users are virtually partitioned on the same server, a digital connexion between the two users is added in both ways, because a potential infection risk exists by cohabitating on the same server. Demilitarized zones can also be modelled specifying WIFI access as explained below.

3.2 The Critical Assets

According to the IDEFØ model (Figs. 1 and 2), some critical assets have been identified. They represent the units which manage core information about patients, activities, employees, etc.

	A1	A2	A31	A32	A33	A34	A35	A36	A41	A42	A43	A44	A45	A5	A6
A1	1	1	0	0	0	0	0	0	0	0	0	0	0	0	0
A2	1	1	0	0	0	0	0	0	1	0	0	0	0	0	0
A31	0	1	1	1	1	0	1	1	0	0	0	0	0	0	0
A32	0	0	1	1	1	0	1	1	0	0	0	0	0	1	1
A33	0	0	1	1	1	0	1	1	0	0	0	0	0	1	1
A34	0	0	0	0	0	1	0	0	0	0	0	0	0	0	0
A35	0	0	1	1	1	0	1	1	0	0	0	0	0	0	0
A36	0	0	1	1	1	0	1	1	0	0	0	0	0	1	1
A41	0	1	0	0	0	0	0	0	1	1	1	1	1	1	1
A42	0	0	0	0	0	0	0	0	1	1	1	1	1	1	1
A43	0	0	0	0	0	0	0	0	1	1	1	1	1	1	1
A44	0	0	0	0	0	0	0	0	1	1	1	1	1	1	1
A45	0	0	0	0	0	0	0	0	1	1	1	1	1	1	1
A5	0	0	0	0	0	0	0	0	0	0	0	0	0	1	0
A6	0	0	0	0	0	0	0	0	0	0	0	0	0	0	1

Fig. 3 The information system mapping

Emergency department and scheduled hospitalization: The patient medical files from emergency and hospitalization wards are collections of personal data about patient diagnosis, patient treatments, patient laboratory tests, medical reports, etc. The patient medical files are accessible by all the care units. They allow deciding: the patient's pathway, the laboratory exams, the drug prescriptions, etc.

Patient registration: The patient administrative files managed by the patient registration unit contain information about personal data (names, addresses, phone number, bank details, etc.), insurance identification, relatives' information … The patient administrative files enable the patient financial support, the communication with patient's relatives, etc. They are at the origin of all patient files. The database is managed by a software suit, which links the medical files, the bed planning and the operating room planning. Patient registration uses websites of health care insurance companies.

Patient registration and central bed booking: The patient registration enables to manage remotely patient admissions by physician and the central bed booking calculates the bed planning which assigns a bed to each patient over a horizon of four weeks. The bed planning allows scheduling the activity of several care units: radiology, surgery, nursing, pharmacy, etc. It is shared by all the care units, and enables to supervise all the nursing and medical activities. The bed planning is also a coordination tool between emergency and hospitalization wards.

Intensive care unit: The patient monitoring system of ICU supervises the oxygen/drug supply to patient requiring intensive cares. The blood pressure, the heart rate, the ECG signal, etc., are vital parameters which are monitored and can trigger nursing/physician alerts. The SCADA system is used to manage the ICU beds [3]. It manages the supervisory control, data acquisition system and pumps' control. The SCADA (Supervisory Control And Data Acquisition) system is potentially vulnerable to disruption of service [4], or to manipulation of data that could result in lack of vital safety for patients.

3.3   The Critical Asset Attractiveness

Based on a brainstorming, the attractiveness of the critical assets has been discussed, by pairing each critical asset and each threat source. For "Scheduled hospitalisation (Elective patient medical files) and Emergency department (Acute patient medical files)", the threats can be:

- Hacktivists: The identification of patients requiring an abortion or of the surgeons practicing abortion, and the diffusion of such information on social networks could be a goal of hacktivists.
- Criminals: The patient medical files contain confidential information about patient test results and patients' treatments. Such information can be sold to malicious financial companies.
- Terrorists: Patients suffering from AIDS, could be identified by terrorists, and could be individually targeted by further criminal actions. Patient's blood types can be changed to mass casualties.

3.4   The Threat Scenarios

The revenge: An employee has been recently fired. He/she approaches a hacker and he convinces him to destroy the hospital network by encrypting the servers. The hacker inserts a worm to get the network node plan, then she/he remotely jeopardises the network introducing crypto-lockers. All the hospital activities are out of order except the ones working with paper and autonomous equipment, i.e. diagnostic devices such as the radiology department. The emergency department has to stay closed and the elective activity must be stopped. The risk of errors is dramatically increased. The whole information system is corrupted. On the financial side, the hospital probably lost $1 million a day in billing.

The murdering: An important politician is in a surgery unit. He/she has just favoured in Parliament the approval of a law in favour of the abortion/euthanasia. A criminal paid by domestic terrorists, infiltrates the hospital information system, and next jeopardizes the patient monitoring system of ICU beds. After the surgery the Politician requires an ICU bed. The politician is killed by disruption of ICU monitoring system that has been compromised (drug dosage and/or ventilation). The SCADA system which manages the supervisory control and data acquisition system of ICU beds was unprotected. Data was not encrypted and could be easily read and modified in order to deliver a lethal dose.

4 The Scenario Assessment with the Information System Mapping

Most of works on information system defence are based on attack graphs to find system breaches and to select countermeasures [5, 6]. Attack graphs have an exponential complexity and are not easily understandable by users [7]. The architecture of the information system is very few questioned [8]. As countermeasures, investments on software are favoured instead of investments on system architecture which require heavier system upgrades due to information system integration. Since this latter is less sensitive in hospital due to the job variety (surgery, nursing, pharmacy, accommodation, etc.) and as the information system network is bounded, we favour information system architecture study. This focus is recommended by the French Network and Information Security Agency (ANSSI: Agence Nationale de la Sécurité des Systèmes d'Information) [9].

4.1 The Strongly Connected Components of the Information System

Figure 3 presents a map of the hospital information system. This map is modelled by a matrix where the N nodes represent the critical assets and the arrows define the digital exchanges between the critical assets. If we calculate the minimum number of strongly connected components of this graph [10] with a computation complexity of N^3, we define sub-graphs which are totally linked and which can be easily contaminated if one of the nodes of the sub-graph is infected. May be, a strongly connected component can be easily compromised but the damages can be limited to the sub-graph if it is physically isolated. The transitive closure gives us an idea of the information system vulnerability.

Our proposal is to create independent sub-graphs to reduce the damage consequences caused by cyber attacks while minimizing the drawback of disconnected sub-graphs [9]. Partitioning data through physically independent sub-systems on one hand reduces the contamination (propagation of viruses by transitivity), also increases the complexity of a potential cyber attack corrupting the whole data, and on the other hand allows the information system to run partially after an attack. Five strongly connected components have been identified in the IDEFØ graph. They lead to define five natural sub-graphs which represent the elective patient activity (A1, A2, A41, A42, A43, A44, A45), the acute patient activity (A31, A32, A33, A35, A36), the pharmacy (A5), the sterilization unit (A6) and a single room (A34: the waiting room of the emergency department which could be demilitarized and enable the patient's relatives to access WIFI). By isolating the information system into five independent sub-systems, we do not enable the sharing of the patient identification between units/services and of the units' requests to the pharmacy/sterilization, by a digital way. Encrypted barcodes identifying patients/products (i.e. only little information) can be an efficient and secure way to exchange data via paper. Information is transmitted by hand.

As mitigation countermeasures, our matrix enables us: to complicate cyber-attacks by partitioning the information system into independent subsystems, to favour the service continuity limiting potential cyber-attacks, to specify attack graphs using simulation [11], to define demilitarized zones for a safe use of internet, and to assess the access rights by units or services, respecting confidentiality and security.

4.2   The K Partitioning Problem

The transitive closure proposes a natural partitioning of the graph into sub-graphs. The strongly connected components represent the critical assets which share the same information and cannot be separated without a heavy lost of digital information exchanges. If we hypothesize that such partitions become physically independent regarding to digital information exchanges, a minimum of information exchanges is lost by partitioning the graph into sub-graphs corresponding to the strongly connected components. To go further, we propose a mixed linear model to partition the graph into at least K components, on one hand with weighted arcs in order to take into account the information traffic level between nodes and on the other hand with weighted critical assets to evaluate the damages of a corrupted partition. The mixed linear model must be used after calculating the transitive closure of the graph, in order to take into account all transitive links.

Data:

- N: number of system components, K: the minimum number of partitions,
- c (i, j): it is equal to 1 if system component i is linked as input with system component j,

- d (i): potential damage of critical asset i when corrupted (disruption of activity, data restoration),
- w (i, j): data traffic from the system component i to the system component j, expressed in working hours to evaluate the potential loss of digital traffic.

Variables:

- X (i, k): this binary variable is equal to 1 if i is assigned to partition k,
- Y (k): this binary variable is equal to 1 if at least one node has been assigned to partition k,
- Loss (i, j): real variable, the link from i to j is deleted and it results a loss of w (i, j) traffic,
- Dmax: real variable, maximum damage of a partition when corrupted.

Objective function:

$$Minimize(Z) = \sum_{i=1}^{N} \sum_{j=1}^{N} Loss(i,j) + Dmax \qquad (1)$$

We minimize the losses of traffic knowing that partitions are not linked, and the maximum damages to a partition. We hypothesize an attack to only one partition instead of the entire information system.

Constraints:

$$\sum_{k=1}^{N} X(i,k) = 1 \quad \forall i = 1, \ldots, N \qquad (2)$$

A system component i must belong to one and only one partition.

$$X(i,k) + X(j,k) - 1 \leq c(i,j) \ \forall i,j,k = 1, \ldots, N \qquad (3)$$

A pair of system components i and j belonging to the same partition, must exchange traffic.

$$\sum_{i=1}^{N} X(i,k) \leq N*Y(k) \ \forall k = 1, \ldots, N \qquad (4)$$

$$\sum_{i=1}^{N} X(i,k) \geq Y(k) \ \forall k = 1, \ldots, N \qquad (5)$$

$$\sum_{k=1}^{N} Y(k) \geq K \qquad (6)$$

A partition must contain at least 1 system component. We count the number of partitions which must be greater than K.

$$Loss(i,j) \geq w(i,j)*X(i,k) - w(i,j)*X(j,k) \ \forall i,j,k = 1, \ldots, N \qquad (7)$$

$$Loss(i,j) \geq w(i,j)*X(j,k) - w(i,j)*X(i,k) \ \forall i,j,k = 1, \ldots, N \qquad (8)$$

The calculation of loss of traffic is done taking into account the links between system components belonging to different partitions.

$$\sum_{i=1}^{N} d(i)*X(i,k) \leq Dmax \quad \forall k = 1, \ldots, N \tag{9}$$

The calculation of the potential damages is done. The maximum damage is sought.

Comments:
We have modelled a mixed linear program to solve a K partitioning problem, where K is greater or equal than a given number, the number of nodes per partition is unconstrained. We minimize on one hand the sum of the weighted arcs between partitions (i.e. lost links), and on the other hand the partitioning risk depending on the critical assets which belong to the same partition. We hypothesis a cyber attack limited to only one sub-system. The resulting K partitioning problem is NP hard [12]. The damages of critical assets are calculated regarding to the activity disruption which is evaluated by the lost turnover and the restoration cost of the information system (i.e. the software loading, the backup recovery, the data update due to the failure period). The traffic losses take into account the working hours to transmit information by hand (data input, barcode scanning, data printing, etc.). All costs are expressed in Euros. The cost horizon reference is one month.

We have solved our mathematical program with the Cplex solver in a few seconds. If K is equal to the number of strongly connected components, the number of partitions is most of the time equal to K and the assignment of components to partitions is the same that was calculated by transitive closure. Only a great variety of traffic costs or a large diversity of damage costs of critical assets introduces changes in the number of partitions. If the graph is a strongly connected graph, than it results just one partition. The mixed linear model will be of a great help to study the partitioning problem with increasing successive values of K.

4.3   Feedback of Partitioning by Scenario

Regarding the revenge scenario, the hacker has to steal the authentication and the identification of five employees who have an authorized access to the five partitions, in order to corrupt the whole system. If the fired employee has still an access to one of these partitions, the ransom-ware attack is limited. A paper kit system can be used to enable the continuity of service of the damaged assets (unit/service) whose data are encrypted [13], and the restoration of the corrupted equipments with an offline redundant hardware can be part of the solution.

The murdering scenario is more complicated because the patient monitoring system of ICU belongs to a partition which includes two kinds of data: SCADA data and patient medical files. Both data could be corrupted to kill the VIP, the medical prescription and/or the SCADA parameters. If we isolate the patient monitoring system of ICU in a 6th partition, the parameters of the intensive care bed will be specified by the anaesthesiologist and the data will not be remotely corrupted. The large information feedback from ICU monitoring which is no more automatically transmitted to the patient medical files, can be transferred with an encrypted USB key. Restrictive access to both systems managing the files, the denial of unauthorized programs, adoption of redundant hardware offline with backups, can be other ways to protect the hacked information sub-system.

Considering both scenarios, internet accesses are necessary: for bed booking (allowing remotely accesses for physicians) patient registration (enabling accesses to the insurance websites) and hospitalization wards, which define the first strongly connected component, for the physicians of the emergency department (which represents the second strongly connected component) and for pharmacy (which is the fourth strongly connected component). However, some restrictions must be recommended: the email access and internet access must be denied as far as possible for the emergency department and the pharmacy on the application server and its connected objects; only physicians/pharmacists might have an email access and internet access on their personal computers by an independent WIFI server; the non encrypted USB keys must be prohibited, and internet accesses must be checked on a daily basis for application servers [14].

5 Conclusion

In this paper we have proposed an approach to assess the vulnerability of hospitals against cyber attacks. After potential hacker identification, we suggest first to define the critical assets of information system in terms of units/services which manage core information about patients, employees, etc. Second, the attractiveness of critical assets is specified per adversaries. The modelling support to find the weaknesses of the hospital information system is a map resulting from an IDEFØ analysis. This map represents the digital exchanges between units/services. Third, some scenarios which represent the most likely critical assets to be attacked with the most potential damages are studied, and the consequences of cyber attacks are analysed thanks to the information system map. By calculating the partitioning of the information system by a transitive closure, we can propose mitigation countermeasures. To go further, a mixed linear program has been modelled to find the best set of sub-systems for the hospital information system, by minimizing losses of digital traffic between the independent sub-systems and contamination damages to components belonging to the same sub-system. This work is the continuation of the Threats project: http://www.threatsproject.eu/project.html.

References

1. IDEFØ, Integration Definition for Function Modelling (IDEFØ) Draft Federal Information Processing Standards Publication (1993). http://www.idef.com/Downloads/pdf/idef0.pdf
2. Bevilacqua, M., Ciarapica, F.E., Paciarotti, C.: Business process reengineering of emergency management procedures: a case study. Saf. Sci. **50**, 1368–1376 (2012)
3. Malavika, S.: Automized intensive care unit (AICU). Int. J. Adv. Electr. Electr. Eng. **1**, 11–14 (2012)
4. Venkateswarlu, G.: Expert and SCADA based centralized patient monitoring and escorting system. Int.J. Adv. Res. Electr. Electr. Instrum. Eng. **3**, 13089–13098 (2014)
5. Phillips, C., Painton-Swiler, L.: A graph-based system for network-vulnerability analysis. In: 9th ACM Conference on Computer and Communications Security, pp. 71–79. ACM Press, Charlottesville, VA, USA (1999)
6. Poolsappasit, N., Dewri, R., Ray, I.: Dynamic security risk management using bayesian attack graphs. IEEE Trans. Depend. Secure Comput. **9**, 61–74 (2012)
7. Yeh, Q.J., Jung-Ting Chang, A.: Threats and countermeasures for information system security: a cross-industry study. Inf. Manag. **44**, 480–491 (2007)
8. Noel, S., Jajodia, S.: Understanding complex network attack graphs through clustered adjacency matrices. In: Proceedings of the 21st Annual Computer Security Applications Conference, IEEE, pp. 1–10 (2006)
9. ANSSI, 40 essential measures for a healthy network. http://www.ssi.gouv.fr/administration/guide/guide-dhygiene-informatique/ (2013)
10. Yelowitz, L.: An efficient algorithm for constructing hierarchical graphs. IEEE Trans. Syst. Man Cybern. **6**, 327–329 (1976)
11. Kuhl, M.E., Sudit, M., Kistner, J., Costantini, K.: Cyber attack modelling and simulation for network security analysis. In: Proceedings of the 2007 Winter Simulation Conference, pp. 1180–1188
12. Kernighan, B.W., Lin, S.: An efficient heuristic procedure for partitioning graphs. Bell Syst. Tech. J. **49**, 291–307 (1970)
13. Grange, H., Leynon, J.: Crisis management plan: preventive measures and lessons learned from a major computer system failure, HCSE 2015. In: Proceedings in Mathematics and Statistics. Springer, Lyon, France, pp. 203–214 (2015)
14. Scott, J., Eftekhari, P.: Hacking Healthcare IT in 2016, Institute for Critical Infrastructure Technology. http://icitech.org/wp-content/uploads/2016/01/ICIT-Brief-Hacking-Healthcare-IT-in-2016.pdf (2016)

Analytical Approaches to Operating Room Management

Projects at Lucile Packard Children's Hospital Stanford

David Scheinker and Margaret L. Brandeau

Abstract In recent decades, healthcare has become increasingly expensive, creating pressure on healthcare providers to cut costs while maintaining or improving quality. Operations research can play an important role in supporting such efforts. A key challenge faced by hospital planners is scheduling and management of operating rooms, as operating rooms typically provide highly specialized care, require significant resources, and contribute significantly to a hospital's bottom line. We describe recent work on hospital operating room management at Lucile Packard Children's Hospital Stanford. We describe preliminary outcomes of three projects aimed at improving the efficiency of the hospital's operating rooms: machine learning to improve surgical case length estimation; queuing analysis to improve operational efficiency; and integer programming to schedule cases to reduce surgical delays.

Keywords Healthcare · Operations management · Optimization
Machine learning · Queueing

1 Introduction

In recent decades, healthcare has become increasingly expensive [17], creating pressure on healthcare providers to cut costs while maintaining or improving quality. The tools of operations research can play a key role in helping to improve the efficiency and effectiveness of healthcare services. Operations research analyses can be used to support high-level decisions such as facility planning (capacity, location, layout and design), public health planning, planning for population health needs, and human resource planning; tactical decisions such as capacity planning, case mix plan-

D. Scheinker
Department of Management Science and Engineering, Stanford University, Lucile Packard
Children's Hospital Stanford, Stanford, CA, USA
e-mail: dscheink@stanford.edu

M. L. Brandeau (✉)
Department of Management Science and Engineering, Stanford University, Stanford, CA, USA
e-mail: brandeau@stanford.edu

© Springer International Publishing AG 2017
P. Cappanera et al. (eds.), *Health Care Systems Engineering*, Springer Proceedings
in Mathematics & Statistics 210, https://doi.org/10.1007/978-3-319-66146-9_2

ning, resource management, patient and resource scheduling, staffing assignment, and quality control and management; and operational decisions such as management of patient flows, waitlists, and staffing levels [1].

A key area of focus in many hospitals is planning and management of the perioperative environment, consisting of the operating rooms and their supporting facilities such as pre- and post-procedure units. Operating rooms typically provide highly specialized care, require significant resources, and contribute significantly to a hospital's bottom line. Even relatively minor delays in an operating room can have a significant impact on quality of patient care, staff satisfaction, and hospital financial stability. Indeed, the average cost of operating room time in the US is approximately $4000 per hour [15, 19] and it is estimated that each procedure that must be cancelled due to operating room delays reduces hospital revenue by approximately $1500 per hour [8].

For these reasons, many operations researchers have focused on developing models to improve the performance of hospital operating rooms. Extensive work has been carried out in areas such as case mix planning and patient and staff scheduling. Work on surgical procedure scheduling for the pediatric environment has combined optimization and simulation (e.g., [2, 4, 5, 23]). For a recent review, see [6]. We note, though, that many such planning models have not been implemented in practice. Additionally, some models that have been implemented in practice are quite specific to the hospital where they were developed and thus cannot be applied in other hospitals. Our goal is to develop methods that can be implemented at our hospital but also generalized to other hospitals.

In this paper we describe preliminary results from three projects in the perioperative environment that we are currently carrying out at Lucile Packard Children's Hospital Stanford (LPCH): using machine learning techniques to improve surgical case length prediction [27]; using queuing analysis to improve the operational efficiency of the perioperative process [10]; and using integer programming to schedule cases to minimize surgical delays [12].

LPCH is a 312-bed hospital that is part of the Stanford University healthcare system. The hospital has 7 operating rooms that are used to perform more than 6000 surgical procedures annually for 23 different services (e.g., cardiology, orthopedics). An expansion that will be completed in 2017 will add 149 beds and 6 additional operating rooms.

Planners at LPCH recently focused their attention on reducing delays in the operating rooms. The elective surgery process can be broken down into three stages: the surgeon sees the patient in clinic and schedules the procedure; the patient prepares for surgery at home, is prepared for surgery in the pre-operating room areas, and has the surgery; and the patient recovers from the procedure in a specialized unit. Three common causes of operating room delays and cancellations associated with these stages at LPCH (and other pediatric hospitals) are: surgical cases are mis-scheduled [3]; surgical preparation resources and processes are managed sub-optimally [13, 26]; and recovery beds for surgical patients are not immediately available [25]. We developed projects to systematically improve performance in each of these stages, as we now describe.

2 Surgical Case Length Prediction

In order to schedule procedures in an operating room, an estimate of the time needed to perform each procedure is required. Creating such estimates is particularly challenging in pediatric hospitals because pediatric patient populations tend to have widely variable needs, even for the same type of procedure [20]. In theory, the operating room time needed for each surgical case at LPCH is estimated by the surgeon after the surgeon examines the patient in clinic. Interviews with surgeons and non-clinician schedulers working at the surgical clinic reveal that, in practice, the scheduler estimates the time needed for surgery based on guidance from the surgeon or based on a historical average. Estimates from the various clinics are then used to manually create a schedule. This schedule is the basis for managing downstream patient flow (e.g., in the post-anesthesia recovery unit).

Predicted procedure durations are often very different from actual durations. When surgeries take less time than expected, operating rooms will be idle and patients may have to wait in the operating room for a recovery bed to become available. When surgeries take longer than expected, delays are incurred for subsequent surgeries, overtime may be required, and in some cases procedures must be cancelled. LPCH planners believed that better estimates of surgical case lengths would lead to improved operating room utilization, fewer delays, and higher patient and staff satisfaction.

We undertook a project to improve the prediction of surgical case lengths. Previous approaches to estimating surgical case length have included not only expert opinion, as in the case of LPCH, but also various types of statistical analysis of historical data (e.g., [11, 14, 21, 22, 24]). We used a prediction approach based on supervised learning, as we describe below, and a classification approach based on support vector machines that we do not describe here. Further details of our models can be found in [27].

We developed tree-based automated models to predict surgical case length: three automated models that use only patient and procedure characteristics and three semi-automated models that additionally use surgeon case length prediction as a feature. The simplest automated model, which we denote by DTR, is a single decision tree regressor. We also use a random forest regressor, denoted by RFR, and a set of gradient-boosted regression trees, denoted by GBR.

We designed and compared the models based on an operationally relevant loss function: the percentage of cases that are significantly mis-scheduled relative to their scheduled duration. Interviews with operating room staff and surgeons revealed that relatively minor differences between actual and scheduled case length do not cause significant disruptions. The impact of mis-scheduling depends on the scheduled length of the procedure. Consider a room in which 10 cases are scheduled, each 1 h long, and a room in which 2 cases are scheduled, each 5 h long. Scheduling errors of 15 min will significantly disrupt the performance of the first room but not of the second. We define a case as mis-scheduled if the actual duration differs from the scheduled duration by more than 25% of the scheduled case length or 15 min.

The automated models develop predictions based on the following features of patients and procedures: sex, weight, age, American Society of Anesthesiologists Physical Status Score (a score ranging from 1–6, indicating a range of health status from normal good health to brain dead), identity of the primary surgeon performing the procedure, location of the procedure (in an operating room or an ambulatory procedures unit), patient class (inpatient or outpatient), and procedure name.

The semi-automated prediction models use the above features and, in addition, use the surgeon's case length estimate as a feature. We denote these semi-automated models as DTR-S, RFR-S, and GBR-S, respectively, corresponding to the automated approach of models DTR, RFR, and GBR.

We tested the prediction models on the 10 most common procedures performed at LPCH from May 2014 through January 2015. The data set had a total of 3426 observations. We divided the data set into a training set and a tuning set of roughly equal size: 1640 observations in the training set and 1846 observations in the tuning set. For each procedure type, we compared the performance of our six prediction methods to two benchmarks: the historical average duration for that procedure and the expert estimate of the procedure duration (i.e., the value currently used when developing the operating room schedules).

As described in [27], our simple DTR model was not better than either benchmark. The other two automated methods, RFR and GBR, outperformed both benchmarks, with GBR performing better than RFR in most cases. The semi-automated prediction models DTR-S and RFR-S performed better than their automated counterparts, while GBR-S had approximately the same performance as GBR. These results suggest that the automated GBR method could be used as an adjunct to expert opinion when estimating surgical case length.

In partnership with LPCH's analytics provider, Qventus, we are now developing a system to implement the results of the work. To minimize disruption, surgical schedulers will continue to submit their case length estimates to EPIC, the LPCH electronic medical record, as they currently do. In real time, the estimate and all relevant information will be transmitted to Qventus for analysis by a variant of our algorithm. If the resulting predicted time differs significantly from the scheduled time, then Qventus will text-message and email an alert to the scheduler with a suggested time. The scheduler can then consult with the surgeon and modify the time appropriately. We will measure the performance of this system using the loss function described above (the percentage of cases that are significantly mis-scheduled relative to their scheduled duration).

Our model is readily generalizable to other hospitals and surgical centers where the relevant patient data, or at least the subset of the most useful features, are collected. For the GBR method, the most important features for prediction were procedure name, patient weight, and primary surgeon identity. For the GBR-S method, the most important features were primary surgeon identity and the surgeon's case length estimate, followed by procedure name and patient weight. The model can be implemented with minimal disruption to operating practices through automated notifications of potential case length mis-estimates.

3 Improving Operational Efficiency of the Perioperative Process

Our second project focuses on the stage between when a surgical procedure is scheduled and when the patient completes surgery and goes to the post-anesthesia care unit (PACU). Numerous interrelated factors contribute to operational inefficiency in this process. These include, for example, variability in patient preferences and clinical needs, lack of patient adherence to guidelines, equipment and supply availability, variability in provider practice, scheduling errors, inefficient resource allocation, and communication errors. In such a complex system, it may be difficult to identify the areas of the process where interventions would most reduce delays. For example, suppose that historical time stamps show that a patient entered the operating room later than scheduled. This may be because the patient came late to the hospital, or came on time and was delayed during preparation, or was prepared on time but delayed by the previous case running late.

We faced three major challenges in this project: identifying the factors that have the most significant impact on efficiency; determining how to address those problems without adverse impact on the wider system; and finding time for staff to implement change while they continue to operate in a very busy environment. We partnered with perioperative leaders, clinicians, and staff to address these challenges systematically.

To identify the factors that have the most significant impact on efficiency we created a detailed queuing representation of patient flows in the system, estimated the utilization and capacity of each step of the process using historical time stamp data, and measured the frequency and magnitude of delays associated with each process. We identified bottlenecks to determine how to make improvements that minimize disruption to the broader system. To achieve change without unduly taxing the perioperative staff, we augmented the current process with several automated notifications powered by data already in the hospital's electronic medical record. Below, we describe the design, implementation, and results of the project. Further details are provided in [10].

We first created a flow chart that maps the process starting from the days before surgery, through the activities in the pre-operative area, to the completion of surgery and the transfer of the patient to the PACU. The process flow is as follows: In the days leading up to surgery, dedicated nurses, nurse practitioners, and physicians from the hospital contact the patient's family to collect relevant information (e.g., allergies to medications) and instruct them to prepare for surgery (e.g., explain NPO guidelines and when to arrive to the hospital). After the patient is admitted and checked in, a nursing assistant takes the patient's vital signs, height and weight, and then brings the patient to a consult room where the patient sees a nurse practitioner and answers a number of questions. Then, depending on various circumstances such as the scheduled time of surgery, the availability of a nurse, or whether the operating room is running late, the nursing team will decide whether the patient needs to be brought back to the waiting area or can immediately see a nurse in the consult room. In both cases, the patient is eventually taken to the holding area, so that the nurse can complete the exam if necessary, and prepare the patient. At some point, the operating room sends a notification that they expect the patient to be ready to see a physician in the holding area within the next 20 min. The patient is then seen by an anesthesiologist and a surgeon for final preparations and then taken to the operating room. When the surgery nears completion, a nurse from the operating room contacts the PACU to reserve a bed.

We interviewed nurses, physicians, and staff members to identify problems and opportunities for improvement in the process flow. We identified several major causes for delays: patients do not follow guidelines to not eat before surgery; on the morning of surgery patients are missing needed paperwork; patients need an interpreter but no interpreter is available; no nurse is available for room turnover; or the PACU is full. In order to quantify the impact of each of these delays and to identify other sources of delays we used a queuing representation. We used two years of historical time stamps to estimate the capacity of each set of resources in the process, the rate of patient arrivals during busy periods, and the frequency of associated delays.

The detailed process mapping revealed that we could relieve the burden of non-clinical work on perioperative staff by using automated communication. We therefore implemented automated text-message reminders to patients to assist with the communication of surgical guidelines such as not eating the morning of a procedure. We redesigned the process for completing the pre-surgical documentation to be electronic rather than paper-based and to allow for automated alerts to notify physicians and staff when documentation was missing. Additionally, we implemented several other, similar interventions, as described in [10].

The queuing representation of the system revealed numerous days in which the PACU was the bottleneck causing surgical delays. Since neither decreasing the number of arrivals to the PACU nor increasing PACU capacity were feasible options, we considered ways to increase the rate of PACU service. A detailed study of the time stamps revealed that the notification from the operating room to the PACU, intended to be made 20 min before the patient is ready to exit the operating room, was frequently premature. When the PACU receives such a notification, a bed is reserved for the patient. Premature notifications thus effectively increase how long a patient occupies a PACU bed.

We implemented our recommended just-in-time operating room notifications to the PACU at the end of June 2016. LPCH's surgical caseload is highest during the summer months, as parents schedule procedures when children are not in school. In the two months following the policy change, the percentage of patients who arrived in the PACU more than 20 min after the notification fell from approximately 54% to approximately 22%. The number of cases with a PACU hold fell from 45 with an average length of 27 min in June, to 6 with an average length of 15 min in July and 16 with an average length of 14 min in August. The improvements were not related to surgical volume, as the average number of weekday cases using the PACU remained constant at approximately 30 per weekday in June, July, and August.

The transition to just-in-time bed requests increased the effective capacity of the PACU. Other institutions that track bed request time stamps and patient arrivals could reproduce our analysis to determine whether a similar intervention is appropriate in their setting, and could explore the potential for implementing automated notifications, using an approach similar to the one we have described.

4 Post-Anesthesia Care Unit Scheduling

Our third project focuses on reducing operating room delays caused by patients waiting for a bed to become available in the PACU. Such delays are a common, extensively studied problem. Research in this area can be categorized into projects that estimate the resources, such as beds or staff, necessary to minimize such delays [7, 18] and projects that develop methods to adjust the order of cases to reduce delays [9, 16]. Results of projects that involve adjusting the order of surgical cases have been largely negative, yielding conclusions such as, "Although effective, such methods can be impractical because of large organizational change required and limited equipment or personnel availability" [9] and "The uncoordinated decision-making of multiple surgeons working in different operating rooms can result in a sufficiently uniform rate of admission of patients into the PACU and holding that the independent sequencing of each surgeon's list of cases would not reduce the incidence of delays in admission or staffing requirements" [16].

We undertook a project to develop an easily implementable surgical procedure scheduling decision support tool that would create a level load of PACU bed and staff utilization. We tested its performance to estimate the resulting improvements and are in the process of implementing it. Below, we describe the current scheduling system at LPCH, the design of the decision support tool, and its implementation.

At LPCH, after a surgical procedure in the operating room, patients are sent to one of 10 recovery beds in the PACU. A patient cannot be assigned to the PACU unless a bed is free and the appropriate staff are available to supervise the patient's recovery. If a bed and needed staff in the PACU are not available when the patient's surgery finishes, the patient must wait in the operating room for a PACU assignment. This means that the next surgical procedure cannot begin and the next patient scheduled for surgery must continue to wait in the pre-operative area. These delays lead to inefficient use of operating rooms and staff as well as lowered patient satisfaction.

The current process to reduce PACU holds is as follows. Each day, starting in the morning, a scheduler 'builds' the operating room case schedule for the following day. Since each operating room is typically reserved for cases performed by a given surgical service, building the schedule consists primarily of determining the order of the cases in each room. The scheduler accounts for special considerations (e.g., patient characteristics, specialized equipment needs, or the need for more than one surgeon for a case) that may require certain cases to be performed at specific times. Each afternoon, by which time a preliminary schedule has been created, a meeting is held to estimate the corresponding demand for PACU and other beds, make changes, and finalize the schedule for the following day. If estimates based on the preliminary schedule suggest that the PACU will reach capacity at a given time of day, then the order of the procedures is shuffled to reduce the number of patients sent to the

PACU during that time. After this process is complete, schedulers call patients to notify them of their surgery time.

Our model uses as input the cases scheduled for the following day, patient information relevant to forecasting PACU length of stay, and the patient and surgeon information relevant to constraining when certain cases must be scheduled. The model uses a random-forest-based method to estimate the likely duration of each patient recovery in the PACU. An integer program is then used to determine the order in which the procedures should be scheduled in each operating room so as to minimize maximum overall PACU occupancy.

We used a discrete event simulation model to test the performance of the optimization. We validated the simulation by reproducing 6 months of historical PACU occupancy based on scheduled order of procedures, procedure durations, and recovery durations. After validation, we used the simulation to compare the historical PACU occupancy to that which would have resulted from scheduling with the optimization. We found that 60% of operating room days finished earlier with the optimized schedule compared to the actual schedule, suggesting that significant operational improvements can be achieved with the optimized scheduling system. We are now in the process of implementing the system at LPCH. For full details of the design, testing, and implementation, see [12].

This model has the potential to improve on the current process in several ways. First, the output of the model is automated; it does not require a scheduler to spend time creating the preliminary schedule. Second, the order of procedures is determined based on case-specific estimates of PACU occupancy whereas the current process assumes equal PACU recovery lengths for all cases. Third, the arrangement of cases is optimal for minimizing maximum PACU occupancy, a combinatorial result not readily achievable with the current manual process. Additionally, implementation of the model is minimally disruptive, as it produces as output a preliminary schedule that can be reviewed and revised at the current afternoon meeting.

Our model is readily generalizable to other healthcare institutions that finalize their operating room schedule after the majority of cases are scheduled. The data used to generate the forecast of PACU length of stay and to determine the order of procedures are routinely tracked by institutions with an electronic medical record. The preliminary schedule is easily modifiable by perioperative staff who can make changes to satisfy ad hoc constraints that are not captured in the model. We are currently working with Stanford Health Care (the adult hospital at Stanford) to explore implementation of our optimization model in their operating rooms.

5   Discussion

Many opportunities exist to improve the efficiency and effectiveness of healthcare services, and operations research can play an important role in supporting such efforts. The projects we are carrying out at Lucile Packard Children's Hospital Stanford demonstrate that a systematic, analytical approach to problems in hospital operating room management can help planners achieve significant operational improvements without expanding resources or unduly taxing hospital staff.

An important goal of our projects is to develop solution approaches that not only can be implemented in the specific setting under study, but that can also be generalized to other hospitals. The models we developed rely on data that are available in almost any electronic medical record system. With the recent expansion of electronic medical record systems, our models have potential usefulness in many settings.

The primary limitation of this work is that the projects described were designed and implemented at a single pediatric hospital. To ensure that these tools are generalizable, future work should implement the tools at a second hospital and report the necessary modifications. We are currently exploring this possibility at the Stanford adult hospital.

Another promising area for further research on improving operating room management is to determine the days on which elective surgery procedures are scheduled so that surgical bed occupancy is balanced. If one knew exactly what the demand for elective surgeries over time would be, then this problem could be solved as an integer program: an assignment problem with the goal of minimizing deviations from an average surgical bed occupancy level. However, future demands for elective surgery cannot all be known when assignments are being made. Thus, the challenge is to develop a prospective algorithm that achieves solutions close to those that would be found with perfect knowledge of future demand and the use of an optimization model.

References

1. Aleman, D., Brandeau, M.L., Carter, M.W., Scheinker, D.: (Draft) Healthcare systems engineering: an analytical approach. Springer Publishers, New York
2. Banditori, C., Cappanera, P., Visintin, F.: A combined optimization-simulation approach to the master surgical scheduling problem. IMA J. Manage. Math. **24**(2), 155–187 (2013)
3. Bravo, F., Levi, R., Ferrari, L.R., McManus, M.L.: The nature and sources of variability in pediatric surgical case duration. Paediatr. Anaesth. **25**(10), 999–1006 (2015)
4. Cappanera, P., Visintin, F., Banditori, C.: Comparing resource balancing criteria in master surgical scheduling: a combined optimisation-simulation approach. Int. J. Prod. Econ. **158**, 179–196 (2014)
5. Cappanera, P., Visintin, F., Banditori, C.: Addressing conflicting stake-holders priorities in surgical scheduling by goal programming. Flex. Serv. Manuf. J. Epub. (2016) (ahead of print)
6. Cardoen, B., Demeulemeester, E., Beliën, J.: Operating room planning and scheduling: a literature review. Eur. J. Oper. Res. **201**(3), 921–932 (2010)
7. Dexter, F., Epstein, R.H., Penning, D.H.: Statistical analysis of postanesthesia care unit staffing at a surgical suite with frequent delays in admission from the operating room: a case study. Anesth. Analg. **92**(4), 947–949 (2001)
8. Dexter, F., Blake, J.T., Penning, D.H., Lubarsky, D.A.: Calculating a potential increase in hospital margin for elective surgery by changing operating room time allocations or increasing nursing staffing to permit completion of more cases: a case study. Anesth. Analg. **94**(1), 138–142 (2002)
9. Dexter, F., Epstein, R.H., Marcon, E., de Matta, R.: Strategies to reduce delays in admission into a postanesthesia care unit from operating rooms. J. Perianesth. Nurs. **20**(2), 92–102 (2005)

10. Durand, A., Kim, H., Pei, F., Petersen, K.: A generalizable, systematic approach to improving perioperative efficiency. Working Paper (2017)
11. Eijkemans, M.J., van Houdenhoven, M., Nguyen, T., Boersma, E., Steyerberg, E.W., Kazemier, G.: Predicting the unpredictable: a new prediction model for operating room times using individual characteristics and the surgeon's estimate. Anesthesiology 112(1), 41–49 (2010)
12. Fairley, M.C., Scheinker, D., Caruso, T.J., Brandeau, M.L.: Improving the efficiency of the operating room environment with a generalizable optimization and machine learning model. Working Paper (2017)
13. Hiltrop, J.: Modeling neuroscience patient flow and inpatient bed management. Ph.D. Thesis, Massachusetts Institute of Technology (2014)
14. Kayış, E., Khaniyev, T.T., Suermondt, J., Sylvester, K.: A robust estimation model for surgery durations with temporal, operational, and surgery team effects. Health Care Manag. Sci. 18(3), 222–233 (2015)
15. Macario, A.: What does one minute of operating room time cost? J. Clin. Anesth. 22(4), 233–236 (2010)
16. Marcon, E., Dexter, F.: An observational study of surgeons' sequencing of cases and its impact on postanesthesia care unit and holding area staffing requirements at hospitals. Anesth. Analg. 105(1), 119–126 (2007)
17. Organisation for Economic Co-operation and Development (OECD): Focus on health spending: OECD health statistics 2015. https://www.oecd.org/health/health-systems/Focus-Health-Spending-2015.pdf (2016)
18. Schoenmeyr, T., Dunn, P.F., Gamarnik, D., Levi, R., Berger, D.L., Daily, B.J., Levine, W.C., Sandberg, W.S.: A model for understanding the impacts of demand and capacity on waiting time to enter a congested recovery room. Anesthesiology 110(6), 1293–1304 (2009)
19. Shippert, R.D.: A study of time-dependent operating room fees and how to save $100 000 by using time-saving products. Am. J. Cosmetic. Surg. 22(1), 25–34 (2005)
20. Smallman, B., Dexter, F.: Optimizing the arrival, waiting, and npo times of children on the day of pediatric endoscopy procedures. Anesth. Analg. 110(3), 879–887 (2010)
21. Stepaniak, P.S., Heij, C., Mannaerts, G.H., de Quelerij, M., de Vries, G.: Modeling procedure and surgical times for current procedural terminology-anesthesia-surgeon combinations and evaluation in terms of case-duration prediction and operating room efficiency: a multicenter study. Anesth. Analg. 109(4), 1232–1245 (2009)
22. Strum, D.P., Sampson, A.R., May, J.H., Vargas, L.G.: Surgeon and type of anesthesia predict variability in surgical procedure times. Anesthesiology 92(5), 1454–1466 (2000)
23. Visintin, F., Cappanera, P., Banditori, C.: Evaluating the impact of flexible practices on the master surgical scheduling process: an empirical analysis. Flex. Serv. Manuf. J. 28(1–2), 182–205 (2016)
24. Wright, I.H., Kooperberg, C., Bonar, B.A., Bashein, G.: Statistical modeling to predict elective surgery time. Comparison with a computer scheduling system and surgeon-provided estimates. Anesthesiology 85(6), 1235–1245 (1996)
25. Zenteno, A.C., Carnes, T., Levi, R., Daily, B.J., Dunn, P.F.: Systematic OR block allocation at a large academic medical center: comprehensive review of a data-driven surgical scheduling strategy. Ann. Surg. 264(6), 973–981 (2016)
26. Zenteno, A.C., Carnes, T., Levi, R., Daily, B.J., Price, D., Moss, S.C., Dunn, P.F.: Pooled open blocks shorten wait times for nonelective surgical cases. Ann. Surg. 262(1), 60–67 (2015)
27. Zhou, Z., Miller, D., Master, N., Scheinker, D., Bambos, N., Glynn, P.: Detecting inaccurate predictions of pediatric surgical durations. In: Data Science and Advanced Analytics (DSAA), 2016 IEEE International Conference. IEEE, pp. 452–457 (2016)

A New Decomposition Approach for the Home Health Care Problem

Nadia Lahrichi, Ettore Lanzarone and Semih Yalçındağ

Abstract Home Health Care (HHC) is a relatively new service that plays an important role to reduce hospitalization costs and improve the life quality for patients. Human resource planning is one of the most important processes in HHC systems, for which service providers have to deal with several operational problems, e.g., the assignment of operators to patients together with their routing process. In the literature, either these problems have been simultaneously solved, or decomposed by first solving the assignment problem and then the routing problem. In this work, we propose an alternative approach, where the decomposition is based on the *First Route and Second Assign* (FRSA) approach. An instance generation mechanism is developed as well, which generates instances inspired from real HHC providers, to test the proposed FRSA approach under different circumstances. Preliminary experiments show the effectiveness of the approach.

Keywords Home health care · Human resource planning · Matheuristic decomposition · First route second assign

1 Introduction

Home Health Care (HHC) is a relatively new service that plays an important role to reduce hospitalization costs and improve the life quality for patients, who receive service at their homes. Due to population ageing and high hospitalization costs, the

N. Lahrichi (✉)
Polytechnique Montreal, Mathematics and Industrial Engineering, Montreal, Canada
e-mail: nadia.lahrichi@polymtl.ca

E. Lanzarone
Istituto di Matematica Applicata E Tecnologie Informatiche (IMATI),
Consiglio Nazionale Delle Ricerche (CNR), Milan, Italy
e-mail: ettore.lanzarone@cnr.it

S. Yalçındağ
Industrial and Systems Engineering Department, Yeditepe University, Istanbul, Turkey
e-mail: semih.yalcindag@yeditepe.edu.tr

© Springer International Publishing AG 2017   27
P. Cappanera et al. (eds.), *Health Care Systems Engineering*, Springer Proceedings
in Mathematics & Statistics 210, https://doi.org/10.1007/978-3-319-66146-9_3

demand for HHC service is increasing all around the world. In 2011, there were about 4.7 million patients in the U.S. and 1 million patients in Canada who were served by several HHC providers [7].

To balance the trade-off between cost and quality of service in HHC, providers need to deal with several optimization problems. Among them, we focus in this work on the patient assignment problem and the nurse (operator) routing problem. The first consists of matching patients with operators, while the latter determines the sequence of visits assigned to each nurse. These problems may be solved for a single period or multiple periods; in this work, we focus on the weekly problem. Both the assignment and the routing problems are handled by taking into account several features, such as patient requirements (frequency of visits), expected duration of each visit, possible visiting schedules (patterns) for patients, continuity of care, and nurse capacities.

We propose a new two-stage approach for the assignment and routing problem, which exploits a new concept for the HHC services, where routing decisions anticipate assignment decisions. In other words, we decompose the problem by deciding the routing at the first stage and the assignments at the second stage. Although this might seem counterintuitive, the variety of contexts in which HHC is provided (e.g., urban vs rural, dense vs sparse) legitimate to investigate the trade-offs for which focusing on the routing at the first stage can be beneficial. To validate the approach, as different HHC providers have different structures and cover different areas, we create a data generation mechanism to generate test instances which are able to mimic several situations of real HHC providers from different countries.

The reminder of this paper is organized as follows. A brief literature review and the problem description are presented in Sect. 2. The proposed methodology is described in Sect. 3. The data generation mechanism with some preliminary results are discussed in Sect. 4. Then, concluding remarks with future perspectives are presented in Sect. 5.

2 Problem Statement and Related Work

As discussed briefly in the previous section, the HHC assignment problem refers to the decision of matching nurses with patients, while the routing problem specifies visiting sequences of patients associated with each nurse. Several works related to these problems have been classified and discussed in a recent literature review [7]. Here, we present a short list of these works and classify them according to the length of the planning period (i.e., Single Period or Multiple Periods) and how the assignment and routing decision are held (i.e., Simultaneously or Sequentially).

The literature is mainly devoted to the simultaneous approach, where assignment and routing decision are obtained together in a single model (Vehicle Routing Problem) both for a single period [1, 6] and multiple periods [3, 9].

Moreover, due to computational complexity and operational flexibility, recent models based on two-stage *First Assign and Second Route* (FASR) approaches have been developed, where the output of the patient assignment problem is integrated as an input to the routing problem of each nurse (Traveling Salesman Problem). Both single planning period [11] and multiple planning period cases [10] have been investigated.

In this work, different from the HHC literature, we develop a new multiple periods two-stage *First Route and Second Assign* (FRSA) approach. Since different providers cover different areas (rural vs urban, different densities, etc.), such a model could be beneficial where travels are the key issue. More generally, with the development of the FRSA approach, a complete analysis can be conducted and the most appropriate approach can be selected depending on the trade-offs and the corresponding key issues of a given provider.

3 Methodology

We use the following notation:

- N: set of nurses, $N = \{1...n\}$;
- P: set of patients, $P = \{1...m\}$;
- D: set of days.
- N_d: subset of nurses available on day d;
- a_i: capacity of nurse i (duration of a workday including service time to provide visits and travel times);
- r_j: total number of visits required by patient j in horizon D;
- H: set of patterns (patterns are defined as the days in which the patient may be visited, e.g., for a frequency of two visits, patterns are: Monday and Thursday, Tuesday and Friday, ...);
- H_j: subset of patterns for patient j;
- t_j: service time for each visit to patient j;
- c_{jk}: cost (time, distance, etc.) between patient j and patient k.

We structure our two-stage FRSA approach as follows:

- **Stage 1: Routing problem** modeled as a Periodic Vehicle Routing Problem (PVRP):

  - total travel time minimization with respect to the overall capacity of nurses is pursued;
  - no nurse-to-patient assignment information is considered;
  - patients are assigned a pattern.

Only the decisions regarding the pattern assignment to patients are kept for the next stage.

- **Stage 2: Nurse-to-patient assignment problem** consists of splitting the giant tours obtained in stage 1, with the objective of minimizing the maximum workload. At this stage, decisions regarding nurse-to-patient assignments are made.

3.1   Stage 1: Route First

We model the HHC as a PVRP in which one vehicle only services all patients. We use
the typical route duration constraint for each day, adding up all nurses' capacities a_i.
Since the number of nurses is not explicitly considered, no attention is paid to the fact
that the nurses' individual travel distances must be minimized after the assignments.
To account for this (and to ensure that the territory is covered every day), we add a
step before solving the PVRP to create *seeds* that will tie each nurse to a physical
territory.

(a) Creation of seeds

The objective is to select n seeds (as many as there are nurses) with a large distance
between them. They are selected using the existing Basic Territory Units (BTUs).
Let $x_s = 1$ if BTU s is chosen to host a seed (and 0 otherwise) and $y_{s_1 s_2} = 1$ if both
s_1 and s_2 are chosen to host a seed (and 0 otherwise). $c_{s_1 s_2}$ is the cost (time, distance,
etc.) between the centers of BTUs s_1 and s_2. We solve the following problem:

$$\max \ \sum_{s_1} \sum_{s_2} c_{s_1 s_2} \, y_{s_1 s_2} \tag{1}$$

s.t.

$$\sum_{s \in S} x_s = n \qquad\qquad \forall s \in S \tag{2}$$

$$\sum_{s_2 \in S} y_{s_1 s_2} \le n x_{s_1} \qquad\qquad \forall s_1 \in S \tag{3}$$

$$y_{s_1 s_2} \le x_{s_2} \qquad\qquad \forall s_1, s_2 \in S \tag{4}$$

$$x_s \in \{0, 1\} \qquad\qquad \forall s \in S$$

$$y_{s_1 s_2} \in \{0, 1\} \qquad\qquad \forall s_1, s_2 \in S$$

Constraints (2) force to choose n BTUs, while constraints (3) and (4) ensure that
$y_{s_1 s_2} = 1$ only if both s_1 and s_2 are chosen. The objective (1) is to maximize the
distance between the chosen BTUs.

Once the n BTUs are selected, we choose in each of them the patient with the
highest number of visits; if there are several equivalent patients, we select the far-
thest from the depot. For this subset of patients $\tilde{P} \subset P$ chosen as seeds, we impose
a frequency of visits equal to $|D|$ for solving the PVRP, while for all others we use
the frequency provided for by parameter r_j.

(b) The PVRP algorithm

We use a tabu search algorithm (sketched in Algorithm 1 and based on [5]) that
proves to perform very well on the PVRP. Two movements are used to explore the
solutions space, i.e., *changing the day combination* or *reinserting on a different route*.

The algorithm allows visiting unfeasible solutions but penalizes the objective function whenever an unfeasible solution is met. These penalties are dynamically adapted during the search. Without loss of generality, the initial solution can be constructed randomly or based on constructive heuristics. Readers are referred to [5] for details.

Algorithm 1 Major lines of the PVRP algorithm

Generate initial solution: patients are assigned a valid pattern of visits and sorted in increasing order of the angle to the depot. Insertion into a route follows this order:
while Stopping criteria not reached **do**
    for each patient **do**
      for each day **do**
        Search for the best insertion using the defined movements;
        Implement best non tabu movement unless aspiration criteria is met;
        Adapt dynamically penalty parameters;
        Update tabu list and statistics.
      end for
    end for
end while

After solving the PVRP, each patient j is assigned to a pattern $p_j \in H_j$.

3.2 Stage 2: Solving the Assignment Problem

After solving the PVRP, we obtain $|D|$ routes (one per day) that need to be *split* into n segments (one for each nurse). The splitting procedure must respect the continuity of care, i.e., each patient is assigned to exactly one nurse; for this purpose, the splitting is performed in parallel for all routes.

The procedure is summarized in Algorithm 2. We denote by x the solution obtained after the splitting procedure, and by $m_u(x)$ the maximum average utilization. x^* and $m_u(x^*)$ refer to the best solution and its value, respectively. Initially, $m_u(x^*)$ and $m_u(x)$ are set to a high value. The idea is inspired by the sweep algorithm where a random angle ω is generated, and patients are collected counter clockwise until a_i is reached. If the solution is infeasible, i.e. the last nurse is overloaded, we introduce parameter β_1 equal to this observed overload divided by n, ensuring to spread the "infeasibility" among all nurses. We repeat this procedure with $a_i + = \beta_1$ until the solution is feasible.

We illustrate how this procedure can be visualized for a given day in Fig. 1. Full line represents the starting point for the collection of patients and the dotted line the point where the capacity of the nurse is reached.

Once the splitting procedure is complete, an assignment of patients to nurses is obtained (i_j denotes the nurse assigned to patient j). We add a step to revisit the assignment of a subset of patients either because the solution is not feasible after the

---

Algorithm 2 Description of the splitting procedure

$m_u(x^*) = m_u(x) = na_i$
for all $|D|$ routes resulting from PVRP: **do**
    Generate ω degrees (points);
    Select a random point;
    Order the nurses (according to: seniority, preferences or randomly);
    for $i = 1$ to n **do**
        for all routes corresponding to the days nurse i works; **do**
            while capacity of nurse is not reached **do**
                collect patients when pattern of visits and days of work coincide.
            end while
            if the problem is infeasible for at least one nurse: increase capacity by parameter β_1 and repeat the collection process starting from nurse 1.
        end for
        If $m_u(x) < m_u(x^*) : x^* \leftarrow x$ and $m_u(x^*) = m_u(x)$.
    end for
end for

Fig. 1 Illustration of the splitting procedure

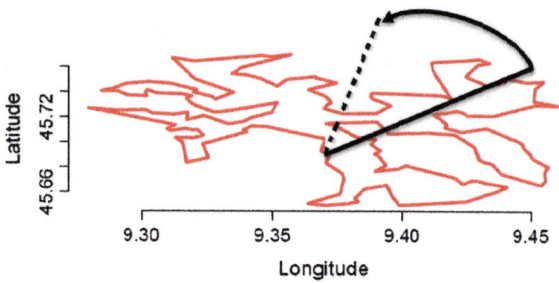

splitting procedure, or as a post-optimization step. We create the set of patients P' to be reassigned as follows:

1. Include in P' γ patients from the beginning and γ patients from the end of each nurse's route. The larger the value of γ is, the more we destroy the routing solution obtained in the previous stage.
2. Include in P' all patients close to the depot.
3. Include in P' all seeds in $\tilde{P}$.

Patients belonging to $P\backslash P'$ generate an initial fixed workload for each nurse.

To solve the assignment model, we use the formulation in [10]. For each patient j, we define τ_j as the average traveling time to reach him. Moreover, five decision variables are defined:

- $u_{ij} = 1$ if nurse i is assigned to patient j (and 0 otherwise);
- $u_{ij}^d = 1$ if nurse i visits patient j on day d (and 0 otherwise);
- $z_{jp} = 1$ if pattern p is assigned to patient j (and 0 otherwise);
- W_{id}: workload of nurse i in day d;
- m_u: maximum of the average utilization over D among the nurses.

The mathematical formulation of the problem is the following:

$$\min \quad m_u \tag{5}$$

s.t.

$$\sum_{i \in N} u_{ij} = 1 \qquad\qquad \forall j \in P \tag{6}$$

$$\sum_{p \in H_j} z_{jp} = 1 \qquad\qquad \forall j \in P \tag{7}$$

$$W_{id} = \sum_{j \in P} (t'_j + \tau_j) \cdot u^d_{ij} \leq a_i \qquad \forall i \in N_d, \forall d \in D \tag{8}$$

$$\sum_{i \in N} u^d_{ij} = \sum_{p:p(d)=1} z_{jp} \qquad \forall j \in P, \forall d \in D \tag{9}$$

$$u^d_{ij} \leq u_{ij} \qquad\qquad \forall i \in N_d, \forall j \in P, \forall d \in D \tag{10}$$

$$\sum_{i \in N} \sum_{d \in D} u^d_{ij} = r_j \qquad\qquad \forall j \in P \tag{11}$$

$$\frac{\sum_{d \in D} W_{id}}{|D| \cdot a_i} \leq m_u \qquad\qquad \forall i \in N \tag{12}$$

$$u_{i,j} = 1 \qquad\qquad \forall j \in P \backslash P' \tag{13}$$

$$z_{jp_j} = 1 \qquad\qquad \forall j \in P \backslash P' \tag{14}$$

$$u_{ij} \in \{0, 1\} \qquad\qquad \forall i \in N, j \in P \tag{15}$$

$$u^d_{ij} \in \{0, 1\} \qquad\qquad \forall i \in N, \forall j \in P, \forall d \in D \tag{16}$$

$$z_{jp} \in \{0, 1\} \qquad\qquad \forall j \in P, \forall p \in H \tag{17}$$

Constraints (6) decide the assignments. Constraints (7) are the scheduling constraints. Constraints (8) control the daily workload of nurses and use τ_j to estimate the travel time to reach patient j. Constraints (9) and (10) link together assignment and scheduling decisions; specifically, constraints (9) state that exactly one nurse per day must visit patient j only if a visit has been scheduled on that day for him/her ($p(d) = 1$ refers to patterns p having a visit on day d), and constraints (10) guarantee that a nurse can visit a patient only if she/he has been assigned to that patient. Constraints (12) link the maximum utilization m_u to the nurses' workloads.

As mentioned earlier, the usual 8 h per day may not allow to visit all patients; thus, parameter a_i in (8) should be carefully fixed. In our context, we solve an optimization problem to determine the smallest increase of this value that allows a feasible solution. We denote this value β_2, and $a_i = 8\,\text{h} + \beta_2$.

Two constraints are added to the general model when used for partial reassignment. Constraints (13) ensure that patients in $P \backslash P'$ are assigned to the nurse from the splitting procedure, and constraints (14) that they are assigned the pattern from the PVRP solution.

3.3   Obtaining the Operational Routes

Once the assignments to nurses and to patterns are obtained, a TSP is run for each day and each nurse to get the actual routes to perform and, thus, the actual daily workload for each nurse. We employ the genetic algorithm of [11], which provides equivalent performance to benchmark solvers, e.g., the *Concorde TSP Solver* [4], for the considered size of instances.

4   Data Generation and Preliminary Experiments

We create a mechanism to generate test instances, for the validation of our approach in a variety of situations inspired from real HHC providers. In particular, we use the characteristics of a large provider operating in the Northern Italy that has been already adopted for other analyses [2, 8].

Let us consider a test instance with m patients. We fix a geographical distribution of some BTUs, each one with a given center (latitude and longitude) and shape. Each patient j is randomly assigned to a BTU according to given probabilities to belong to each BTU; then, his/her coordinates are uniformly generated within the assigned BTU. A Care Profile (CP) is also assigned to each patient j, according to given probabilities to belong to each CP. Each CP is characterized by a range for the number of visits and an associated discrete distribution; once a patient is assigned to a CP, his/her demand is drawn from such distribution. Then, from the number of visits, the list H_j of all possible patterns is derived. The duration of each visit is finally generated from a uniform distribution within the interval [35, 45] minutes.

This mechanism is versatile and allows us to generate, for example, urban/rural instances by imposing high/low number of patients in a small/large territory. For example, we consider the average travel time $\bar{\tau}$ between patients as metrics to characterize an instance: $\bar{\tau} = 7$ min refers to an urban instance and $\bar{\tau} = 25$ min to a rural context.

An illustration of patient scattering in the territory is presented in Fig. 2.

We test the approach on the instances reported in Table 1. Details about their features are given in the first three columns. The following two columns show the solution obtained with the classical FASR decomposition in terms of maximum average utilization m_u and maximum workload W_{max} observed over the horizon. Finally, the last columns give the improvement (in %) when using the FRSA approach in two settings: only using the splitting, or using the splitting and the partial reassignment.

FRSA generally outperforms FASR, and in particular FRSA with partial reassignment outperforms FASR in all instances when comparing m_u. Moreover, results show that the partial reassignment step after the FRSA is very important when small distances are involved; however, the reassignment deteriorates the solutions when the distances are large. With partial reassignment, the improvement ranges between 0.37% and 1.61% when the distances are small $(\bar{\tau}_j = 7)$, while the improvement is

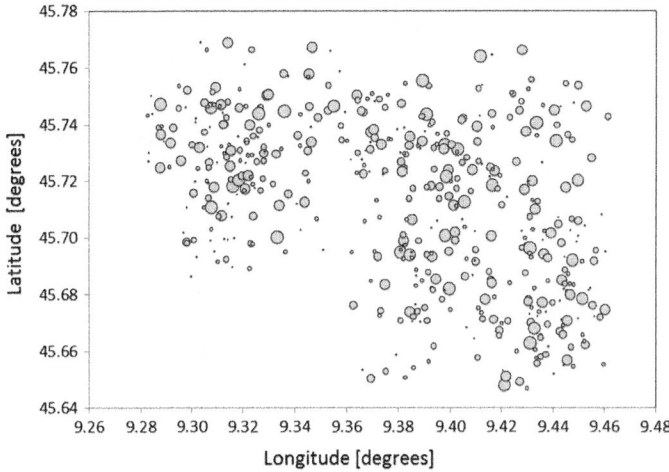

Fig. 2 Example of tested instance; each point represents a patient, whose size is proportional to the number of visits required in the time horizon

Table 1 Comparison of utilization rates and workloads (in %) obtained with FASR and FRSA

Instance			Classical FASR		FRSA: only splitting		FRSA: splitting & partial reassignment	
m	n	$\overline{\tau}$	m_u	W_{max}	m_u (%)	W_{max} (%)	m_u (%)	W_{max} (%)
300	18	7	90.51	95.26	+12.28	+9.14	−1.25	+6.60
			87.34	93.82	+7.53	+10.93	−0.72	+8.31
			94.71	96.03	+3.12	+10.39	−1.61	+6.04
300	20	7	82.63	94.91	+11.01	+9.87	−0.37	+8.64
			79.40	92.40	+22.85	+12.34	−0.93	+6.97
			85.60	94.64	+16.27	+12.06	−0.76	+6.35
300	18	25	113.60	117.44	−4.64	+1.61	−2.34	+1.00
			122.58	128.13	−9.48	−7.51	−2.95	+0.79
			118.27	121.07	−11.54	−3.37	−4.39	+0.91
300	20	25	106.26	111.04	−4.41	0.00	−3.05	+0.66
			114.31	117.35	−9.04	−4.45	−2.70	−0.13
			105.78	111.21	−4.40	+2.04	−2.15	−0.27

higher, up to the 11.54% without partial reassignments, when the distances are large ($\overline{\tau}_j = 25$). We remark that, in absolute terms, 1% improvement corresponds to 5 min approximately. As for W_{max}, we only observe an improvement with FRSA when the distances are large.

Finally, we may observe that m_u is higher than the 100% in some instances; in fact, for $\overline{\tau}_j = 25$, we considered cases in which the staff is overloaded to test both underloaded and overloaded situations.

5 Conclusion and Perspectives

In this work we propose an alternative decomposition for the HHC problem. While the literature usually proposes to first assign patients to nurses and then solve the routing problem, we first consider the routing problem and then solve the assignment problem. Moreover, to evaluate such approach in a variety of realistic cases, we define a versatile mechanism to generate instances. Preliminary experiments are promising, and seem to assess the appropriateness of our alternative decomposition.

In our future work, we will test additional instances considering in particular large settings. Finally, we will further compare our FRSA approach with the classical FASR decomposition available in the literature [10, 11], to deeply investigate for which types of instances our alternative decomposition performs better.

References

1. Akjiratikarl, C., Yenradee, P., Drake, P.R.: PSO-based algorithm for home care worker scheduling in the UK. Comput. Ind. Eng. **53**, 559–583 (2007)
2. Carello, G., Lanzarone, E.: A cardinality-constrained robust model for the assignment problem in home care services. Eur. J. Oper. Res. **236**, 748–762 (2014)
3. Cappanera, P., Scutella, M.G.: Joint assignment scheduling and routing models to home care optimization: a pattern-based approach. Trans. Sci. **49**, 830–852 (2015)
4. Concorde TSP Solver. http://www.math.uwaterloo.ca/tsp/concorde
5. Cordeau, J.-F., Laporte, G., Mercier, A.: A unified tabu search heuristic for vehicle routing problems with time windows. J. Oper. Res. Soc. **52**, 928–936 (2001)
6. Eveborn, P., Flisberg, P., Ronnqvist, M.: LapsCare-an operational system for staff planning of home care. Eur. J. Oper. Res. **171**, 962–976 (2006)
7. Fikar, C., Hirsch, P.: Home health care routing and scheduling: a review. Comput. Oper. Res. **77**, 86–95 (2017)
8. Lanzarone, E., Matta, A., Sahin, E.: Operations management applied to home care services: the problem of assigning human resources to patients. IEEE Trans. Syst. Man Cybern. A **42**, 1346–1363 (2012)
9. Nickel, S., Schroder, M., Steeg, J.: Mid-term and short-term planning support for home health care services. Eur. J. Oper. Res. **219**, 574–587 (2012)
10. Yalçındağ, S., Cappanera, P., Matta, A., Scutellá, M.G., Sahin, E.: Pattern-based decompositions for human resource planning in home health care services. Comput. Oper. Res. **73**, 12–26 (2016)
11. Yalçındağ, S., Matta, A., Sahin, S., Shanthikumar, J.: The patient assignment problem in home health care: using a data-driven method to estimate the travel times of care givers. Flex. Serv. Manuf. **281**, 304–335 (2016)

Strategic Operations Management in Healthcare: A Reference Model for Cardiac Rehabilitation

Barbara Resta, Vittorio Giudici, Sergio Cavalieri, Wei Deng Solvang, Stefano Dotti and Paolo Gaiardelli

Abstract Although operations strategy has been recognised as a relevant topic for the healthcare sector, scarce attention has been devoted to understand how internal and external operations characteristics may interact each other affecting strategic decision-making process. On this premise, this paper introduces a framework built on current literature and further validated through a case study carried out in a Cardiac Rehabilitation department. It aims at identifying the key characteristics of operations for healthcare providers, linking them to the context where a hospital operates. The final framework suggests that internal characteristics are differently influenced by external factors and generate mutual relationships. In addition, they are affected by other internal factors, including variables describing hospital clinical context and hospital operations strategy.

Keywords Healthcare operations management · Operations characteristics Strategic decision-making · Cardiac Rehabilitation · Strategic framework Hospital operations strategy

B. Resta (✉) · S. Cavalieri · S. Dotti · P. Gaiardelli
CELS - Research Group on Industrial Engineering, Logistics and Service Operations,
Department of Management, Information and Production Engineering,
Università degli Studi di Bergamo, Bergamo, Italy
e-mail: barbara.resta@unibg.it

V. Giudici
Bolognini Hospital, Seriate, Italy

W. D. Solvang
Faculty of Engineering Science and Technology, Department of Industrial Engineering,
UiT – the Arctic University of Norway, Tromsø, Norway

© Springer International Publishing AG 2017   37
P. Cappanera et al. (eds.), *Health Care Systems Engineering*, Springer Proceedings
in Mathematics & Statistics 210, https://doi.org/10.1007/978-3-319-66146-9_4

1 Introduction

In the healthcare sector, governments around the world are facing significant expenditure constraints, while at the same time trying to improve quality, customization and access to healthcare services for population [1]. Although the quality of medical care is improving for most types of illness, the attention to detail in business systems and processes that transform resources into healthcare services has not kept pace [2], resulting in long waiting times, patients lying on stretchers in hallways, overburdened and stressed medical staff [3]. In this context, while the appropriate use of traditional operations management (OM) concepts could be potentially of great help relying on the hypothesis that the healthcare sector can benefit from the lessons learned in the industrial sector [4], the use of methods, approaches and tools initially established for manufacturing companies has been proved to be challenging because of the differences between manufacturing and healthcare operations [5].

Existing studies in this area offer a collection of frameworks, models and classifications, providing some guidance on how to configure an operations strategy for healthcare providers [6, 7], but are insufficient to provide a complete and detailed picture of medical care delivery. In particular, most of the operations-oriented studies focus narrowly on specific issues [8, 9], with a few examples integrating long-term structural and intermediate operations decisions [4]. However, focusing on one decision area is not enough; on the contrary, developing a congruent operations strategy is the key to improve healthcare providers' performance [2]. Moreover, though many organizations flounder on internal factors, failure of execution can also be related to environmental features and their evolution [10].

On these premises, this study will address the following research questions: *Which are the main variables characterising healthcare operations management? How are the variables connected?*

Therefore, a framework providing a set of leverages for strategic Operations Management (OM) in healthcare will be proposed. The framework aims at capturing the key characteristics of operations for healthcare providers, linking such aspects to the overall healthcare system in which they function, including drivers for change and factors that influence decision-making.

In the next section, the research methodology is described. Section 3 presents the initial theoretical model. Section 4 is focused on the discussion of results from an Italian case study and on the novel model for Healthcare Strategic OM in the cardiac rehabilitation area. Section 5 closes the paper with contributions to theory and practice, outlining future research directions.

2 Methodology

Given the explanatory nature of the research ("how"-type of questions), the originality of the field, as well as the criticality of the context, a case study methodology was adopted [11]. In the first phase of the research, a conceptual framework was developed based on a literature review in the following research field: Healthcare, Service and Manufacturing Operations Management and Strategy.

Research issues and interview questions were then developed and included in a research protocol. Then, an in-depth case study was carried out. The case, the cardiac rehabilitation (CR) unit of the "Ospedale Bolognini" (ASST Bergamo Est) in Seriate (Bergamo, Lombardy region, Italy), with 500 patients/year, supported in refining the theoretical model and in exploring how the context influence healthcare operations management characteristics. In particular, the cardiac rehabilitation services, defined as "comprehensive, long term programmes involving medical evaluation, prescribed exercise, cardiac risk factor modification, education, and counselling. These programmes are designed to limit the physiological and psychological effects of cardiac illness, reduce the risk for sudden death or re-infarction, control cardiac symptoms, stabilise or reverse the atherosclerotic process, and enhance the psychosocial and vocational status of selected patients" [12]. Multiple investigators and respondents have been used as a form of triangulation to handle the richness of the contextual data and provide more confidence in research findings. During two visits, six informants from the CR unit have been interviewed: head of the unit, a physician, the head of the physiotherapists, the head-nurse. In the third and final stage the theoretical framework was redefined and relationships between variables characterising healthcare strategic operations management were identified.

3 Literature Review and Definition of the Theoretical Model

The theoretical model is composed of two main dimensions, internal and external. It is intended to capture the general characteristics of healthcare operations strategy.

3.1 Internal Factors

Internal factors represent the structural and infrastructural characteristics that describe how operations are configured. Internal factors have been developed based on the framework proposed by Baines et al. [13], further tailored on healthcare distinctive features. Such framework represents a comprehensive model obtained from the analysis and synthesis of a wide literature on strategic operations management (Table 1).

3.2 External Factors

The external environment is defined as all of the political, economic, social and regulatory forces that influence on the organisation [20]. Drawing on the model developed by Swayne et al. [16], five main areas are included in the framework:

Table 1 Internal factors of the theoretical model

Process and technology	A process can be defined as the chain of operations that need to be performed to produce a particular service [14]. Vissers and Beech [14] provide an overview of the most important characteristics of care processes from an operations management point of view: emergency versus elective; level of urgency; length of the process (number of operations that constitute the chain); complexity (number of diagnostic and therapeutic procedures, necessity to consult another specialist, cyclic character); predictability (number of operations, durations, routing); decoupling point. Moreover, this factor entails also the range of technologies supporting the delivery of service
Capacity	This factor covers decisions related to the definition and allocation of key resources, such as personnel/specialist time (labour availability of physicians, nurses and other providers), accommodation/facility space (number of beds, treatment or examination rooms and clinics), supply and other resources (cafeteria, parking and support services), equipment and material (key medical technology and equipment) [2, 14, 15]. Such resources can be classified as: dedicated/shared; leading/following; bottleneck; continuous/intermittently available [14]. Capacity can also be distinguished between different types [14]: potential capacity (total amount of resources available of one resource type when all resources are used for production); available capacity (total amount of capacity available, in principle, for production); usable capacity (capacity normally available for production); utilised capacity (available capacity actually used for production). It includes both productive and non-productive purposes, such as set-up capacity
Facilities	In the healthcare field, "facilities" represent the physical environment in which health care is delivered [16]. Decisions concerning this category refer to facility location (in terms of accessibility, adaptability and availability) to achieve equitable healthcare access and efficient delivery [15]
Supply chain positioning	In healthcare, two main supply chain areas can be identified: care development (upstream) and care delivery (downstream). With this category, a healthcare actor can be positioned within the supply chain and its vertical integration degree can be described, intended as "those parts of the value network that belong to the company" [17]
Planning and control	As defined by Hans et al. [18], healthcare operations planning and control involves four key areas: medical planning (development of new medical treatments), resource capacity planning (dimensioning, planning, scheduling, monitoring and control of renewable resources), materials planning (acquisition, storage, distribution and retrieval of consumable resources), financial planning (how to manage costs and revenue to achieve organizations' objectives.)
Human resources	Healthcare organizations must be able "to gather resources, skills and knowledge in a unique way, coordinate diverse operational skills and integrate multiple streams of technologies" [16]. Therefore, this category describes the composition of the workforce, as well as how the workforce will be scheduled and paid. In particular: personnel mix (physicians, doctors, nurses, therapists, administration staff), competences (technical and non-technical), workforce planning recruitment, payment, training and rewarding [16]

(continued)

Table 1 (continued)

Quality control	Donabedian [19] proposes a framework for quality assessment and evaluation, that considers the quality of care into three fundamental parts: structure, process, and outcome. The variable 'structure' describes the setting in which care is delivered including hospital facilities and equipment, human resources and organizational structure. The variable 'process' refers to the transactions between patients and providers in which healthcare services and treatments are provided. Finally, the dimension 'outcomes' refers to the health status of patients and populations
Service range and new service introduction	The range of healthcare delivery service includes inpatient service, outpatient service, other community oriented activities (wellness program, fitness centres, pharmacy information systems, health promotion programs), and day care centres, as well as its specialization degree (generalist service, market specialist, service specialist, super specialist) [16]
Performance measurement	In healthcare, three categories can be useful for evaluating performance [20]: structural measures (assess the features of operations management in delivery organizations); process measures (assess the activities carried out to deliver healthcare services); outcome measures (assess the impact of the healthcare service on the health status of patients)
Stakeholder relations	Different needs of different stakeholders can affect the delivery of the final service [16]. More specifically, three main stakeholders' categories can be identified: internal stakeholders (actors who operate primarily within the organization, such as managers and employees, medical staff and corporate officers), interface stakeholders (who function both internally and externally) and external stakeholders (community, third-party payers, competitors, clinics which operate external the organization)

- *Legislative/Political/Regulatory*—defining how governments and legislation bodies intervene in the healthcare economy;
- *Economic*—related to healthcare economics;
- *Social and demographic*—describing gender, ethnicity, religious and cultural aspects, as well as health consciousness, population growth rate, age distribution, living standards and income level;
- *Technological*—including R&D activity, automation, technology incentives and the rate of technological change;
- *Competitive*—representing the strategic behaviour of competing actors.

4 Results from the Case Study

The description of the case using the theoretical model previously defined is presented in Table 2.

Table 2 Case description

External factors	
Legislative/Political/Regulatory	Italian National Health Service: principles of universal coverage and non-discriminatory access to the health care services. Beveridge model (healthcare is provided and financed by the government through tax payments). "Livelli essenziali di assistenza" (basic assistance level) defined at national level. Clinical aspects defined by Italian and European guidelines. Process standards defined by the Lombardy Region. Promotion of the integration between hospitals and the territory (general practitioner) (Regional law 23/2015)
Economic	For CR services provided by public hospitals (or by licensed private healthcare facilities), service delivery is reimbursed by the regional healthcare system based on bed occupancy
Social and demographic	Catchment area: 400.000 citizens (foreign citizen: 11%)/7 hospitals. Gender is not discriminating
Technological	Major technological trend of the last years: telemedicine for rehabilitation and remote monitoring
Competitive	Other cardiac rehabilitation unit in the territory (catchment area): Romano di Lombardia (public service/hospital for acute patients—no cardiology department); San Pellegrino (no acute patients); Gavazzeni (private structure—not specific for cardiac rehabilitation only); Trescore Balneario (private structure)
Internal factors	
Process and technology	Minimum duration: 7 days (defined by legislation). The duration is not deterministic (given the required minimum duration) because it depends on patient's pathology, clinical trend; general complexity level. Each day the therapy can be re-defined, based on patient's conditions. Therefore, the service delivery process is personalised. In order to reduce variability, similar patients are clustered into groups. After patient's stabilization (in average 1 week), the process continues with outpatient activities. If patients live far away from the hospital, they continue with inpatient activities. Telemedicine is extensively used for remote monitoring of patients' conditions (supporting function). Telemedicine technology provides physicians with a large amount and range of data (not related to cardiac aspects only). Such data could be used by general practitioners to manage the patients. To this end, the pilot project "Respirare" was launched a few years ago, involving 10 general practitioners with access to a platform containing patients' data from telemedicine (for selected relevant parameters). In 2015 "Respirare II" has been approved to further develop such delivery service, promoting integration between hospital and territory. Internal protocols (defined by the Quality Department) outline the service delivery process (clinical aspects defined by Italian and European guidelines)

(continued)

Table 2 (continued)

External factors	
Capacity	18 beds for inpatients; outpatient departments; two gyms; nursery room with tele-medicine control unit (12/18 patients can be tele-monitored). Additional 8 beds and 10 tele-monitoring units in another hospital controlled by Ospedale Bolognini (Gazzaniga). An additional outpatient department is going to be opened in Calcinate for patients' joint-management between the hospital and the general practitioner (actual number of beds: 10—Expected number of beds: 30)
Facilities	Activities are carried out in 2 different buildings, located in the hospitals (lower degree of synergies). Ospedale Bolognini is an urban hospital
Supply chain positioning	The cardiac rehabilitation unit moved to the Bolognini hospital (from Gazzaniga) 8 years ago as a strategic decision to have acute and post-acute treatments in the same hospital. Emergency medicine is also present to handle emergency situations. Moreover, there is a novel tendency to integrate CR activities carried out by the hospital with follow-up activities supervised by general practitioners. Therefore, a new unit is going to be opened soon (Calcinate—see Capacity) where the patients will be jointly managed
Planning and control	Planned access to the department: max 2 patients/day in the afternoon (from 2 p.m. to 5.30 p.m.)—1 additional patient/day in exceptional situations (in the morning). The head-nurse receives the requests (from cardiac surgery—mostly from the Bolognini cardiac surgery department and from Bergamo hospital; from general medicine and cardiology departments—from all the Bergamo province, with some patients coming also from the Brescia province) by phone or by fax. Planning meetings: –8.30 a.m.: physiotherapists with their head to organise morning activities; 2 physicians meet the night-shift nurse for updates. Physicians starts their inpatient visits; –12.00 a.m.: meeting with physicians, head nurse and nurses to plan acceptance/discharge activities; Planning the discharge (not only the acceptance) is very important: the patients need to set-up its social and living environment before being able to return to its home
Human resources	5 cardiologists; 5 physiotherapists (3 full-time and 2 part-time); 12 nurses (shared with 10 beds from the medicine department—usually patients having cardiac problems)—daily schedule: 4 part-time operators (2 for inpatient and 2 for outpatient departments) and 1 full-time operator (outpatient department); 6 supporting operators (4 full-time and 2

(continued)

Table 2 (continued)

External factors	
	part-time). Three shifts per day. All the workforce (including doctors for nutrition and psychological consulting) is employed by the hospital. There is a reward system for personnel based on parameters defined by the Quality Department. A cultural intermediary is also available (shared with other departments and made available by the hospital) to support foreigner patients
Quality control	Quality control parameters focus on process delivery aspects, based on hospital's certifications. Parameters cover inpatient activities only. Long-term quality control (post-discharge) is not implemented
Service range and new service introduction	The cardiac rehabilitation unit is responsible for managing a wide range of medical interventions, including nutrition and psychological consultancy. Moreover, support services to general practitioners will be available in the near future. Cultural intermediary service is provided to support communication between foreigner patients and institutional staff
Performance measurement	Measures related to process quality, defined by the Quality Department, related to economic incentives for the personnel
Stakeholder relations	The relationships with patients' relatives and community is fundamental (see planning and control). Integration of the hospital with the territory (general practitioners) is a key aspect introduced by legislation

5 Towards a Reference Model for Strategic Operations Management in Healthcare

The discussion on how both internal and external factors are connected (Table 3) was the key step for the definition of a reference model for Strategic Operations Management in Healthcare, with reference to cardiac rehabilitation.

From the analysis of the case studies, the following changes have been introduced in the initial model:

- "Capacity" has been split into two variables: *key resources* (types of resources for service delivery) and *capacity utilisation* (utilisation level of the resources);
- Internal context: including variables describing hospital clinical context (availability of health departments) and hospital operations strategy characteristics. The internal context influences: CR human/key resources, facilities, supply chain positioning and quality control.

Table 3 Relationships between internal and external factors

	Link with internal factors
External factors	
Legislative/Political/Regulatory	*Process&Technology:* for the definition of the minimum length of stay for CR services and the promotion of the use of innovative technologies. *Human resources*: for the definition of the number of nursing and supporting personnel based on several parameters (defined at national/regional level). *Stakeholder relationships*: for the integration of hospitals with the territory (general practitioners). *Quality control*: for the definition of quality standards (at regional/national level)
Economic	*Planning&Control*: related to the reimbursement of service delivery by the regional healthcare system. *Performance measurement*: for the definition of proper key performance indicators
Social and demographic	*Process&Technology*: patient's social and demographic characteristics influence the features of the delivery process and the use of the technology. *Stakeholder relationships*: patient's community/relatives characteristics have an influence on its relationships with the hospital
Technological	*Process&Technology:* the availability of new technologies on the market is a necessary condition for their adoption. *Service range and new service introduction*: the availability of new technologies on the market enables the introduction of new services or the process improvement of existing health solutions
Competitive	Not directly connected with internal factors. It is linked to the operations strategy of the hospital
Internal factors	
Process and technology	*Planning&Control*: process delivery characteristics and the adopted technologies influence planning and control choices. *Human/Key resources (capacity)*: process delivery characteristics and the adopted technologies influence the required human/key resources for its implementation. *Facilities*: in terms of decisions on facility location and types. *Supply chain positioning*: in terms of make or buy decisions
Capacity—Key resources	*Facilities*: where the key resources are allocated/installed. *Planning&Control*: related to the availability of key resources than can be allocated during planning activities
Capacity—Utilisation	It does not influence other internal factors
Facilities	*Human resources*: in terms of possible synergies within the same unit and also with other hospital departments
Supply chain positioning	*Human/Key resources (capacity)*: in terms of make or buy decisions for human/key resources

(continued)

Table 3 (continued)

	Link with internal factors
Planning and control	*Capacity utilization*: planning activities determine the level of utilisation of internal capacity
Human resources	*Planning&Control*: related to the availability of human resources than can be allocated during planning activities
Quality control	*Performance measurement*: quality control aspects drive the definition of proper key performance indicators
Service range and new service introduction	*Process&Technology:* the choice of service offering involves the definition of a proper delivery process and its characteristics (including technological aspects). *Supply chain positioning*: in terms of actors involved in service delivery
Performance measurement	*Human resources*: economic incentives are connected with performance
Stakeholder relations	Service range and new service introduction: services supporting the establishment of relationships with stakeholders (community and territory). *Planning&Control:* for discharge planning, patient's community and relatives should be ready for its homecoming

6 Conclusion

Nowadays, healthcare delivery actors have to face an increasing number of challenges to create operational efficiency being at the same time financially viable. Therefore, they must apply sound business and operations management, ensuring improvements in clinical and organisational performance. In this context, an integrated framework for strategic operations management was developed and then relationships between variables were defined through an in-depth case study of an Italian cardiac rehabilitation unit. The final framework is composed of internal operations strategy factors as well as variables describing the context, both external and internal. Moreover, it provides relationships between the identified variables and how they influence each other. Therefore, the framework can be used as a tool helping healthcare managers engineer and configure service delivery processes, considering both the external and internal context. It also supports decision-making processes, simulating how acting upon a specific variable or changes in the context affect the other components of the operations strategy. Nevertheless, several limitations and further developments can be outlined. Firstly, more case studies should be carried out to validate the model (selected variables and their links), both for other CR departments and other hospital units operating in different health systems. Then, a system thinking model can be developed for running simulations and scenario analysis to test and optimise configuration choices.

Acknowledgements The authors would like to thank the staff of the cardiac rehabilitation unit of the Bolognini Hospital (Seriate, Italy) for their kind and active participation in the interviews, as well as their constructive feedback and suggestions.

References

1. Deloitte: 2017 Global Healthcare Outlook—Making Progress Against Persistent Challenges. Deloitte Consulting LLP, US (2017)
2. Langabeer II, J.R., Helton, J.: Health Care Operations Management—A System Perspective. Jones & Bartlett Learning, Burlington, MA (2016)
3. Steinke, C.: Examining the role of service climate in health care: an empirical study of emergency departments. Int. J. Serv. Ind. Manag. **19**(2), 188–209 (2008)
4. Li, L.X., Benton, W.C., Leong, G.K.: The impact of strategic operations management decisions on community hospital performance. J. Oper. Manag. **20**(4), 389–408 (2002)
5. Bertrand, J.W., Vries, D.G.: Lessons to be learned from operations management. In: Vissers, J., Beech, R. (eds.) Health Operations Management: Patient Flow Logistics in Health Care. Routlede, Oxon New York (2005)
6. Mahdavi, M., Malmström, T., van de Klundert, J., Elkhuizen, S., Vissers, J.: Generic operational models in health service operations management: a systematic review. Soc. Econ. Plan. Sci. **47**(4), 271–280 (2013)
7. de Vries, J., Huijsman, R.: Supply chain management in health services: an overview. Supply Chain Manag. Int. J. **16**(3), 159–165 (2011)
8. Dobrzykowski, D., Deilami, V.S., Hong, P., Kim, S.C.: A structured analysis of operations and supply chain management research in healthcare (1982–2011). Int. J. Prod. Econ. **147**, 514–530 (2014)
9. Jha, R.K., Sahay, B.S., Charan, P.: Healthcare operations management: a structured literature review. Decision **43**(3), 259–279 (2016)
10. McLaughlin, D.B., Hays, J.M.: Healthcare Operations Management. Health Administration Press (2008)
11. Yin, R.K.: Case Study Research: Design and Methods. Sage, Newbury Park, CA (1994)
12. Wenger, N.K.: Current status of cardiac rehabilitation. J. Am. Coll. Cardiol. **51**(17), 1619–1631 (2008)
13. Baines, T., Lightfoot, H., Peppard, J., Johnson, M., Tiwari, A., Shehab, E., Swink, M.: Towards an operations strategy for product-centric servitization. Int. J. Oper. Prod. Manag. **29**(5), 494–519 (2009)
14. Vissers, J., Beech, R. (eds.): Health Operations Management: Patient Flow Logistics in Health Care. Routledge, Oxon New York (2005)
15. Brandeau, M.L., Sainfort, F., Pierskalla, W.P. (eds.): Operations Research and Health Care: A Handbook of Methods and Applications, vol. 70. Springer Science & Business Media (2004)
16. Swayne, L.E., Duncan, W.J., Ginter, P.M.: Strategic Management of Health Care Organizations. Wiley, Malden Oxford Victoria, Oxford (2012)
17. Hayes, R.H., Wheelwright, S.C.: Restoring Our Competitive Edge. Wiley, New York (1984)
18. Hans, E.W., Van Houdenhoven, M., Hulshof, P.J.: A framework for healthcare planning and control. In: Hall, R. (ed.) Handbook of Healthcare System Scheduling. Springer, New York Heidelberg Dordrecht London (2012)
19. Donabedian, A.: Evaluating the quality of medical care. Milbank Quart. **83**(4), 691–729 (2005)
20. Shortell, S.M., Kaluzny, A.D.: Health Care Management: Organization, Design, and Behavior. Cengage Learning, Delmar (2000)

Mining the Patient Flow Through an Emergency Department to Deal with Overcrowding

Davide Duma and Roberto Aringhieri

Abstract The Emergency Department (ED) management presents a really high complexity due to the admissions of patients with a wide variety of diseases and different urgency, which require the execution of different activities involving human and medical resources. This have an impact on ED overcrowding that may affect the quality and access of health care. In this paper we apply Process Mining techniques to a real case study: from the ED database, discovery techniques identify the possible paths of a patient on the basis of the information available at the triage. Our purpose is to obtain precise process models for replicating and predicting the patient paths.

Keywords Emergency department · Overcrowding · Process mining · Patient flow

1 Introduction

The Emergency Department (ED) is a sub-unit of the hospital that operates 24 h per day, 365 days per year, providing immediate treatments to patients. Although a substantial number of patients is self-referred, but does not need emergency care, EDs must ensure treatments to all of them. The ED management presents a really high complexity due to the admissions of patients with a wide variety of diseases and different urgency, which require the execution of different activities involving human and medical resources. The uncertainty determined by the high heterogeneity of cases brings about different problems, such as the extension of waiting times and the inefficient use of the available resources. For all these reasons, the ED overcrowding represents an international phenomenon [8], which may affect the quality and access of health care because of medical errors, delays in treatments, risks to

D. Duma (✉) · R. Aringhieri
Dipartimento di Informatica, Università degli Studi di Torino,
Corso Svizzera 185, 10149 Torino, Italy
e-mail: davide.duma@unito.it

R. Aringhieri
e-mail: roberto.aringhieri@unito.it

© Springer International Publishing AG 2017   49
P. Cappanera et al. (eds.), *Health Care Systems Engineering*, Springer Proceedings
in Mathematics & Statistics 210, https://doi.org/10.1007/978-3-319-66146-9_5

patient safety and poor patient outcomes [3], but also high levels of stress, impairment of the staff morale and increased costs [1].

The ED overcrowding is manifested through an excessive number of patients in the ED, patients being treated in hallways, ambulance diversions, long patient waiting times and patients leaving without treatment [5]. Since the perception of the crowding level from the staff is subjective [9] and because of the need to adequately prevent the phenomenon, several indices for the real-time measurement of overcrowding have been introduced and studied. They are based on different indices about the current operating status: the amount of available resources, the number of patients in the ED involved in some activities or waiting for a resource, their waiting times, the patient outcome and the predicted arrivals. However, the analysis in [4] shown that none of most popular overcrowding measures is capable of providing an adequate forewarning. Simulation has been widely used to understand causes of the overcrowding and to analyse the impact of several interventions (patient flow analysis, the bottlenecks detection, resource allocation) to alleviate its effects.

After interviewing the ED-staff operating on a real case study, we observed how the patient paths can be different and intricate, since they depend on specific needs that are not easily identifiable at the time of their arrival, taking into account many variables that affect medical conditions, shifts of the ED staff, crowding of the ED but also availability of beds for hospitalisation within the other wards. Although the flowchart of the ED process can be easily designed interviewing the management staff, rules defined by physicians are usually subjective and not sufficient precise due to their complexity, the practical sense of physicians taking decisions and the flexibility required by urgent actions. Then the replication of the patient flow is difficult without making significant assumptions, which do not allow us to act in real time on the bottlenecks to prevent or relieve overcrowding.

Nowadays huge amounts of data are collected by EDs, recording diagnosis and treatments of patients. Process Mining can exploit such data and provide an accurate view on health care processes [7], ensuring their understanding in order to generate benefits associated with efficiency [10]. In literature there are several discovery techniques that use specialised data-mining algorithms to extract knowledge from datasets, creating a process model that takes into account dependency, order and frequency of events, but also decision criteria and durations. However, the ED process we would to mine has the characteristics of a *Spaghetti process*, that is an unstructured process in which the huge variety of sequences of events affects the trade-off between simplicity and precision discovering the process.

In this paper we deal with the discovery of the ED process in a real case study. We use several Process Mining techniques to identify the possible paths of each patient on the basis of the only information known at the access of the patient. Our purpose is to obtain precise process models for replicating and predicting patient paths. An accurate replication of the patients paths allows us to simulate the process, generating a resource demand that depends on decisions taken in the past activities. Then the prediction of the future patient activities can be used to implement online approaches for the resource allocation to act on bottlenecks.

The paper is structured as follows. In Sect. 2 we describe a case study of a medium-size Italian ED, providing an analysis of the patient population. The process mining approach is presented in Sect. 3, discussing some preliminary results. Section 4 closes the paper.

2 A Case Study

We present a real case study concerning the ED sited at *Ospedale Sant'Antonio Abate di Cantù*, which is a medium size hospital in the region of Lombardy, Italy. The ED serves about 30000 patients per year, the urgency of which is classified by a code from 1 (most urgent) to 5 (less urgent) assigned by the triage-nurse at the time of their arrival, in accordance with Table 1. After the triage, the patient is visited in one of the visit rooms by a physician, which can prescribe X-ray examinations, several laboratory tests or therapies and a Short-Stay Observation (SSO). Certain patients are visited in other special rooms, such as the shock-room that is properly equipped for severely urgent interventions, and the Minor Codes Ambulatory (MCA), provided by the ED from Monday to Friday in the time slot 8:00–16:00 for adult patients with low urgency codes and good ambulation ability. In addition to all these ED tasks, there are activities that can be performed outside the ED, that is several specialist visits and the paediatric visit that is provided for non-urgent young patients instead of the medical visit. After examinations, treatments and specialist visits, the patient is revalued again by a physician of the ED, which establishes how to continue the treatments, the need of hospitalisation or the discharge. In case of hospitalisation, patients are observed in the SSO units until a bed is available within the assigned hospital ward.

Table 1 Urgency codes: description and frequency over 2013–2015

Number	Color	Description	Frequency (%)
1	Red	Immediate danger of Death	1.5
2	Yellow	Need of a timely medical visit	15.8
3	Green	Need of treatments or investigations	61.6
4	Blue	Symptoms that could be treated as primary care	13.8
5	White	Symptoms that could be treated as primary care	7.3

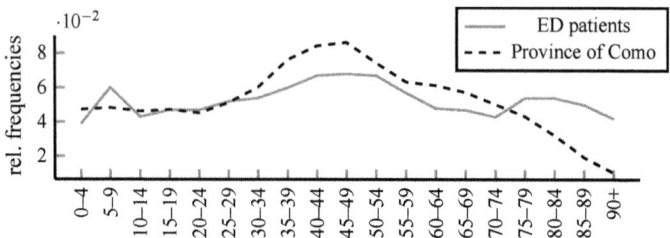

Fig. 1 Comparison between territorial and patient age distributions

The resources available within the ED are: 4 beds for the medical visits placed in 3 different visit rooms, in addition to one bed within the shock-room and another one in the MCA, one X-ray machine, 5 SSO units, 10 stretchers and 10 wheelchairs to transport patients with walking difficulties. The medical staff is composed of 4–6 nurses and 1–3 physician(s), depending on the time of day and the day of week, in addition to the X-ray technician.

2.1 Patient Population

Thanks to the collaboration with the ED, we have available data concerning all accesses made in the years 2013–2015. Such data contains sex (male 52.7% or female 47.3%) and age of the patient, type of access (autonomously 79.9% or with a rescue vehicle 20.1%), the urgency code (1–5), the main symptom (undefined 35.2%, trauma 30.7%, abdominal pain 6.9%, flue 4.5%, chest pain 3.8%, dyspnea 3.4%, and other 25 options), timestamps and resources used during the activities, and type of discharge (ordinary 82.0%, hospitalisation 8.0%, abandonment 6.9%, transfer to another facility 2.6%, death 0.3% or hospitalisation refusal 0.2%).

The patient population is quite uniformly distributed across the different ages, with slight peaks for the age groups 5–9 and 35–54. To motivate this fact we compared the access frequencies of the five-year age classes with the demographic distribution. As shown in Fig. 1, the almost uniform distribution of accesses among the age classes is due to the balance between the lower percentage of children and older people in the territorial area and the higher percentage of adults, which have a lower number of accesses per person. For the comparison, we used ISTAT data about 2014 in the province of Como, in which Cantù is located, observing that Lombardy Region and Italian territory have very similar distributions.

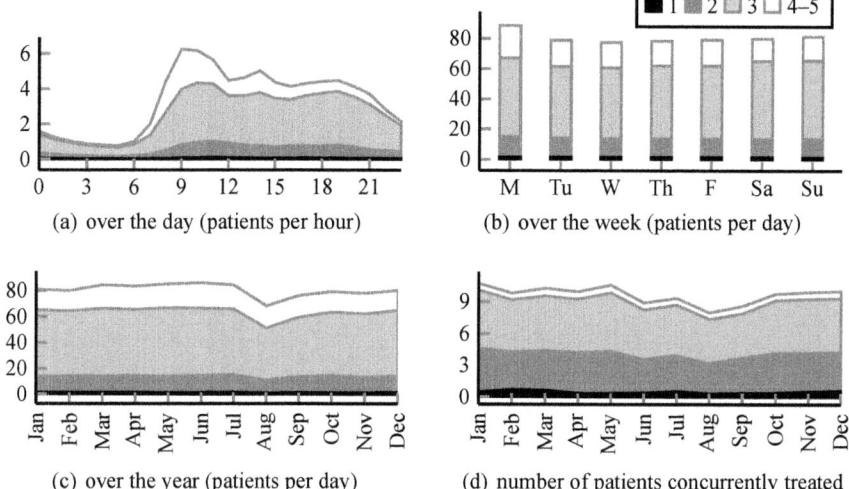

Fig. 2 Patient accesses divided for urgency code

2.2 Retrospective Analysis

The ED of Cantù performed a retrospective analysis using the National Emergency Department Overcrowding Scale (NEDOCS) [12] in the aftermath of several management changes, such as the introduction of the MCA or a new staff rostering. In addition to inadequacy of this measure, proved in [4], the analysis performed by the ED of Cantù has been affected by the lack of several information that has been dealt with approximations. For all these reasons, we omit the NEDOCS results, focusing on a brief retrospective analysis that describes the variability of demand over time.

The accesses have different fluctuations over the day, among the days of the week and among the seasons, but also among the urgency classes. The higher arrival rate fluctuations occur during the business hours of the day, as shown Fig. 2a, especially for the minor codes, which usually go to the ED instead of relying on primary care. For the same reason, a higher number of non-urgent arrivals has been registered on Monday, as shown in Fig. 2b. Conversely, the urgency class 1 has the highest coefficient of variation among the different months of the week, because of medical and epidemiological reasons that cause more arrivals in winter. Nevertheless, from Fig. 2c a uniform workload over the year (except for August) could be deducted, in fact, the workload do not depend directly of the number of accesses. Then, we report in Fig. 2d the average number of patients concurrently treated (including all the activities between the first visit and the discharge), that is a more consistent indicator with respect to the ED staff perception. The statistics in Table 2 justify this fact, indeed more urgent patients have a longer average EDLOS (Emergency Department Length-Of-Stay). Such a difference is due to the higher frequency of SSO for patients

Table 2 Waiting times, LWBS and statistics on the treatment of patients

Urgency code	Average wait time (min)	Percentage of LWBS (%)	Average EDLOS (h)	Percentage of SSO (%)	Average SSO duration (h)	Percentage of hospitalisation (%)
1	15	0.2	9	28.4	19	60.3
2	34	0.5	7	19.5	20	28.9
3	65	3.3	2.5	5.4	19	7.8
4–5	68	11.1	1	0.5	17	1.1

with urgency codes 1 and 2, caused by a higher percentage of hospitalisations. The average waiting times confirm us that the priority among urgency codes is respected. Finally, lower urgency codes also have an higher rate of patients Leaving Without Been Seen (LWBS).

3 Process Modelling

In order to use discovery mining techniques, we need to preprocess the ED database in order to create an event log, that consists of a set of traces (i.e. ordered sequences of events of a single case), their multiplicity and other information about the single events, such as timestamps and/or durations, resources, case attributes and event attributes. In our case the events correspond to the activities concerning the patient treatments recorded for each access within the ED of Cantù's database, while each trace identifies a patient path. Then, our objective is to discover a simple process model that allows us: (i) to simulate accurately patient paths, resource allocation and workload, and (ii) to predict the next activities and the required resources of patients on the basis of their characteristics and their partial paths. All the process mining techniques cited in this Section have been used as plug-ins of *ProM 6.6*.

3.1 Preprocessing

The event log has been generated taking into account the accesses of the year 2015, removing all the accesses interrupted by death or a voluntary abandonment, because in this cases we do not know what would be the continuation of the treatments. However, such data can be used to study the patient LWBS phenomenon and their impact on performance. Each case of the event log consists in an access and events consists in activities, which has been classified into 15 event classes: *T* (*triage*), *V* (*medical visit*), *K* (*shock-room treatment*), *A* (*MCA visit*), *P* (*paediatric visit*), *S* (*specialist visit*), *R* (*revaluation visit*), *X* (*X-ray examinations*), *E* (*laboratory examinations*),

Y (*therapy*), *O* (*SSO*), *H* (*hospitalisation*), *Z* (*hospitalisation refusal*), *F* (*transfer to facility*) and *D* (*discharge*). Consecutive events of the same event class were merged because of the irrelevance from a control-flow perspective and because this allow us to simplify the process models.

The event log is composed by 141 202 events regarding by 27 039 cases, which generated 3 986 different traces. On average traces are formed by 5 events belonging to distinct classes, but the number of events per trace ranges between 3 and 25. The high number of traces with a low frequency is caused by three different factors, that is medical reasons (i.e. patients need very different treatments), incorrect recordings (i.e. noise) and incomplete data. For instance: some activities has been recorded later and not always immediately at the end of their execution because of the need to deal with an urgency; for technical reasons, revaluation visits are not recorded every time; some traces contains both medical visit and paediatric visit but they refer to the same activity.

3.2 Process Discovery

The huge number of traces suggests the use of discovery techniques that deal with low frequent behaviour and noise. We used two different process miners, the *HeuristicMiner* (HM) [11] and the *Inductive Miner – infrequent* (IMi) [6], both based on the control-flow perspective. The HM takes into account the order and the causal dependencies among the events within a trace, generating a model that uses the Heuristic Net (HN) notation, which is flexible because it can be easily converted in other notations, for instance a Petri Net (PN). The IMi is an extension of the *Inductive Miner* (IM), that is a divide-and-conquer approach based on dividing the events into disjoint sets taking into account their consecutiveness within traces, then the event log

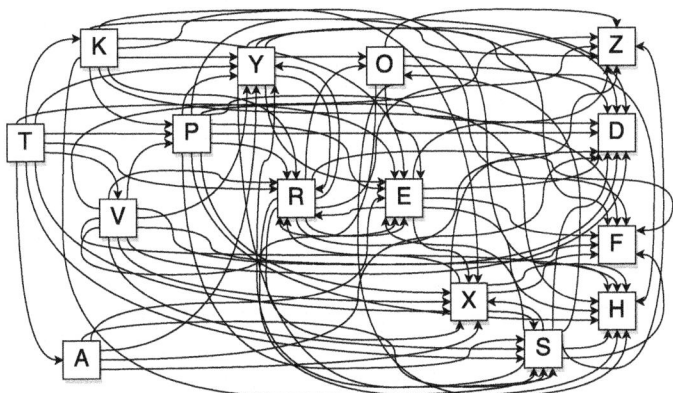

Fig. 3 Process model mined with the HM: model $\mathcal{H}$ (heuristic net)

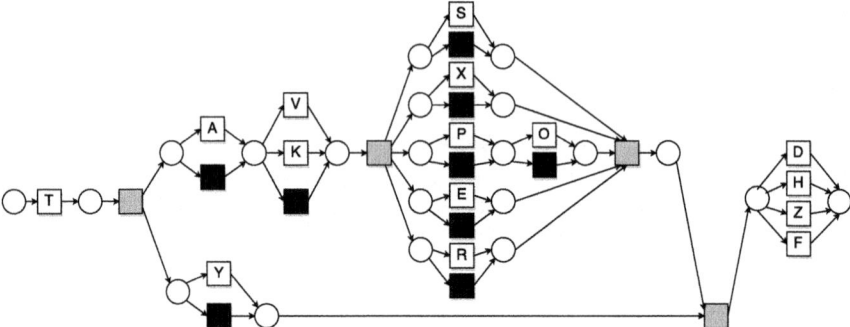

Fig. 4 Process model mined with the IMi: model $\mathscr{I}$ (Petri net)

is splitted into sub-logs using these sets. The IMi uses the same approach but filters a fixed percentage of traces representing infrequent behaviour to create a PN. Both the techniques require low computational time, that is an important requirement because of the dimension of our event log. We discovered from the event log two different process models using the HM and the IMi, that are shown in Figs. 3 and 4, called $\mathscr{H}$ and $\mathscr{I}$.

The model $\mathscr{H}$ has been generated varying the parameters *dependency* and *relative-to-best* of the HM in such a way to reach the best fitness, that is an index of the capacity to reproduce the behaviour recorded in the event log, equal to 62%. The obtained model $\mathscr{H}$, as well all the other generated varying the parameters, is a so-called *Spaghetti process* that is not sufficient simple to understand the whole process.

The model $\mathscr{I}$ has been obtained varying the noise parameter of the IMi in order to have a good precision avoiding or limiting infrequent behaviour. However, we observed very slight deviations among the models ranging the noise percentage, that has been fixed to 20%. Contrariwise to $\mathscr{H}$, this model is very simple but not precise: the parallelisms among activities (represented by the grey boxes in the Fig. 4) allowed by $\mathscr{I}$ implies additional behaviour that is not present in the event log.

3.3 Patient Clustering

In order to have a better balance of the four most important quality criteria of the process modelling (i.e. fitness, precision, generalisation and simplicity), we use the *Decision-Tree Miner* (DTM) described in [2], which analyses data flow to find rules explaining why individual cases take a particular path. Although the approach that is most straightforward and common in literature is the *Trace Clustering* (TC) [10], we opt for the DTM because of its computational efficiency and the ability to determine the most influential variables by itself. We execute the DTM using the event log, which contains the patient attributes, and the PN of the model $\mathscr{I}$ that is easier to

Table 3 Guards detected by the DTM on $\mathscr{I}$ (variables: age a, sex x, main symptom s, arrival mode m and urgency code c)

Id	Transition	Predicate	F_1-score
G_1	*Paediatric visit*	$a \leq 17$	0.954
G_2	*X-rays*	$s =$ trauma $\vee$ $(a \geq 66 \wedge m = 118$-ambulance$)$	0.703
G_3	*Examinations*	$s =$ abdominal-pain $\vee$ $(c \leq 2 \wedge a \geq 61)$	0.655
G_4	*Therapy*	$(c \leq 2 \wedge a \geq 19) \vee (c = 3 \wedge a \in [19,70] \wedge s \neq$ trauma $\wedge x = $ F$)$	0.624
G_5	*Medical visit*	$a \geq 17$	0.562

handle of $\mathscr{H}$. We selected only the variables known at the access of the patient, because we would classify the patients at their arrival and to predict their paths on the basis of his/her cluster.

Several guards have been determined on a subset of the transitions of $\mathscr{I}$. Such guards are predicates that indicates a criterion to fire a transition, using the F_1-*score* as accuracy index. We denoted with white block transitions the performing of a certain activity, while black block transitions fire when activities represented by their parallel transitions are not executed.

The guards detected by the DTM are reported in Table 3. We remark that such guards are more objective and accurate than the indications given by the medical staff, for instance we are now able to determine exact boundary ages to classify the patient behaviour. Furthermore, the guards involve the most relevant variables from a control-flow perspective.

We clustered the patients in accordance with the characteristics determined by the guards. We started from the guard G_1 that have the highest F_1-score and we split the set of all the patients into 2 clusters on the basis of the guard satisfaction. We repeated this operation on the clusters for all the other guards in decreasing order of their F_1-score, avoiding the splits when one of the two sub-cluster had less than 270 cases, that is 1% of all patients, in order to ensure significant samples.

The patient clustering determined 10 clusters that we used to create the same number of event logs. The cases of the new event logs ranges between 355 and 6 493, that is between the 1.3% and 24% of the initial event log. The number of different traces within the same event log is significantly decreased, obtaining on average 585 and maximum 1 128 different traces per cluster. All the corresponding process models mined through the HM are simpler than that in Fig. 3 because they have less transactions between events, but they are also more precise than the model $\mathscr{I}$ in Fig. 4 because of the absence of parallelisms. We report in Fig. 5 the model $\mathscr{C}$ obtained using the event log of a single cluster and the HM. We observe that the

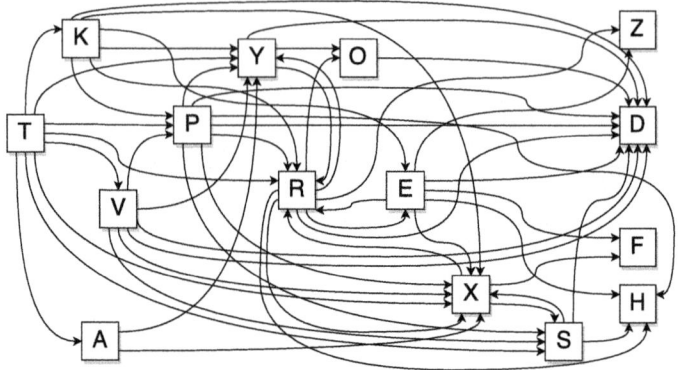

Fig. 5 Model $\mathscr{C}$ (heuristic net) for the cluster that satisfies $G_1 \wedge G_2$

model $\mathscr{C}$ is considerably simpler than $\mathscr{H}$ but gives information about the behaviour that is not deductible by $\mathscr{H}$, such as sequential order of activities or frequencies of the transactions between the same couple of events in different clusters.

The patient clustering may be further continued converting the new process models in PNs and run the DTM to discover new guards. Additional event attributes can be taken into account in order to discover different decisions that depends on several time variables (e.g. MCA visits are never performed over the weekend) or the system state (e.g. degree of crowding). Furthermore, the HM compute frequencies of transition between two events, that could be very useful to estimate the probability of a patient to follow a certain path and consequently to compute the expected waiting and execution times.

4   Conclusions

In this paper we apply Process Mining techniques to a real case study in order to obtain precise process models for replicating and predicting the patient paths. This is an important starting point on the perspective of implementing online algorithms for the patient flow optimisation with the aim of reducing overcrowding. Finally, the computational inefficiency of some procedures, such as the TC or the ILP-Based Process Discovery, suggests ad-hoc implementations of the process modelling techniques.

Acknowledgements The authors wish to thank Alessandra Farina, Elena Scola and Filippo Marconcini of the ED at *Ospedale Sant'Antonio Abate di Cantù* for the fruitful collaboration and for providing us the data set and allowing their use in this paper.

References

1. Anantharaman, V., Seth, P.: Emergency department overcrowding. In: Emergency Department Leadership and Management: Best Principles and Practice, pp. 257–269. Cambridge University Press (2014)
2. De Leoni, M., van der Aalst, W.: Data-aware process mining: Discovering decisions in processes using alignments. In: Proceedings of the ACM Symposium on Applied Computing, pp. 1454–1461 (2013)
3. George, F., Evridiki, K.: The effect of emergency department crowding on patient outcomes. Health Sci. J. **9**(1), 1–6 (2015)
4. Hoot, N., Zhou, C., Jones, I., Aronsky, D.: Measuring and forecasting emergency department crowding in real time. Ann. of Emerg. Med. **49**(6), 747–755 (2007)
5. Hwang, U., Concato, J.: Care in the emergency department: how crowded is overcrowded? Acad. Emerg. Med. **11**(10), 1097–1101 (2004)
6. Leemans, S., Fahland, D., van der Aalst, W.: Discovering block-structured process models from event logs containing infrequent behaviour. Lect. Notes Bus. Inf. Process. **171 LNBIP**, 66–78 (2014)
7. Mans, R., van der Aalst, W., Vanwersch, R., Moleman, A.: Process mining in healthcare: data challenges when answering frequently posed questions. Lect. Notes Comput. Sci. (including subseries Lecture Notes in Artificial Intelligence and Lecture Notes in Bioinformatics) **7738 LNAI**, 140–153 (2013)
8. Paul, S., Reddy, M., Deflitch, C.: A systematic review of simulation studies investigating emergency department overcrowding. Simulation **86**(8–9), 559–571 (2010)
9. Reeder, T., Burleson, D., Garrison, H.: The overcrowded emergency department: a comparison of staff perceptions. Acad. Emerg. Med. **10**(10), 1059–1064 (2003)
10. Rojas, E., Munoz-Gama, J., Seplveda, M., Capurro, D.: Process mining in healthcare: a literature review. J. Biomed. Inf. **61**, 224–236 (2016)
11. Weijters, A., Ribeiro, J.: Flexible heuristics miner (FHM). In: IEEE SSCI 2011: Symposium Series on Computational Intelligence—CIDM 2011: 2011 IEEE Symposium on Computational Intelligence and Data Mining, pp. 310–317 (2011)
12. Weiss, S., Derlet, R., Arndahl, J., Ernst, A., Richards, J., Fernndez-Frankelton, M., Schwab, R., Stair, T., Vicellio, P., Levy, D., Brautigan, M., Johnson, A., Nick, T.: Estimating the degree of emergency department overcrowding in academic medical centers: Results of the national ed overcrowding study (NEDOCS). Acad. Emerg. Med. **11**(1), 38–50 (2004)

A Location Problem for Medically Under-Served Areas in Korea

Hoon Jang, Kyosang Hwang, Taeho Lee, Minji Kim, Hansu Shin
and Taesik Lee

Abstract Satisfying the needs for healthcare services in medically under-served areas (MUAs) is an important task to address in public health policies in many countries. Motivated by the Korean government's MUA program, we develop a covering location model to optimally establish a fixed number of medical facilities to best serve MUAs. A unique feature of the proposed model lies in the definition of service coverage, which considers both potential and realized accessibility to healthcare services. In this paper, we first review the operational definition for MUA the Korean government currently uses, describe the proposed location model in detail, and present the results of location solutions using the actual MUA data in Korea for its obstetrics service. In particular, the effects of different MUA definitions and the policy maker's specific objectives on the location solutions are highlighted.

Keywords Location problem · Optimization · Potential accessibility · Realized accessibility · Medically under-served area

H. Jang
Georgia Institute of Technology, 755 Ferst Drive, Atlanta, GA 30332, USA
e-mail: hoon.jang@isye.gatech.edu

K. Hwang · T. Lee (✉)
KAIST, 291 Daekak-ro, Yuseong-gu, Deajeon 34141, Republic of Korea
e-mail: taesik.lee@kaist.edu

K. Hwang
e-mail: kyosanghwang@kaist.ac.kr

T. Lee · M. Kim · H. Shin
National Medical Center, 245 Euljiro, Jung-gu, Seoul, Republic of Korea
e-mail: leeth@nmc.or.kr

M. Kim
e-mail: dwamjy@nmc.or.kr

H. Shin
e-mail: hansu@nmc.or.kr

© Springer International Publishing AG 2017
P. Cappanera et al. (eds.), *Health Care Systems Engineering*, Springer Proceedings in Mathematics & Statistics 210, https://doi.org/10.1007/978-3-319-66146-9_6

1   Introduction

A medically under-served area (MUA) refers to a region where its residents do not have a sufficient level of access to health care services. Satisfying the needs for health care services in MUAs is an important task for public health policies in many countries. As such, there are many examples of MUA support programs around the world that aim to filling the void in health care service provision for the population in MUAs [1].

When the support for MUAs takes the form of establishing new care facilities (or creating new capabilities in the existing care facilities), the decision problem in MUA programs can be formulated as a facility location problem, a covering location problem in particular. While the well-researched standard location models provide a prescription for an approach to solve these problems (see for example [2]), what is required beyond tailoring a standard covering location model is that the definition of MUAs unique to the specific program in question must be considered.

In our problem, which is motivated by the Korean government's MUA support program, the unique feature comes from the definition of MUA adopted by the Korean government. Two different aspects of accessibility to health care services are considered, namely potential and realized accessibility [3]. The former measures the accessibility by reflecting the service provider's point of view. In this view, what is achieved by the health care providers is the availability of health care services, and it is assumed that the increased availability would lead to improved access to health care services. It is thus referred to as the *potential* accessibility. The latter, on the other hand, reflects the service consumer's point of view; given the provided availability, it is through the patients' choices that the actual use of health care service is realized—hence, it is referred to as the *realized* accessibility.

To incorporate the MUA definition described above, we develop a covering location model in which the definition of coverage is specifically constructed per the MUA criteria the Korean government uses. To address the MUA definition by the potential accessibility criteria, we newly introduce a constraint that determines the service coverage. To address the realized accessibility criteria, we use a simple choice model to represent the patients' hospital selection behavior, thereby predicting the actual volume of health care service use by the patients.

This paper is structured as follows. In Sect. 2.1, we first review the operational definition for MUAs currently used by the Korean government, with a particular emphasis on the difference between the potential and realized accessibility. Section 2.2 then presents the proposed location model and how MUA definitions have been translated into the model is described in detail. In Sect. 3, we illustrate the characteristics of the proposed model by presenting results of location solutions using actual MUA data in Korea for its obstetrics service. In particular, the effects of different MUA definitions and the policy maker's specific objectives for the location solutions are highlighted. Section 4 concludes the paper by highlighting the key contributions of this research.

2 Problem Description

2.1 Operational Definitions for MUA

The Korean government's MUA definition addresses two aspects of accessibility: potential and realized accessibility. Potential accessibility is related to the geographical coverage for a district provided by nearby care facilities. Specifically, a district is considered to have insufficient access to health care services if more than 30% of its residents cannot reach a care facility within 60 min. For later use, let us define the fraction of under-served population, λ^p, as follows:

$$\lambda^p = \frac{\text{Number of residents for whom no hospital is open within 60 min}}{\text{Total number of residents}}$$

Thus, if $\lambda^p \geq 30\%$, then the district is considered to be lacking sufficient access to health care services. It is worth noting that it ignores the patient's choice and only indicates the presence of a nearby care provider regardless of its desirability to the patients.

The second aspect of accessibility in the definition of MUA concerns the actual use of health care services by MUA residents. If more than 70% of the health care service use cases by its residents involve a trip to a care provider longer than 60 min, then the access to health care services in this district is considered insufficient. That is, a district has insufficient access if $\lambda^r \geq 70\%$, where λ^r denotes the fraction of hospital visits that involved an excessively-long travel:

$$\lambda^r = \frac{\text{Hospital visits of longer than 60 min by residents}}{\text{Total number of hospital visits by residents}}$$

Motivation for the government to use the potential accessibility criteria is quite clear. From the policy planning purpose, the concept of potential accessibility is intuitive and easy to measure—'people in this town simply don't have hospitals nearby'. Also, perhaps more importantly, the lack of potential accessibility is directly addressed by a policy intervention, i.e., by establishing a care provider in the right location. On the other hand, the government knows that it has to respect the patients' choice of care providers. Low realized accessibility may signal that the district may suffer from accessibility issues other than the geographical accessibility issue. For example, if a district has enough care providers to ensure sufficient potential accessibility and yet shows poor realized accessibility, it most likely indicates the need for other types of policy intervention than establishing additional capacities.

2.2 Location Model

The primary objective of the MUA support program of the Korean government is to establish a number of care facilities at locations that will maximally *cover* the current MUAs. Given the MUA criteria by the Korean government, an MUA is said to be covered in our problem if either λ^p is below 30% or λ^r is below 70%. This section introduces a location model to determine optimal hospital locations to cover MUAs. The novelty of the proposed model lies in how the MUA criteria has been incorporated into a location model framework.

We use index k to denote a district, and it is at this level that the government designates MUA. The set of MUAs is K, where $k \in K$. We let i be an index for a demand region and I be the set of all demand regions. Practically, i represents the smallest geographical unit for which necessary demand-related data (e.g., service use records, demographics, etc.) is available. The size of demand which is interpreted as the number of residents in region i is denoted by ρ_i. I_k is the set of demand regions that belong to district k. For hospitals, we use j as an index for a candidate site to establish a hospital, with J denoting the set of all such sites. Finally, we let I_j represent the set of demand regions that are within a specified distance standard, e.g., 60 min, from hospital site j; likewise, J_i denotes the set of hospital sites that are within the specified distance standard from demand region i.

Now, let us introduce six types of principal decision variables used in our model:

- y_j, binary variable that indicates whether a hospital is open at candidate site j ($y_j = 1$ if it is open and 0 otherwise);
- x_{ij}, continuous, non-negative variable that represents the fraction of patients at region i who (are expected to) choose the hospital located at j ($x_{ij} \geq 0$);
- h_i, binary variable that indicates whether there exists at least one hospital that can be reached from region i within the time standard ($h_i = 1$ if there is one or more hospitals within the time standard and 0 otherwise);
- s_k^p, binary variable that indicates whether the potential accessibility of district k is at an acceptable level ($s_k^p = 1$ if $\lambda^p < 30\%$);
- s_k^r, binary variable that indicates whether the realized accessibility of district k is at an acceptable level ($s_k^r = 1$ if $\lambda^r < 70\%$);
- s_k, binary variable that indicates the MUA status of district k ($s_k = 0$ if district k is MUA).

To express s_k^p and s_k^r in the math program, we define functions g_k^p and g_k^r as follows:

$$g_k^p = \Lambda^p - \lambda_k^p = \Lambda^p - (1 - \frac{\sum_{i \in I_k} \rho_i h_i}{\sum_{i \in I_k} \rho_i}) \tag{1}$$

$$g_k^r = \Lambda^r - \lambda_k^r = \Lambda^r - (1 - \frac{\sum_{i \in I_k, j \in J_i} \rho_i x_{ij}}{\sum_{i \in I_k} \rho_i}) \tag{2}$$

where we use Λ^p and Λ^r to denote the threshold value for the two accessibility criteria, 0.3 and 0.7, respectively. If g_k^p is positive, the fraction of the under-served population is less than the threshold value, and thus, district k has sufficient potential accessibility. Similarly, a positive g_k^r indicates that district k has sufficient realized accessibility. Note that λ_k^p is computed by using decision variable h_i; the numerator in the last term in (1) is the number of residents in district k for whom there is at least one hospital within the time standard. To compute λ_k^r, we use decision variable x_{ij}. Recall that x_{ij} represents the fraction of patients at region i who are expected to choose the hospital located at j. Then, the numerator in the last term in (2) is the fraction of patients in district k who are expected to seek care at a hospital located within the time standard.[1]

By using the above definitions and terms, we now formulate a mathematical program P_{MUA} for the location problem as follows:

$$(P_{MUA}) \qquad \max \sum_{k \in K} s_k \qquad (3)$$

$$\text{s.t.} \sum_{j \in J_i} x_{ij} \leq 1 \quad \forall i \in I \qquad (4)$$

$$x_{ij} \leq p_{ij} y_j \quad \forall i \in I, \forall j \in J_i \qquad (5)$$

$$h_i \leq \sum_{j \in J_i} y_j \quad \forall i \in I \qquad (6)$$

$$\sum_{i \in I_j} \rho_i x_{ij} \leq c_j y_j \quad \forall j \in J \qquad (7)$$

$$\sum_{j \in J} y_j \leq n \qquad (8)$$

$$g_k^p s_k^p \geq 0 \quad \forall k \in K \qquad (9)$$

$$g_k^r s_k^r \geq 0 \quad \forall k \in K \qquad (10)$$

$$s_k^p + s_k^r \geq s_k \quad \forall k \in K \qquad (11)$$

Objective function (3) maximizes the number of MUAs that are covered—i.e., changes its status to non-MUAs ($s_k = 1$)—by the hospitals to be established. Constraint (4) requires that the sum of the fraction of patients in region i to choose hospital j should be no greater than 1 when summed over all reachable hospitals, J_i. Constraint (5) limits the fraction of patients in i to choose hospital j at its pre-determined choice probability, p_{ij}. Note that p_{ij} is the probability that a patient in region i will choose a hospital at j. A detailed explanation of p_{ij} will be given in Sect. 3. This constraint also ensures that the fraction is set to zero if there is no hospital open at j. Constraint (6) specifies that demand region i can only be served if there exists at least one hospital within the time standard. Constraint (7) imposes a capacity limit, c_j, for each hospital. Constraint (8) limits the total number of hospitals to be established.

---

[1]When we let $\lambda_k^r = \sum_{i \in I_k, j \in J_i} \rho_i x_{ij} / \sum_{i \in I_k} \rho_i$, we are making an assumption that one patient equals one hospital visit.

Constraints (9)–(11) jointly specify that district k is considered covered, i.e., $s_k = 1$, if it has sufficient access in terms of either potential ($s_k^p = 1$) or realized ($s_k^r = 1$) accessibility.

Note that P_{MUA} contains nonlinear constraints (9) and (10). By introducing four auxiliary variables ($u_k^p, v_k^p, u_k^r, v_k^r$), we linearize these constraints following techniques suggested in [4]. Linearization of constraint (9) can be done as shown below:

$$u_k^p = \frac{\sum_{i \in I_k} \rho_i h_i}{\sum_{i \in I_k} \rho_i} \quad \forall k \in K \tag{12}$$

$$v_k^p \geq (1 - \Lambda_k^p)s_k^p \quad \forall k \in K \tag{13}$$

$$v_k^p \geq u_k^p + M(s_k^p - 1) \quad \forall k \in K \tag{14}$$

$$v_k^p \leq u_k^p \quad \forall k \in K \tag{15}$$

$$v_k^p \leq Ms_k^p \quad \forall k \in K \tag{16}$$

Similarly, constraint (10) is linearized in the exactly same fashion.

In next section, we apply the above model to solve an obstetric care facility location problem for Korea's obstetrics MUA support problem.

3   Case: MUA Support Program in Korea for Obstetrics Care

3.1   Case Illustration

To address the country's low fertility rate issue, the Korean government is developing various policies and programs. One of the strategies is to foster a more favorable childbirth environment by establishing obstetrics care units especially in the MUAs. Motivated by this program, we apply the proposed model described in previous Sect. 2.2 to the problem of locating obstetrics hospitals. For this study, we use actual nationwide data and examine the characteristics of the location solutions from the model.

Our case study uses data sets gathered from the prior nationwide study in which the authors participated [5].[2] For the demand side, we use the residential locations and obstetrics care records of women aged 15–49 years. By the MUA criteria of the Korean government, there are 34 MUAs in 2015, and 209,933 women in the age group live in these areas. On the supply side, there are 66 candidate sites to establish an obstetric care capacity. Most of these sites are at a secondary care hospital level (55 candidates), two are at a tertiary-care level, and nine sites are at a primary-care

---

[2]This study was approved by the Institutional Review Board of National Medical Center in Korea.

level. The maximum number of delivery cases that each hospital can handle is set at 20 cases per day.

Recall that our model addresses the realized accessibility by using variable x_{ij}, which is the fraction of patients at region i who are *expected* to choose the hospital j. In the planning stage, we do not know for sure how patients will respond to a newly established care providers, and thus x_{ij} can only be guessed with some uncertainties. Accurately predicting x_{ij} and properly incorporating it into the location model requires significant research on its own, which is beyond the scope of this paper. Here we adopt a simplifying approach by using an exogenously specified choice parameter, p_{ij}. In our study, we set p_{ij} based on the actual obstetrics service use records from the data set. In doing so, we made an assumption that travel time is the only factor for patient's choice of a hospital. From the care use records of a region i, if, for example, a hospital j is 15 min from i, we approximate p_{ij} by counting the fraction of visits by residents in i to hospitals that are between 10 to 20 min away from i. While a better model based on rigorous analyses should be developed in future work, this simple, distance-based model is still able to display the bypassing behavior of patients, which is good enough to illustrate how our model works to capture the realized accessibility.

Using these experimental settings, we conduct experiments to obtain location solutions under various decision scenarios. Specifically, we examine the following two questions:

(1) How significant is it to incorporate the realized accessibility in computing the location solution?

Our model computes allocation decision variable x_{ij} by using the extrinsically estimated parameter p_{ij}. x_{ij} represents the expected number of visits to hospital j from region i, with which the (expected) realized accessibility is measured. We conjecture that using realized accessibility in the location model would provide more accurate prediction for how many MUAs will be *actually* covered by newly established hospitals than a model that only looks at the potential accessibility. To investigate this question, we compare solutions from P_{MUA} with solutions obtained from its variant P'_{MUA}. In P'_{MUA}, we replace constraint (5) with a new constraint that enforces that patients choose a hospital closest to their residence.[3] This is a common assumption in most of the standard location models. With the modification, solutions from P'_{MUA} ignore the patient choice aspect of accessibility and represent location solutions purely from the potential accessibility aspect.

(2) How much difference exists between the two sets of solutions when our objective is (a) to minimize the number of remaining MUAs versus (b) to minimize the size of under-served *population* by the investment?

To the government, the number of MUAs relieved by the investment made in the program would be a direct measure of the program's success in terms of the investment efficiency, and as such P_{MUA} uses the measure as its objective function.

---

[3] $\sum_{k \in J_i} t_{ik} x_{ik} \leq t_{ij} + M(1 - y_j)$ $\quad \forall i \in I, j \in J_i$ (t_{ij} denotes the travel time between i and j).

However, such solutions, while minimizing the number of remaining MUAs, would not necessarily minimize the under-served population. We develop a variant from P_{MUA} where $\sum \rho_{ij}x_{ij}$ replaces the current objective function and constraints (9)–(11) are removed. The modified program P''_{MUA} has the form of a capacitated maximum covering location problem [6]. Considering that minimizing the volume of the under-served population would also be a equally plausible goal for the policy makers, we examine the difference between two models.

3.2 Results

We solve P_{MUA} to see how many hospitals need to be established to serve the 34 obstetric MUAs in Korea. Figure 1 shows that establishing the first hospital, if optimally located, can serve four MUAs providing sufficient accessibility to relieve them from their MUA status. After the first hospital, the number of MUAs covered by the established hospitals increases in proportion to the number of hospital, and 30 hospitals will relieve all MUAs. The size of Medically Under-served Population (MUP) also increases more or less proportional to the number of hospitals. Note that opening more hospitals beyond 30 continues to contribute to further reducing MUP. Opening hospitals in all 66 candidate sites will provide access to 138 thousand women, reducing MUP to 35% of the initial size.

Now we discuss each of the two questions posed in Sect. 3.1. The first question is on the significance of realized accessibility in obtaining the location solutions. Recall that, to examine this question, we obtain location solutions from the original program P_{MUA} and its modified version P'_{MUA} which makes people always choose the nearest hospital from them. The results show that for all scenarios, the objective function

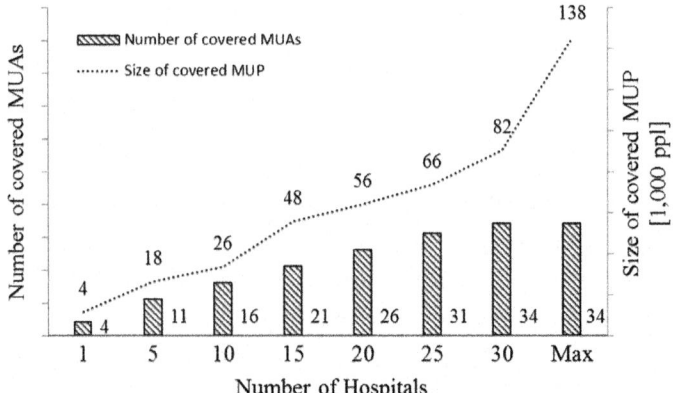

Fig. 1 Number of covered MUAs and size of MUP (Medically Under-served Population) computed from P_{MUA}

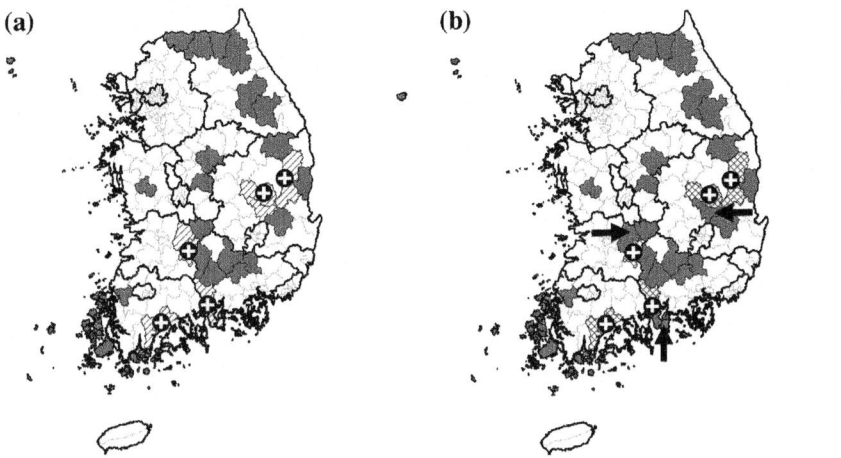

Fig. 2 Locations of the five hospitals obtained from P'_{MUA} and MUA regions (grey-colored): **a** 12 MUAs are *expected* to be relieved (hatched) **b** three (areas where the arrow points to) among the 12 MUAs turn out to remain MUAs

value, $\sum s_k$, is higher in P'_{MUA} than in P_{MUA}. With the same number of hospitals, two more MUAs, on average, are relieved under P'_{MUA} compared with P_{MUA}. Although it is not meaningful to compare the objective values of two different optimization models, there is an important implication suggested by the observation: effectiveness of investment to support MUAs can be overestimated if the expected, actual use of new care capacity is not properly accounted for. Figure 2 shows an example of such possible overestimation.

Figure 2a shows the location solution from P'_{MUA} for $n = 5$. It shows locations of the five hospitals and the respective MUAs that these hospitals are expected to relieve. As indicated in the figure as hatched areas, there are 12 MUAs that are expected to become non-MUAs due to new service capacity provided by these hospitals. To see if this expectation is realized, we take the location solution into P_{MUA} by fixing y_j, and we solve P_{MUA} to obtain s_k. Then, s_k indicates which MUAs have been relieved by these five hospitals. This is shown in Fig. 2b. In the figure, we see three districts that are expected to be relieved by the five hospitals, but it turns out that they will remain as MUAs when we take into account their choices as dictated by p_{ij}. The government invests its resources in new capacity with an *expectation* that provision of the capacity will lead to its consumption by the residents in the target MUAs. However, this expectation, when based solely on the potential accessibility aspect, is only prognostic from the view point of a service supplier. To narrow the gap between prognostic expectations and actual, realized use of service, it should be pro-actively modeled so that location solutions are derived with a model that takes into account the service consumers' reaction to new capacity.

The second question concerns the choice of objective function for P_{MUA}. Again, we compare solutions from P_{MUA} with solutions obtained from another modified

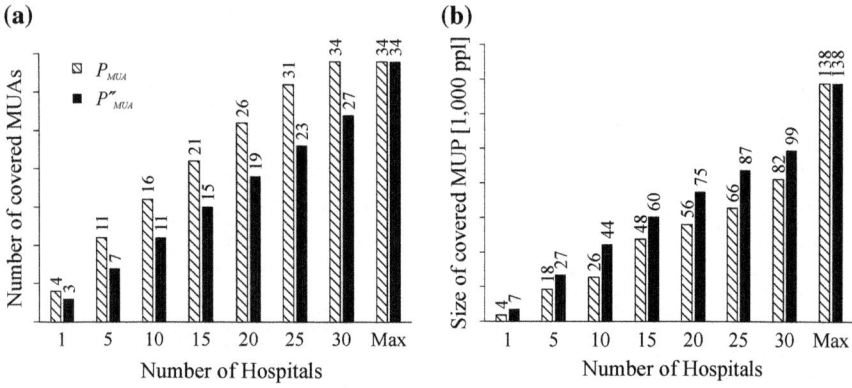

Fig. 3 Pursuing the number of MUAs (P_{MUA}) versus the size of MUP (P''_{MUA}) produces very different location solutions

version, P''_{MUA}. Potential and realized accessibility are reflected in both models, but the two models differ in the objective function. P_{MUA} maximizes the number of MUAs to be relieved, whereas P''_{MUA} maximizes the size of the under-served population (MUP). Figure 3 clearly shows the contrast between the two sets of solutions. In Fig. 3a, we see that solutions from P_{MUA} relieve more MUAs, as many as eight, (See the case of n = 25) than P''_{MUA}. On the other hand, in terms of the number of women benefited by the new capacity, solutions from P''_{MUA} dominate P_{MUA}, as shown in Fig. 3b. Although this contrast is expected, the magnitude of the difference in the size of covered MUP is simply too large to ignore. In case of $n = 25$, the difference is as large as 21,000 women. In a sense, this 21,000 women can be thought of as the cost of regional equity pursued in policy making. It is understandable that the government treats all districts equally and decides to relieve as many MUAs as possible, but the cost of equity is potentially very significant such that it certainly warrants a serious discussion.

4 Conclusion

Motivated by Korean government's MUA support program, we develop a location model to determine optimal locations to establish obstetric hospitals. Novel features of the developed model are twofold. First, the government's operational definition of MUA requires to determine the service coverage at the aggregate geographical unit (district). Individual regions are examined for their accessibility, and then the results are rolled up at the district level to compute the size of low accessibility population. This was achieved by defining auxiliary indicator functions—g_k^p and g_k^r—and using them to constrain the decision variable for accessibility at the district level, s_k^p and s_k^r. The second feature is that we take expected responses of the MUA residents into consideration within the model. This was needed to properly describe the two accessibility concepts as specified in the government's MUA definition. While potential accessibility is represented in standard location models, realized accessibility is rarely considered. We incorporate parameters p_{ij} to model the expected responses of the MUA residents—how likely they will choose to use hospital at j.

As a case study, we apply the model to Korean government's MUA support program for obstetric care. The results have been provided to the government to aid in their decision making for the program. In addition to the location solution, we found important implications from the study. Effectiveness of the program's investment may be overestimated if the actual use of new care capacity is not properly predicted and accounted for in the decision model. In the worst case, the government opens a hospital that ends up being unused by the target residents. We also found that there is a significant difference when the support program focuses on the number of MUAs versus the size of underserved population. Pursuing the number of MUAs as the policy objective is certainly justifiable since it is based on the premise of regional equity. But the cost of regional equity is potentially very significant, and thus it is worthwhile to discuss if it is possible and necessary to seek balance between the two objectives.

References

1. Steinhaeuser, J., Otto, P., Goetz, K., Szecsenyi, J., Joos, S.: Rural areas in a European country from a health care point of view: an adaptation of the rural ranking scale. BMC Health Serv. Res. **14**, 147 (2014)
2. Ahmadi-Javid, A., Seyedi, P., Syam, S.S.: A survey of healthcare facility location. Comput. Oper. Res. **79**, 223–263 (2017)
3. Khan, A.A., Bhardwaj, S.M.: Access to health care a conceptual framework and its relevance to health care planning. Eval. Health Prof. **17**(1), 60–76 (1994)
4. Peterson, C.C.: A note on transforming the product of variables to linear form in linear programs. Working paper. Purdue University (1971)
5. Lee, T.H., Kwak, M.Y., Kim, M.J., Kim, J.S., Lee, T., Hwang, K.S., Hwang, J.Y., Kim, M.S.: Monitoring Medically Under-served Areas in Korea. The Korean Ministry of Health and Welfare Report. http://www.ppm.or.kr/board/thumbnailList.do?MENUID=A04030000 (2015). Accessed 22 Jan 2016
6. Pirkul, H., Schilling, D.: The capacitated maximal covering location problem with backup service. Ann. Oper. Res. **18**, 141–154 (1989)

A Choice Model for Estimating Realized Accessibility: Case Study for Obstetrics Care in Korea

Kyosang Hwang, Hoon Jang, Taeho Lee, Minji Kim, Hansu Shin
and Taesik Lee

Abstract Improving accessibility to care in medically under-served areas(MUAs) is the goal of MUA support program in public health policy. In the planning phase of such programs, we often use geographical proximity of care facilities as a measure of accessibility, and the programs resource is used to maximize the geographical accessibility. While it is easy to assess the geographical accessibility, this is not always an accurate assessment of the actual, realized accessibility because true accessibility is realized by the actual service use by patients. The choice of a specific care provider by patients is made not just by a physical distance, but by many other factors including the size of a care provider, physician's demographic, etc. Predicting true accessibility thus requires a model that considers various factors in the patients decision making, and in this paper, we use a choice model known as the conditional logit model. We use the actual health insurance data from Korea to identify factors affecting patients choice of care providers and model the provider choice behavior of patients by using the MNL model. To validate the proposed model, we compare the actual patient volumes for care providers with the model prediction, and the results show a good agreement suggesting the MNL model is a promising approach to assess true accessibility to care.

K. Hwang · T. Lee (✉)
KAIST, 291 Daekak-ro, Yuseong-gu, Deajeon 34141, Republic of Korea
e-mail: taesik.lee@kaist.edu

K. Hwang
e-mail: kyosanghwang@kaist.ac.kr

H. Jang
Georgia Institute of Technology, 755 Ferst Drive NW, Atlanta, GA 30332, USA
e-mail: hoon.jang@isye.gatech.edu

T. Lee · M. Kim · H. Shin
National Medical Center, 245 Euljiro, Jung-gu, Seoul, Republic of Korea
e-mail: leeth@nmc.or.kr

M. Kim
e-mail: dwamjy@nmc.or.kr

H. Shin
e-mail: hansu@nmc.or.kr

© Springer International Publishing AG 2017
P. Cappanera et al. (eds.), *Health Care Systems Engineering*, Springer Proceedings
in Mathematics & Statistics 210, https://doi.org/10.1007/978-3-319-66146-9_7

Keywords Healthcare · Choice model · Conditional logit model · Under-served areas · Obstetrics care

1 Introduction

A medically under-served area (MUA) is a region where access to health care services is limited. This is often caused by the lack of health care service capacity, hospitals in most contexts. Often, public health authority installs policy interventions, for example by establishing new care capacity, to improve accessibility to health care for people residing in MUAs [1].

Accessibility has two components: potential accessibility and realized accessibility [2]. Potential accessibility is measured by the existence of a hospital within reasonable proximity, e.g. 60-min for obstetrics care. The term *potential* indicates that the mere existence of a hospital does not necessarily mean it provides sufficient access to the patients due to various factors; they may travel long distance to get care from a hospital with higher quality of service or from a hospital with lower expenses. Realized accessibility, as its name suggests, measures accessibility in terms of actual, realized use of health care services.

While potential accessibility is prognostic and easy-to-measure in the planning stage, realized accessibility can only be guessed at the time of planning. In planning an intervention strategy, policy makers need to have a good understanding as to how health service consumers would respond to newly established capacity. Without it, they may invest in capacity that ends up being unused, yielding less-than-desired outcomes.

A key to the accurate prediction is the understanding of health service consumers' choice of care providers, and this type of problem has been analyzed by using what is known as a discrete choice model [3, 4]. Discrete choice models, originally developed in economics, describe choices between two or more discrete alternatives. For example, we can construct a model to describe patients' decision regarding a choice between available hospitals to visit to receive care. Specifically, the attributes of each of the patient and the attributes of the available hospitals are statistically related to the particular choice made by the patient.

In this paper, we develop a discrete choice model to capture patients' choice behavior for health care service providers. We use nation-wide data of MUAs for obstetrics care in Korea. Using the data, we construct a conditional logit model, which is one of the popular discrete choice models in economics and other relevant studies. In the context of the MUA support program, such a model would offer an opportunity to develop a location model that takes into account the future (expected) consumer responses to the care providers to establish. If we develop an accurate choice model with manageable degree of complexity, the expected responses can be endogenized into a location model.

Remainder of this paper is structured as follows. Section 2 provides a brief overview on discrete choice models. In Sect. 3 discusses the conditional logit model and its specification, along with the data used in our study. Section 4 summarizes the results and discusses the major findings from the model. Finally, Sect. 5 concludes with a few issues to address in future research.

2 Discrete Choice Model

Originated in psychology and economics, discrete choice models describe and explain choices between two or more discrete alternatives. Specifically, discrete choice models specify the probability that an individual decision maker chooses an option among a set of alternatives. For example, we can construct a discrete choice model to estimate the probability that a patient will choose to receive a tonsillectomy or not, or the probabilities for each of five available hospitals that the patient will choose to receive the surgery from. Such prediction is made by statistically relating the choice made by a decision maker to the attributes of the decision maker and the attributes of the alternatives.

Discrete choice models assume that a decision maker chooses an alternative that returns the greatest utility. Let U_{ni} denote utility that decision maker n obtains from alternative i. Then, the probability that decision maker n chooses alternative i is

$$P_{ni} = Prob(U_{ni} > U_{nj}, \forall j \neq i) \tag{1}$$

In discrete choice models, U_{ni} is assumed to consist of two parts: $U_{ni} = V_{ni} + \varepsilon_{ni}$ where V_{ni} is observable utility and ε_{ni} is unobservable utility. V_{ni} depends on the attributes that the modeller observes—the attributes of the decision maker n and the attributes of alternative i faced by n. ε_{ni} represents the random effects of all factors that are not observable to the modeller. It is assumed to follow some probabilistic distribution, giving the stochastic nature to discrete choice models. With V_{ni} and ε_{ni}, (1) can be rewritten as follows:

$$P_{ni} = Prob(V_{ni} + \varepsilon_{ni} > V_{nj} + \varepsilon_{nj}, \forall j \neq i) \tag{2}$$
$$= Prob(\varepsilon_{nj} - \varepsilon_{ni} < V_{ni} - V_{nj}, \forall j \neq i)$$

Different choices for a stochastic distribution of ε_{ni} give rise to different choice models. For example, in the case of binary choices, assuming the extreme value distribution for ε_{ni} yields the *logit* model. If ε_{ni} is assumed to follow the standard normal distribution, then the resulting choice model is the *probit* model. A wide range of choice models have been developed, and the reader is referred to [5, 6] for a full account of the discrete choice models.

In our application, we use the conditional logit model to describe hospital choices by patients. The conditional logit model was introduced by McFadden [7], and has been extensively used to estimate choice behaviors in numerous applications. In the logit model, a decision maker faces more than two alternatives, hence multinomial choices, and its unobserved utility ε_{ni} is assumed to follow the extreme value distribution—the Gumbel distribution in particular. It is similar to the multinomial logit model, a more basic version of the multinomial choice models, but differs in its modeling scope. A key benefits of the conditional logit model is that *a choice among alternatives is treated as a function of the characteristics of the alternatives, rather than (or in addition to) the characteristics of the individual making the choice* [8]. In the conditional logit model, we have a closed-form expression for the choice probability that decision maker n chooses alternative i among the alternatives, J:

$$P_{ni} = \frac{\exp(V_{ni})}{\sum_{k \in J} \exp(V_{nk})} \tag{3}$$

Since Eq. (3) is a function of only the observable utilities, it can be easily calculated as long as we have identified V_{ni} in a functional form. Recall that the main motivation for us to develop a choice model is to use it in the context of location models. The fact that P_{ni} is easily computed under the conditional logit model enables us to conveniently incorporate the choice probability, hence the expected responses from the patients, into a location model.

3 Choice Model for Obstetrics Patients in Korea

This section presents the choice model that we construct to understand the hospital choice behavior of obstetrics patients in Korea. We first describe two modeling components in Sect. 3.1. Specifically, we define two major modeling components: a choice set J and decision maker n's utility V_{ni}. Then, we discuss the obstetrics care data we used to construct the model in Sect. 3.2.

3.1 Model Components

The choice set is a set of alternatives available to the decision maker. For our problem, the choice set consists of obstetric care providers in Korea. Korea runs its health care system under the national health insurance program, which is mandated for every resident to subscribe. Under the national health insurance program, individual health service consumers—pregnant women in our study—are given complete freedom to choose any obstetrics care provider.[1] In this sense, we may include all obstetric care providers in Korea in the choice set for each pregnant woman. However, using the entire set of obstetric care providers in Korea as the choice set for all decision maker

---

[1]In general, health service consumers in Korea has "almost" complete freedom in a sense that there exists a mechanism to induce appropriate use of health care resources. For example, the government implements price differentiation between care providers in different tiers to prevent over- and unwarranted use of high-tier care providers.

Table 1 Selected factors for the patients' hospital selection model

Group	Factors	Group	Factors
Accessibility easy access by transport (e.g., public or own, parking)	Travel time	Staff	Medical qualification; Specialization/interest; No. of staff per patient
Availability (e.g., open hr)	Language incentivizing by insurers	Organization of care	Convenience (e.g., open hours)
Type/Size of the institution	Provider ownership (e.g., public, for-profit, private non-profit) quality of facilities provider size (e.g., no. of beds)	Physician's demographic; Cost	Patient experience; Gender; Age; Out-of-pocket expenses

does not seem very sensible. First, we conjecture that many of the cases of deliveries at a hospital far away from their residence are possibly due to incomplete data, for example mismatch between actual residence at the time of delivery and the address on the national residence registry. Second, the actual data shows that majority of women have chosen a hospital not too far away from them. It turns out that 91.6% of the total delivery cases are handled by a hospital located within 120 minutes of travel distance from the mother's residence. Thus, we define the choice set for pregnant woman n as a set of hospital located within 120-min of n's residence.

Next, to define decision maker n's utility, V_{ni}, we need to identify the variables that characterize the attributes of alternative i faced by decision maker n. Simply put, we need to identify factors affecting the decision maker's choice of a hospital. In this paper, we follow the framework proposed by Victor et al. [9]. It identifies various factors that are found to affect patients' hospital selection. These factors are classified into seven groups. See Table 1.

We examine each of the factors shown in Table 1 for their relevance to our problem and availability of necessary data. Factors in the organization of care, cost groups, and physician's demographic are excluded due to data unavailability. Language and insurance factors in the availability group are excluded as they are irrelevant to Korean heath care environment. We adopt travel time factor in the Accessibility group, while the easy-access-by-transport factors are excluded due to the data issue. For the institution factors, we use two alternative factors that are available for our study: level of a hospital and the degree of urbanization of the town a hospital is located. For the Staff category, we adopt the number of obstetrics specialists as a representative alternative for the factors identified above. We assume, albeit without empirical evidence, these three factors are reasonable surrogates to capture the attributes that the institution and staff factors intend to represent in [9].

Let us use t_{ni}, Lv_i, Urb_i, and Num_i denote the travel time, hospital level, urbanization of the hospital location, and the number of obstetrics specialists, respectively. Their operational definition used in our model is as follows.

- Level of hospital, Lv_i

The level of a hospital is an important factor when patients choose the hospital to receive care. Like the health care systems in many other countries, medical institutions in Korea are hierarchically structured—primary, secondary, and tertiary.[2] For the purpose of our discussion here, we can further simplify it into a two-tier system as there is not much difference between the primary and secondary hospitals when it comes to obstetric care provision. Then, Lv_i enters the model as a dummy variable:

$LvH_i = 1$ if i is a tertiary hospital, and 0 otherwise;
$LvL_i = 1$ if i is an either primary or secondary hospital, and 0 otherwise.

- Urbanization of the hospital location, Urb_i

It is expected that whether a hospital is located in a metropolitan area or a rural county influences patient's perception, hence their choice, of the hospital. In Korea, for many types of health care services, it is believed that patients strongly prefer hospitals located in metropolitan areas. Due to its categorical nature, this variable enters the model as a dummy variable as well:

$UrbMetro_i = 1$ if i is in a metropolitan region, and 0 otherwise;
$UrbCity_i = 1$ if i is in a city region, and 0 otherwise;
$UrbRural_i = 1$ if i is in a rural region, and 0 otherwise.

- Number of obstetrics specialists, Num_i

The number of obstetrics specialists is another variable relevant to the size and quality of a hospital. It is also correlated to the number of female obstetrics physicians, which is another presumably important factor when pregnant women choose a hospital for their delivery. According to our survey through obstetrics care providers' web page, the correlation coefficient between the number of physicians and the number of female physicians is 0.82.

- Travel time, t_{ni}

Presumably, travel time is important consideration when choosing a hospital as it is a primary determinant of physical accessibility. Note that, unlike the other variables, t_{ni} depends on a patient-hospital pair, indicated by its double-index ni. We use the hospital addresses and residential addresses to compute travel times for all $\{ni\}$ pairs, assuming automotive transportation.

---

[2]The hierarchical structure in Korea's healthcare system is a little more complicated than that, but for the purpose of our discussion it suffices to use a three-tier (primary, secondary, and tertiary) classification.

With these four variables, our specification for V_{ni} is as follows:

$$V_{ni} = \beta_{Lv} * Lv_i + \beta_{Urb} * Urb_i + \beta_{Num} * Num_i + \beta_t * t_{ni} \tag{4}$$

Coefficients in Eq. (4) are estimated by the maximum likelihood estimation method, and we use the MDC Procedure in SAS.

3.2 Data

We use the data for actual uses of obstetrics care providers by pregnant women for their deliveries in Korea in year 2015. Given the model specification (4), we need data, for each delivery case, on the mother's residential address and identification of the hospital she gave birth in. Then, for these hospitals, we need information on their designation level, address, and the number of obstetrics specialists. These sets of data have been obtained from three sources: National Health Insurance Service (NHIS), Health Insurance Review & Assessment Service (HIRA), and National Transportation DB center. Using these data sources, we define the choice set (Fig. 1).

Birth data from NHIS consists of mothers' residential addresses and the id code of the hospitals they used. After excluding invalid entries, 291,126 delivery cases have been obtained from the database. It should be noted that due to the NHIS privacy requirement, we were not given an individual mother's residential address; individual birth records have been compiled at an aggregate geographical unit before being made available for our analysis.

Hospital information is obtained from HIRA. There are 576 medical institutions that has at least one case of delivery during year 2015. Some of these institutions have less than 50 cases of deliveries in one year. Also there are some institutions for which HIRA data shows no obstetricians. Our discussion with the government officials and public health experts suggests that these institutions do not provide delivery services under nominal circumstances or as part of their regular services, and thus we exclude these institutions from our analysis. This leaves us 480 hospitals across the country. The HIRA database contains information on each hospital for their designation, address, number of beds and labor beds, number of obstetrics specialists and physicians, number of nurses and other equipment.

The travel time between each geographical unit and hospitals is obtained by using the network analysis tool offered in ArcGIS 10.0. For each unit, we use a population-weighted centroid as its center point from which we measure the travel distance to each of the hospitals. Traffic analysis network data from the National Transportation DB Center is used to provide information on the road network and average vehicle speed, etc.

Motivated by the MUA support program for obstetrics care in Korea, our objective of constructing the choice model is to understand the choice behavior of prospective mothers in MUAs. We conjectured that the choice behavior of prospective mothers will be largely influenced by where they live. In particular, people who live in and

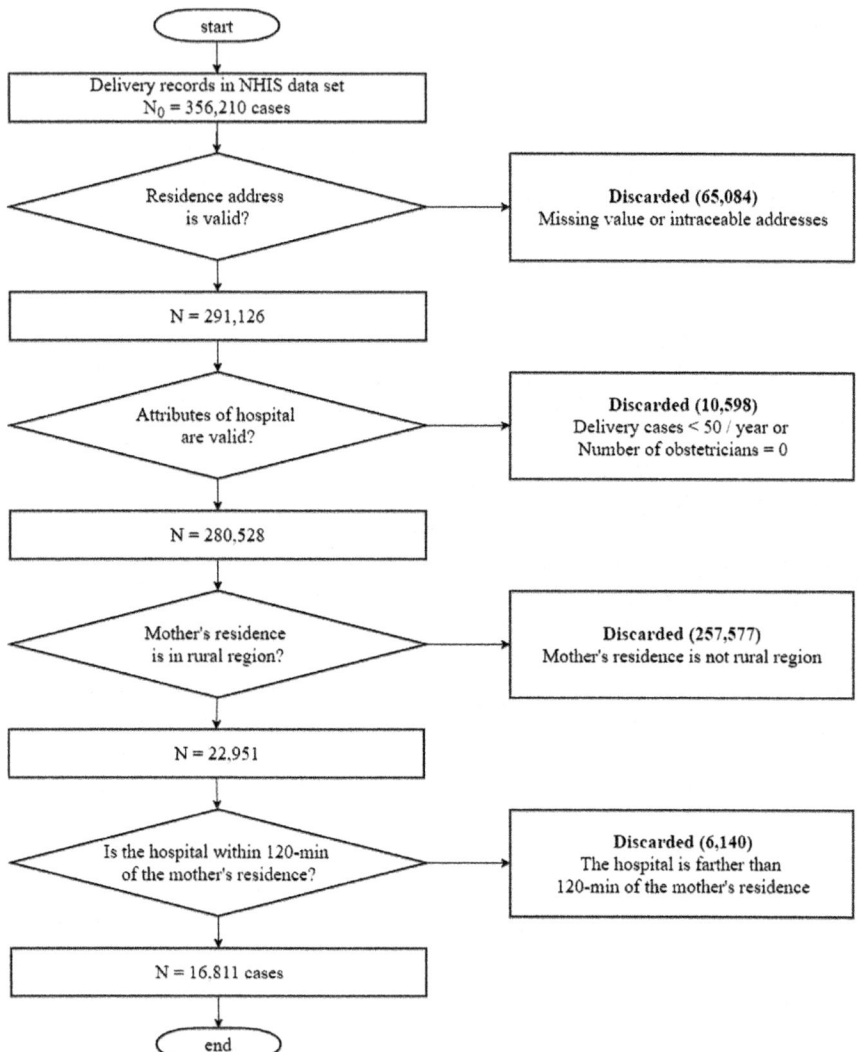

Fig. 1 Flow chart to define the final data set

around MUAs are likely to choose a hospital on the different rationale than those who have access to abundant alternatives. Thus from the entire records of delivery cases in Korea, we include delivery cases from the rural regions. Note that 33 out of 34 MUAs are rural region and that 33 out of 82 rural regions in Korea are MUAs. The final data used in our analysis is, for each of 82 rural regions, a list of hospitals and the number of cases each hospital served. The descriptive statistics of hospitals are shown in Table 2.

Table 2 Descriptive statistics of alternatives

Categorical factor		N		Ratio (%)
Hospital level		468		100.0
LvL		370		79.1
LvH		98		20.9
Hospital location		468		100.0
UrbMetro		311		66.5
UrbCity		147		31.4
UrbRural		10		2.1
Numerical factor	Mean	Std.	Min	Max
Number of obstetricians				
Num	5.42	4.61	1	38

4 Result

Observable utility V_{ni} of hospital i faced by pregnant women in region n is

$$V_{ni} = -1.0722 * LvH_i - 0.1738 * UrbCity_i - 0.8619 * UrbRural_i \qquad (5)$$
$$+ 0.1511 * Num_i - 0.0637 * t_{ni}$$

The detailed results for the conditional logit model is shown in Table 3. All coefficients are statistically significant with their p-value very small. The last column of

Table 3 Results of the patients' hospital choice model

Factor	DoF	Estimate	Relative risk	Standard error	t-value	p-value	VIF
Hospital level LvL (reference) LvH	1	−1.0722	0.34	0.0287	−37.32	≤ 0.0001	1.06
Hospital location UrbMetro (reference) UrbCity	1	−0.1738	0.84	0.0305	−5.7	≤ 0.0001	1.11
UrbRural	1	−0.8619	0.42	0.0499	−17.28	≤ 0.0001	1.06
Number of obstetricians Num	1	0.1511	1.16	0.0023	65.22	≤ 0.0001	1.17
Travel time (min.) t	1	−0.0637	0.94	0.0001	−116.92	≤ 0.0001	1.02

McFadden's pseudo R-squared = 0.2748

Table 3 shows the VIF values for the factor variables, which suggests they are not correlated with each other.[3] Note that the model's McFadden's pseudo R squared is 0.2748, and thus the goodness of fit of the model is deemed appropriate [10]. In addition to the goodness of fit, we examine the validity of the model for its prediction. We validated our model by 5-fold cross-validation [12]. Since the purpose of the model is to predict the number of visits for each hospitals, we examine R^2 between the actual and the predicted number of visits in the test data set. R^2 value for the entire dataset is 0.7326. R^2 is decreased to an average of 0.7195 (0.7389, 0.7062, 0.7006, 0.7319, 0.7199) in the 5-fold cross-validation, which is still acceptable.

The coefficient of each variable is interpreted via its relative risk as follows:

$$\frac{P_{ni|LvH_i=1}}{P_{nj|LvH_j=0}} = \frac{\frac{e^{V_{ni}}}{\sum_k e^{V_{nk}}}}{\frac{e^{V_{nj}}}{\sum_k e^{V_{nk}}}} = \frac{e^{\beta_{Lv} \star (LvH_i=1)}}{e^{\beta_{Lv} \star (LvH_j=0)}} = e^{\beta_{Lv}} = e^{-1.0722} = 0.34 \quad (6)$$

- Probability of choosing a high-level (tertiary) hospital is, with everything else being equal, 0.34 times the probability of choosing a low-level hospital (primary and secondary);
- Probability of choosing a hospital in a city region is 0.84 times the probability of choosing a hospital in a metropolitan region;
- Probability of choosing a hospital in a rural region is 0.42 times than the probability of choosing a hospital in a metropolitan region;
- Probability of choosing a hospital in a city region is 2.0 times the probability of choosing a hospital in a rural region;
- Probability of choosing a hospital that has one more obstetricians over the probability of choosing the hospital with one less obstetrician is 1.16;
- Probability of choosing a hospital that takes one additional minute of travel is 0.94 times the probability of choosing the hospital that takes one less minute to reach.

In the above results, there are a few notable findings from the results. First, pregnant women living in rural regions in Korea would choose a hospital of the lower level (primary and secondary) over a tertiary hospital. This is probably due to the fact that, under current obstetrics practices in Korea, those lower level hospitals provide care with reasonable quality. Also, getting care at a tertiary hospital generally accompanies significant overhead and additional cost to patients, which can only be justified for high risk delivery cases. Second, the results show that hospitals located in a large, urban environment are favored over the ones in a more local and rural setting. This will be a concern in the MUA support program for its possible implication; newly established obstetrics hospitals in MUAs may not effectively attract and serve the target population. Third, the number of obstetricians influences the choice of obstetric hospital. This can also be a concern since it will be practically

---

[3]The multicollinearity between variables is judged by Variance Inflation Factor(VIF); if any of the VIF values exceeds 5 (or 10), it implies that the associated regression coefficients are poorly estimated because of multicollinearity. VIF of all variables in Table 3 is less than 1.2, so there is no multicollinearity problem in our model.

Table 4 Confusion matrix for the prediction from the choice model

		Model prediction	
		MUA	non-MUA
Actual	MUA	28	28
	non-MUA	2	194

very difficult to operate a hospital in MUAs—rural environment in general—with many obstetrics specialists.

With the coefficients estimated from the data as shown in Eq. (5), now we can evaluate Eq. (3) to determine the choice probability and assess the fidelity of the derived choice model. We compare the model's prediction on the choice of hospitals with the actual choices in the data. Specifically, from the actual data we examine the number of delivery cases in each region that were served by a hospital farther than 60-min travel distance. From the model, we obtain P_{ni} as given in Eq. (3) and multiply the number of women between age 15–49 in regional unit n to compute the counts of corresponding cases. Note that 60-min is the travel time standard used by the Korean government to measure accessibility to obstetrics care. When the fraction of delivery cases served beyond the 60-min travel distance is higher than 70%, then the region is considered as an MUA from the realized accessibility criteria [11].[4] Thus, we obtain the model's prediction on the MUA status for each region and compare the prediction with the actual MUA status for the region. This is a relevant test for us as the choice model will be used in answering the question of whether a new obstetrics capacity will relieve the MUA status of the region it is intended to serve.

Table 4 shows that the overall accuracy of the prediction is 88%. Its sensitivity is rather low at 50%, while specificity is very high at 99%; thus the MUA prediction based on our model is highly specific but not sensitive. Our model rarely mistake a region as an MUA when it actually is not (few false positives). On the other hand, there is a good chance of overlooking many regions that are actually MUAs (many false negatives). In the context of MUA support program, this can be overly conservative, and its sensitivity should be enhanced.

While the results shown in Table 4 certainly suggest the need to improve the underlying choice model, we would like to emphasize it is still much better than the current practice. In the current practice of the MUA support program, location decisions for new hospitals are made based on the potential accessibility. That is, whether a region will be relieved from its MUA status is predicted purely by geographic proximity criteria. To examine the accuracy of the prediction based on such simple model—i.e., using the geographic proximity as a rule for allocating patients to hospitals, we conduct the same test and evaluate the prediction outcome. The results

---

[4]The other criteria is the potential accessibility, which concerns geographical accessibility due to the existence of care provider within the time standard. If more than 30% of a region's patients do not have a hospital within 60-min, then the region is MUA from the potential accessibility criteria [11].

Table 5 Confusion matrix for the prediction using the proximity-rule

		Model prediction	
		MUA	non-MUA
Actual	MUA	18	38
	non-MUA	0	196

are shown in Table 5. It turns out that the overall accuracy drops to 85%, primarily due to its poor sensitivity at 32%.

5 Conclusion

In this paper, we construct a choice model to describe and predict the hospital choice decision for obstetrics care in Korea. Specifically, we use the actual records of delivery cases during year 2015, and the data is fitted by a conditional logit model. The resulting model confirms a prior notion that the distance to the hospital is an important consideration, that the size of the hospital matters as well (measured by the number of obstetrics specialists), and that the hospitals in a more urban region are preferred. On the other hand, being a tertiary hospital does not translate into an attractive attribute as the data shows lower-level hospitals are in fact preferred.

These findings cast some implications for designing and implementing the MUA support program in Korea. In the MUA support program, new obstetrics hospitals are established within or near MUA regions with an expectation that this hospitals will absorb the demand from the region, hence relieving its MUA status. But then in most cases MUAs are rural regions, and practically it is difficult to expect more than one obstetrics specialist in those hospitals. These newly established hospitals may not be able to attract as many pregnant women for their deliveries as expected in the planning phase. Our model allows to predict the obstetrics care consumers' response to new hospitals in a way that the prediction can be incorporated into a location decision model.

There are a few aspects in which the choice model developed in this study can be improved. First, even though we adopt a conditional logit model, we use primarily the hospital attributes. We need to refine and expand the current model by including more attributes for the decision makers. Second, other than the distance variable, the attributes in V_{ni} are assumed to be independent of individual decision makers. Certainly there are possibility that other multinomial choice models provide a better description for the choice behavior, and further exploration into alternative models is warranted.

References

1. MacKinney, A.: Access to Rural Health Care—A Literature Review and New Synthesis. RUPRI, Iowa City, IA (2014)
2. Khan, A.A., Bhardwaj, S.M.: Access to health care a conceptual framework and its relevance to health care planning. Eval. Health Prof. **17**(1), 60–76 (1994)
3. Brekke, K.R., Gravelle, H., Siciliani, L., Straume, O.R.: Patient choice, mobility and competition among health care providers. In: Levaggi, R., Montefiori, M. (eds.) Health Care Provision and Patient Mobility, pp. 1–26. Springer, Milan (2014)
4. Pellegrini, P.A., Fotheringham, A.S.: Modelling spatial choice: a review and synthesis in a migration context. Prog. Hum. Geogr. **26**(4), 487–510 (2002)
5. Garrow, L.A.: Discrete Choice Modelling and Air Travel Demand: Theory and Applications. Ashgate Publishing, Aldershot, UK (2010)
6. Train, K.: Discrete Choice Methods with Simulation. Cambridge University Press, Cambridge (2003)
7. McFadden, D.: Conditional logit analysis of qualitative choice behavior. In: Zarembka, P. (ed.) Frontiers in Econometrics, pp. 105–142. Academic Press, New York (1973)
8. Hoffman, S.D., Duncan, G.J.: Multinomial and conditional logit discrete-choice models in demography. Demography **25**(3), 415–427 (1988)
9. Victoor, A., Delnoij, D.M., Friele, R.D., Rademakers, J.J.: Determinants of patient choice of healthcare providers: a scoping review. BMC Health Serv. Res. **12**(1), 272–288 (2012)
10. McFadden, D.: Quantitative methods for analyzing travel behaviour on individuals: some recent developments. In: Hensher, D.A., Stopher, P.R. (eds.) Bahvioural Travel Modelling, pp. 278–318. Croom Helm, London (1978)
11. Lee, T.: A Study on Monitoring of Underserved Area for Medical Care. National Medical Center, Seoul (2015)
12. Kohavi, R.: A study of cross-validation and bootstrap for accuracy estimation and model selection. Ijcai **14**(2), 1137–1145 (1995)

Handling Time-Related Demands in the Home Care Nurse-to-Patient Assignment Problem with the Implementor-Adversarial Approach

Giuliana Carello, Ettore Lanzarone, Daniele Laricini and Mara Servilio

Abstract The nurse-to-patient assignment is one of the main decisions in planning Home Care (HC) services under continuity of care. In the literature, this problem has been tackled with several approaches to take demand variability into account. However, patient's demands at different time periods have been always assumed as independent, while they are highly correlated in practice. In this work, we propose a robust assignment model that includes the time-dependency of the demands in the HC nurse-to-patient assignment problem, based on the implementor-adversarial framework. Results from a relevant test case show the appropriateness of the approach and the capability to contain costs while respecting the continuity of care constraints.

Keywords Home care · Nurse-to-patient assignments · Continuity of care Time related demands · Implementor-adversarial approach

1 Introduction

Parameter uncertainty is common to several health care-related optimization problems, where patients evolve along with time and their future demands are affected by high variability. This is particularly critical in some health care services, e.g. in the Home Care (HC) service, where patients are usually assisted for a long time.

G. Carello · D. Laricini
Politecnico di Milano, DEIB, Milan, Italy
e-mail: giuliana.carello@polimi.it

D. Laricini
e-mail: daniele.laricini@mail.polimi.it

E. Lanzarone (✉)
CNR–IMATI, Milan, Italy
e-mail: ettore.lanzarone@cnr.it

M. Servilio
CNR–IASI, Rome, Italy
e-mail: mara.servilio@cnr.it

© Springer International Publishing AG 2017
P. Cappanera et al. (eds.), *Health Care Systems Engineering*, Springer Proceedings in Mathematics & Statistics 210, https://doi.org/10.1007/978-3-319-66146-9_8

A crucial task in HC is to assign nurses to patients over a planning horizon while taking into account such variability and meeting all requirements from both operators and patients. One of the main requirements, which is crucial for a good quality of service, is the *continuity of care*, i.e. the assignment of a patient to the same *reference nurse* during the entire care period. Thus, HC provider managers must assign nurses to patients in order to satisfy the continuity of care, given the available operators and the set of patients in the charge with their uncertain demands.

The HC nurse-to-patient assignment problem is solved over a planning horizon usually divided into time slots. In the literature, this problem has been already tackled under uncertain demands either applying the stochastic programming or the robust optimization. However, to the best of our knowledge, the proposed solutions do not take into account the correlation of each patient's demands over the time periods. On the contrary, in this work, we solve the problem assuming that the demands are correlated over the periods.

Starting from a deterministic formulation in which three different continuity of care requirements are considered at the same time, we propose a robust model for the time-correlated demands, based on the so-called *implementor-adversarial* approach introduced by Bienstock for portfolio selection [5]. Briefly, the robust model is viewed as a two-stage game: the implementor computes the nurse-to-patient assignments before the realization of the uncertainty, while the adversary generates the worst demand evolution for that assignment.

The paper is structured as follows. A literature analysis of the HC nurse-to-patient assignment problem is presented in Sect. 1.1. The deterministic assignment model considered in this paper is described in Sect. 2. Then, our robust optimization approach is detailed in Sect. 3 including the formalization of the uncertainty set. The computational tests and the results are presented in Sect. 4. A final discussion is drawn in Sect. 5.

1.1 Related Works

Different features can be considered while solving the HC nurse-to-patient assignment problem. If continuity of care is not required, the assignment problem turns out to be an assignment of operators to visits rather than patients, and the different time periods can be independently considered. In this case, the aim is usually to jointly optimize the assignment of operators to visits and the scheduling and routing problems [12, 13]. On the contrary, when continuity is considered, the decisions related to a time period affect the following ones.

Uncertainty inherently arises in HC due to changes in patients' conditions and needs, and several models have been developed to predict the uncertain demands in order to support the management of HC services [2, 3, 8].

The nurse-to-patient assignment problem including both continuity of care and uncertain demands has been addressed in the literature with stochastic programming, robust approaches, and also heuristic policies.

The problem has been tackled with stochastic programming based on scenario generation in [9] where, due to the high number of demand scenarios, a limited number of them has been considered for a computationally feasible solution; as a consequence, a high expected value of perfect information and a low value of the stochastic solution were found. Analytical policies have been developed in [10, 11], based on strict assumptions regarding the workload probability density functions, the number of new patients (one at a time), and the number of periods in the planning horizon (equal to only one).

To overcome the drawbacks of the stochastic programming and the analytical policies, a robust model based on the cardinality-constrained approach [4] has been proposed in [6]. This approach avoids the scenario generation and at the same time does not require the strict assumptions of the policies. More in general, the cardinality-constrained approach has been recognized to be a useful tool for health care problems [1]. However, the cardinality-constrained approach assumes that the uncertain parameters in the different constraints are independent of each other. Thus, it is not suitable for dealing with time-related demands in the HC nurse-to-patient assignment problem.

The correlation can be considered with the implementor-adversarial approach [5]. However, such an approach has been never applied to HC and only once in health care, to the hospital master scheduling problem [7].

2 Nurse-to-Patient Assignment Problem

We consider a set of patients $\mathscr{P} = \{1, \ldots, P\}$ who require care in a discrete planning horizon $\mathscr{T} = \{1, \ldots, T\}$, and we denote by r_{jt} the demand required from patient $j \in \mathscr{P}$ at period $t \in \mathscr{T}$. Set $\mathscr{P}$ is partitioned in three different subsets:

- $\mathscr{P}_{hc}$: patients requiring *hard continuity* of care, who must be assigned to a single reference nurse during the entire planning horizon.
- $\mathscr{P}_{pc}$: patients requiring *partial continuity* of care, who must be assigned to a single nurse in each period t but can be assigned to different nurses in different periods. Each time a patient $j \in \mathscr{P}_{pc}$ is reassigned to a different nurse from period to period, a penalty β affects the total assignment cost.
- $\mathscr{P}_{nc}$: patients having *no-continuity* of care, who can be assigned to different operators even in the same period.

Finally, $\mathscr{I} = \{1, \ldots, I\}$ is the set of all available nurses, and v_i the working hours associated with nurse i in each period of the planning horizon. When nurse i works over v_i, an overtime cost c_i has to be paid for each extra hour.

The goal is to determine the minimum assignment cost (function of nurses' overtimes and reassignment penalties) while assigning all patients in the charge to one or more available nurses, according to their continuity of care requirements. The following decision variables are used:

- x_{ij}: equal to 1 if $j \in \mathscr{P}_{hc}$ is assigned to $i \in \mathscr{I}$, 0 otherwise;
- ξ_{ij}^t: equal to 1 if $j \in \mathscr{P}_{pc}$ is assigned to $i \in \mathscr{I}$ at period t, 0 otherwise;
- y_j^t: equal to 1 if $j \in \mathscr{P}_{pc}$ is reassigned from period $t-1$ to t, 0 otherwise;
- $\chi_{ij}^t \in [0, 1]$: fraction of r_{jt} for $j \in \mathscr{P}_{nc}$ assigned to $i \in \mathscr{I}$ at period t;
- $w_i^t \geq 0$: extra time of $i \in \mathscr{I}$ at period t;
- $u_i^t \geq 0$: idle time of $i \in \mathscr{I}$ at period t;
- η_i^t: equal to 1 if nurse i has positive overtime at period t, 0 otherwise.

The deterministic model (based on [6]) is:

$$\min \left\{ \sum_{i \in \mathscr{I}} c_i \sum_{t \in \mathscr{T}} w_i^t + \beta \sum_{t \in \mathscr{T} \setminus \{1\}} \sum_{j \in \mathscr{P}_{pc}} y_j^t \right\} \tag{1}$$

s.t.

$$\sum_{i \in \mathscr{I}} x_{ij} = 1 \qquad\qquad\qquad j \in \mathscr{P}_{hc} \tag{2}$$

$$\sum_{i \in \mathscr{I}} \xi_{ij}^t = 1 \qquad\qquad\qquad j \in \mathscr{P}_{pc}, t \in \mathscr{T} \tag{3}$$

$$\sum_{i \in \mathscr{I}} \chi_{ij}^t = 1 \qquad\qquad\qquad j \in \mathscr{P}_{nc}, t \in \mathscr{T} \tag{4}$$

$$\xi_{ij}^t - \xi_{ij}^{t-1} \leq y_j^t \qquad\qquad j \in \mathscr{P}_{pc}, i \in \mathscr{I}, 2 \leq t \leq T \tag{5}$$

$$\sum_{j \in \mathscr{P}_{hc}} r_{jt} x_{ij} + \sum_{j \in \mathscr{P}_{pc}} r_{jt} \xi_{ij}^t + \sum_{j \in \mathscr{P}_{nc}} r_{jt} \chi_{ij}^t - w_i^t + u_i^t = v_i \quad i \in \mathscr{I}, t \in \mathscr{T} \tag{6}$$

$$w_i^t \leq v_i \eta_i^t \qquad\qquad\qquad i \in \mathscr{I}, t \in \mathscr{T} \tag{7}$$

$$u_i^t \leq v_i \left(1 - \eta_i^t\right) \qquad\qquad i \in \mathscr{I}, t \in \mathscr{T} \tag{8}$$

Constraints (2)–(4) force the assignments of patients according to their continuity requirements. Constraints (5) keep track of the reassignments for patients in $\mathscr{P}_{pc}$. Constraints (6) compute nurses' overtimes and idle times based on the assigned patients and the capacities v_i. Constraints (7) establish the maximum overtime at each time period; without losing generality this maximum value is set equal to the capacity v_i. Finally, constraints (8) prevent w_i^t and u_i^t from being both positive in the optimal solution. Objective (1) minimizes the total assignment cost due to overtime and reassignments of patients in $\mathscr{P}_{pc}$. A reassignment is accepted if the overtime cost would increase too much avoiding the reassignment; β is the maximum overtime cost we are willing to accept for preserving the continuity of care for one patient and for one visit.

3 The Robust Optimization Approach

According to the implementor-adversarial approach, the model is divided in two stages, which are iteratively solved. In the first stage, the service manager (i.e. the *implementor*) assigns the nurses to the patients according to the current demand and respecting the requirements; in the second stage, the *adversary* chooses the worst demand pattern for the current assignments selected by the implementor.

We assume that each patient $j \in \mathscr{P}$ is associated, at each time period t, with a probability distribution describing his/her demand. Such distributions are divided into H equiprobable bands, and a value r_{jt}^h is associated to each band (e.g. the upper level of the interval). $\mathscr{H} = \{1, \ldots, H\}$ is the set of these bands. In this way, each patient j is characterized by a set of values r_{jt}^h ($h \in \mathscr{H}$) for each period t. At the beginning of the planning process, we consider that each demand belongs to band h^*, thus defining the nominal demands $\bar{r}_{jt} = r_{jt}^{h^*}$.

The evolution of the demand for each patient is then modeled considering that the actual demand may move from the nominal band h^* to another band h in any time period t. For this purpose, we define $\delta_{jt}^h = r_{jt}^h - \bar{r}_{jt}$ as the deviation affecting the demand when it moves towards band h. By constraining the deviations, we can include the two following aspects in the robust model:

1. *cardinality*: deviations occur for a limited number of patients;
2. *correlation*: deviations at consecutive time periods are not independent.

The second one, in particular, is the innovative contribution of this work to the HC nurse-to-patient assignment problem.

We remark that, in our framework, patients' demands may evolve for two reasons. First, demands at two different periods can be different for a given h ($r_{jt_1}^h \neq r_{jt_2}^h$, $t_1 \neq t_2$) because the probability distribution of the class which the patient belongs to is moving over the periods. Second, the band can h may vary, meaning that the patient is evolving in a different way than the global behavior of his/her class.

3.1 Uncertainty Set

We formalize the uncertainty set U as $\left\{ \tilde{r}_j^t = \bar{r}_{jt} + \delta_{jt}^h, j \in \mathscr{P}, t \in \mathscr{T}, h \in \mathscr{H} \right\}$, where $\bar{r}_{jt}$ represents the *true* demand of patient j at period t. Further restrictions are taken into account to model the conditions *cardinality* and *correlation* mentioned above. To this end, we define the following variables:

- p_{jt}^h: equal to 1 if demand $\tilde{r}_{jt}$ belongs to band h, 0 otherwise;
- z_{jt}^{hd}: equal to 1 if demand $\tilde{r}_{jt}$ moves from band h towards band $h + d$ from period $t - 1$ to period t, and 0 otherwise.

The *cardinality* is modeled by adding to U the constraints:

$$\sum_{j\in\mathcal{P}}\sum_{h\in\mathcal{H}} z_{jt}^{hd} \leq \alpha_d \quad t\in\mathcal{T}\setminus 1, 0\leq d\leq H-h \tag{9}$$

where α_d is the maximum number of patients whose demand is allowed to move from the actual band to another one at distance d. Observe that, as robust approaches search for the worst realization, only forward jumps are of interest.

To model the *correlation*, we define θ as the maximum distance between two demand bands at two consecutive time periods. Indeed, given $\delta_{j,t-1}^h$ and δ_{jt}^k for patient j at periods $t-1$ and t, then $|k-h|\leq\theta$. This requirement is expressed through the following constraints:

$$\sum_{k\in\{h+\theta+1,\ldots,H\}} p_{jt}^k + p_{j,t-1}^h \leq 1 \quad j\in\mathcal{P}, t\in\mathcal{T}\setminus\{1\}, h\in\mathcal{H} \tag{10}$$

$$\sum_{h\in\mathcal{H}} p_{jt}^h = 1 \quad j\in\mathcal{P}, t\in\mathcal{T} \tag{11}$$

$$p_{jt}^{h+d} + p_{j,t-1}^h \leq 1 + z_{jt}^{hd} \quad j\in\mathcal{P}, t\in\mathcal{T}\setminus\{1\}, h\in\mathcal{H}, 0\leq d\leq\theta \tag{12}$$

$$z_{jt}^{hd} \leq p_{j,t-1}^h \quad j\in\mathcal{P}, t\in\mathcal{T}\setminus\{1\}, h\in\mathcal{H}, 0\leq d\leq\theta \tag{13}$$

$$z_{jt}^{hd} \leq p_{jt}^{h+d} \quad j\in\mathcal{P}, t\in\mathcal{T}, h\in\mathcal{H}, 0\leq d\leq\theta \tag{14}$$

Observe that, with the definition of θ, constraint (9) can be rewritten as:

$$\sum_{j\in\mathcal{P}}\sum_{h\in\mathcal{H}} z_{jt}^{hd} \leq \alpha_d \quad t\in\mathcal{T}\setminus 1, 0\leq d\leq\min\{\theta, H-h\} \tag{15}$$

Summing up, the uncertainty set U reads:

$$U = \{\tilde{r}_j^t = \bar{r}_{jt} + \delta_{jt}^h, \quad (10)-(15),$$
$$p_{jt}^h \in\{0,1\}, \ z_{jt}^{hd}\in\{0,1\}, \ j\in\mathcal{P}, \ t\in\mathcal{T}, \ h\in\mathcal{H}\} \tag{16}$$

3.2 Robust Model

Assuming that demands range in U, the robust assignment problem is:

$$\min_{\{x_{ij}, \xi_{ij}^t, \chi_{ij}^t, y_j^t\}\in X} \left\{ W(U) + \beta \sum_{t\in\mathcal{T}\setminus\{1\}}\sum_{j\in\mathcal{P}_{pc}} y_j^t \right\} \tag{17}$$

where X defines the feasible assignments induced by constraints (2)–(6) plus the integrality clauses, and $W(U)$ is the maximum (worst-case) overtime cost over U for a given assignment solution $\left\{x_{ij}, \xi_{ij}^t, \chi_{ij}^t, y_j^t\right\}$, which is determined as follows:

$$\max \sum_{i \in \mathscr{I}} c_i \sum_{t \in \mathscr{T}} w_i^t \tag{18}$$

s.t.

$$\sum_{j \in \mathscr{P}_{hc}} \tilde{r}_{jt} x_{ij} + \sum_{j \in \mathscr{P}_{pc}} \tilde{r}_{jt} \xi_{ij}^t + \sum_{j \in \mathscr{P}_{nc}} \tilde{r}_{jt} \chi_{ij}^t - w_i^t + u_i^t = v_i \qquad i \in \mathscr{I}, t \in \mathscr{T}$$

$$w_i^t \le v_i \eta_i^t \qquad\qquad i \in \mathscr{I}, t \in \mathscr{T}$$

$$u_i^t \le v_i (1 - \eta_i^t) \qquad\qquad i \in \mathscr{I}, t \in \mathscr{T}$$

$$\eta_i^t \in \{0, 1\} \qquad\qquad i \in \mathscr{I}, t \in \mathscr{T}$$

$$w_i^t, u_i^t \ge 0 \qquad\qquad i \in \mathscr{I}, t \in \mathscr{T}$$

$$\{\tilde{r}_{jt}\} \in U$$

Thus, problem (17) can be rewritten as:

$$\min \quad \gamma + \beta \sum_{t \in \mathscr{T} \setminus \{1\}} \sum_{j \in \mathscr{P}_{pc}} y_j^t \tag{19}$$

s.t.

$$\gamma \ge \sum_{i \in \mathscr{I}} c_i \sum_{t \in \mathscr{T}} w_i^t$$

$$\left\{ x_{ij}, \xi_{ij}^t, \chi_{ij}^t, y_j^t \right\} \in X, \quad \left\{ w_i^t, u_i^t, \eta_i^t \right\} \in V, \quad \left\{ \tilde{r}_{jt} \right\} \in U$$

where V defines the workloads that generate feasible overtimes, according to constraints (6)–(8) plus the integrality clauses.

3.3 Implementor-Adversarial Algorithm

As introduced above, the robust problem can be interpreted as an implementor-adversarial iterative game. The implementor solves the following problem

$$\min \quad \gamma + \beta \sum_{t \in \mathscr{T} \setminus \{1\}} \sum_{j \in \mathscr{P}_{pc}} y_j^t$$

$$\left\{ x_{ij}, \xi_{ij}^t, \chi_{ij}^t, y_j^t \right\} \in X, \quad \left\{ w_i^t, u_i^t, \eta_i^t \right\} \in V$$

$$\left\{ \tilde{r}_{jt} \right\} \in \text{realizations generated so far} \tag{20}$$

for deciding the assignments x_{ij}^*, ξ_{ij}^{t*}, χ_{ij}^{t*} and y_j^{t*} that minimize the costs over all demand realizations generated so far.

Table 1 Adopted implementor-adversarial algorithm

Initialization:	D = constraints (6) associated with $\{\bar{r}_{jt}\}$
Iterate:	
1. Implementor problem:	solve problem (20) with solution $\left\{x_{ij}^{*}, \xi_{ij}^{t*}, \chi_{ij}^{t*}, y_{j}^{t*}\right\}$
2. Adversarial problem:	solve problem (21) with solution $\left\{r_{jt}^{*}, w_{i}^{t*}, u_{i}^{t*}, \eta_{i}^{t*}\right\}$
3. Test	**if** $\gamma \geq \sum_{i \in \mathscr{I}} c_{i} \sum_{t \in \mathscr{T}} w_{i}^{t*}$
	then exit
	else add (6) associated with $\left\{r_{jt}^{*}\right\}$ to D; go to 1

The adversary solves the following problem:

$$\max \sum_{i \in \mathscr{I}} c_{i} \sum_{t \in \mathscr{T}} w_{i}^{t}$$

$$\left\{w_{i}^{t}, u_{i}^{t}, \eta_{i}^{t}\right\} \in V, \quad \left\{\tilde{r}_{jt}\right\} \in U$$

$$\text{last assignments } \left\{x_{ij}, \xi_{ij}^{t}, \chi_{ij}^{t}, y_{j}^{t}\right\} \text{ from the implementor} \qquad (21)$$

for choosing the demands r_{jt}^{*} that maximize the cost with respect to the last assignments just selected by the implementor.

Both problems take into account constraints (6); however, the implementor satisfies them by choosing the assignments for fixed values of demand, while the adversary does the very opposite. Let D be the set of constraints (6). We adapt the basic template of the implementor/adversarial algorithm [5] as follows. Each run of the adversarial problem provides a realization of the demand $\left\{r_{jt}^{*}\right\} \in U$ and the corresponding workload variables $\left\{w_{i}^{t*}, u_{i}^{t*}, \eta_{i}^{t*}\right\} \in V$. Either $\sum_{i \in \mathscr{I}} c_{i} \sum_{t \in \mathscr{T}} w_{i}^{t*} > \gamma$ or γ is already the maximum cost, where γ is the last cost given by the implementor problem. In the former case, an equation of type

$$\sum_{j \in \mathscr{P}_{hc}} r_{jt}^{*} x_{ij} + \sum_{j \in \mathscr{P}_{pc}} r_{jt}^{*} \xi_{ij}^{t} + \sum_{j \in \mathscr{P}_{nc}} r_{jt}^{*} \chi_{ij}^{t} - w_{i}^{t} + u_{i}^{t} = v_{i}$$

is added to the implementor formulation, which is consequently reoptimized. A sketch of the algorithm is illustrated in Table 1.

4 Computational Tests and Results

We test our robust approach on several instances generated by assuming several mixes of patients, whose characteristics and demand evolution follow those in [8], with either $P = 70$ or $P = 98$. The number of bands H is chosen either equal to

6 or 10. Parameter α_d takes the same value α for any $d = 1, \ldots, H - 1$, and we assume α either equal to 1, 2 or 4. Moreover, we set $\alpha_0 = P$ in such way that all patients have the opportunity to remain in their current band without varying their actual demand. Finally, we set the $\beta = 1$, $c_i = 1$ for each nurse i, and $\theta = 3$.

The commercial framework IBM Cplex 12.5 and the language AMPL have been used to solve and implement the model. Instances have been run on a 2-core Linux processor clocked at 2 GHz with 16 GB of RAM. Time limits of 1200 and 300 s have been imposed for the implementor and the adversarial problem, respectively. On the one hand, we evaluate the choice of the parameters in terms of the computational time spent for finding the optimal robust solution. On the other hand, we measure the quality of the assignments with the aim of finding the best values for all of the parameters involved.

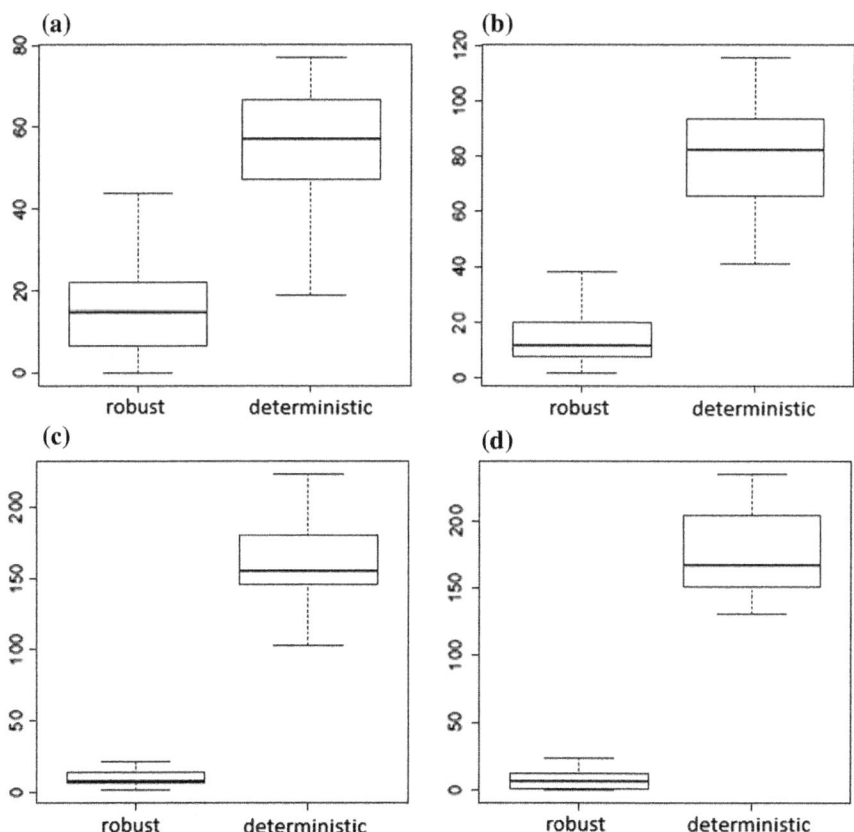

Fig. 1 Boxplots of the robust vs the deterministic solution. Case 98 patients with $T = 5$, $\theta = 4$; $\alpha_d = 4$ (**a**); case 98 patients with $T = 5$, $\theta = 4$; $\alpha_d = 1$ (**b**); case 70 patients with $T = 5$, $\theta = 5$; $\alpha_d = 2$ (**c**); case 70 patients with $T = 8$, $\theta = 5$; $\alpha_d = 1$ (**d**)

As for the computational times, we found that the maximum time spent for solving a robust problem over the tested instances is 11,777 s, which is acceptable for a weekly solution of the assignment problem in practice. We observed that the length of $\mathcal{T}$ directly affects the number of the iterations, thus increasing the computational time. Moreover, it seems that there is no correlation between the computational time and the values of H and α.

As for the quality of the robust assignments, we have compared the deterministic and the robust solutions when executed on a set of simulated scenarios. In particular, we have generated 20 simulated scenarios by drawing the demands from their probability density functions with a Monte Carlo approach. Then, we have compared the boxplots of the costs over the 20 scenarios from both solutions. Results are provided in Fig. 1 for four tested configurations. We observed that, in all cases, the robust solutions have a lower cost than the deterministic ones. The robust model is able to decrease the cost needed to cope with the worst demand realization and, thus, it allows to properly represent the real context.

Looking at the parameters, we observed that the best robust results occur when both H and α take small values.

5   Conclusions

In this work we consider, for the first time in the literature, time-related demands in the HC nurse-to-patient assignment problem under continuity of care and uncertain demands. The adopted implementor-adversarial approach [5] has proved to adequately address the time-dependency of demands, and the results show good quality solutions in terms of costs when executed in several scenarios.

Future work will extend the computational analyses. Moreover, we will improve the approach by postponing the assignment decisions for patients without continuity of care, to adapt the solution based on the actual demands, as in the real practice.

References

1. Addis, B., Carello, G., Grosso, A., Lanzarone, E., Mattia, S., Tànfani, E.: Handling uncertainty in health care management using the cardinality-constrained approach: advantages and remarks. Oper. Res. Health Care **4**, 1–4 (2015)
2. Argiento, R., Guglielmi, A., Lanzarone, E., Nawajah, I.: A Bayesian framework for describing and predicting the stochastic demand of home care patients. Flex. Serv. Manuf. J. **28**(1–2), 254–79 (2016)
3. Argiento, R., Guglielmi, A., Lanzarone, E., Nawajah, I.: Bayesian joint modeling of the health profile and demand of home care patients. IMA. J. Manag. Math. (2017). https://doi.org/10.1093/imaman/dpw001
4. Bertsimas, D., Sim, M.: The price of robustness. Oper. Res. **52**, 35–53 (2004)
5. Bienstock, D.: Histogram models for robust portfolio optimization. J. Comput. Financ. **11**, 1–64 (2007)

6. Carello, G., Lanzarone, E.: A cardinality-constrained robust model for the assignment problem in home care services. Eur. J. Oper. Res. **236**, 748–762 (2014)
7. Holte, M., Mannino, C.: The implementor/adversary algorithm for the cyclic and robust scheduling problem in health-care. Eur. J. Oper. Res. **226**, 551–559 (2013)
8. Lanzarone, E., Matta, A., Scaccabarozzi, G.: A patient stochastic model to support human resource panning in home care. Prod. Plan. Control **21**, 3–25 (2010)
9. Lanzarone, E., Matta, A., Sahin, E.: Operations management applied to home care services: the problem of assigning human resources to patients. IEEE Trans. Syst. Man Cybern. A **42**, 1346–1363 (2012)
10. Lanzarone, E., Matta, A.: A cost assignment policy for home care patients. Flex. Serv. Manuf. J. **24**, 465–495 (2012)
11. Lanzarone, E., Matta, A.: Robust nurse-to-patient assignment in home care services to minimize overtimes under continuity of care. Oper. Res. Health Care **3**, 48–58 (2014)
12. Rasmussen, M.S., Justesen, T., Dohn, A., Larsen, J.: The home care crew scheduling problem: preference-based visit clustering and temporal dependencies. Eur. J. Oper. Res. **219**, 598–610 (2012)
13. Trautsamwieser, A., Hirsch, P.: Optimization of daily scheduling for home health care services. J. Appl. Oper. Res. **3**, 124–136 (2011)

A Cardinality-Constrained Robust Approach for the Ambulance Location and Dispatching Problem

Vittorio Nicoletta, Ettore Lanzarone, Valérie Bélanger and Angel Ruiz

Abstract Emergency Medical Services (EMS) systems aim to provide immediate care in case of emergency. A careful planning is a major prerequisite for the success of an EMS system, in particular to reduce the response time. Unfortunately, the demand for emergency services is highly variable and uncertainty should not be neglected while planning the activities. Several optimization models have been proposed in the literature to deal with EMS planning-related problems, e.g. the Ambulance Location and Dispatching Problem (ALDP). However, most of the models are deterministic and neglect demand uncertainty. In this paper, we formulate and validate a robust counterpart of the ALDP to deal with demand uncertainty, exploiting the cardinality-constrained approach. Numerical experiments inspired by a real case show promising results and prove the practical applicability of the approach.

Keywords Emergency medical services · Demand uncertainty · Robust optimization · Cardinality-constrained approach

V. Nicoletta · A. Ruiz (✉)
Department of Operations and Decision Systems, Université Laval,
Québec, Laval, Canada
e-mail: angel.ruiz@osd.ulaval.ca; angel.ruiz@fsa.ulaval.ca

V. Nicoletta
e-mail: vittorio.nicoletta.1@ulaval.ca

E. Lanzarone
Istituto di Matematica Applicatae Tecnologie Informatiche (IMATI),
Consiglio Nazionale delle Ricerche (CNR), Milan, Italy
e-mail: ettore.lanzarone@cnr.it

V. Bélanger
Department of Logistics and Operations Management, HEC Montréal,
Montréal, Laval, Canada
e-mail: valerie.belanger@cirrelt.ca

© Springer International Publishing AG 2017
P. Cappanera et al. (eds.), *Health Care Systems Engineering*, Springer Proceedings in Mathematics & Statistics 210, https://doi.org/10.1007/978-3-319-66146-9_9

1   The Ambulance Location and Dispatching Problem

Emergency Medical Services (EMS) systems consist of the clinical activities and the
ambulance transportation in response to an emergency call. EMS plays an important
role in modern health care systems, as an adequate response to distress calls may have
a crucial impact on patients' health conditions. In particular, one of the main issues
is to reduce, or to keep under a given threshold, the time between the distress call and
the arrival of the ambulance to the emergency site. The body of literature devoted to
EMS design and management is huge; we refer the interested reader to [2] and [6],
which report classifications of EMS problems, formulations, and solving approaches.
Unsurprisingly, both reviews identify uncertainty on the demand and availability of
ambulances upon a call arrival as open and important issues to address in future
research.

This paper deals with the Ambulance Location and Dispatching Problem (ALDP),
which aims to choose at the same time the location for the available ambulances and
a dispatch policy, which decides which ambulance should answer any arriving call.
Although most of location problems assume that any call is answered by the closest
available ambulance, this policy is not always optimal [3, 11]. Thus, we consider the
formulation proposed in [4, 5], where the dispatch policy takes the form of a list of
ambulances for each demand zone, in which the ambulances are sorted according to
their priority to answer the calls in the zone.

To deal with stochastic demands, we introduce a robust counterpart of the deterministic ALDP, based on the cardinality-constrained approach [8]. Such an approach
has been recognized to be very effective to cope with uncertainty in health care problems [1] and has been successfully applied to other health care facilities [9], but it
has never been applied to EMS. In fact, although several works deal with uncertainty
in the context of ambulance location [7, 12], none of them has previously used this
approach.

The paper is organized as follows. The ALDP formulation and the proposed
robust counterpart are presented in Sect. 2. Computational experiments are detailed
in Sect. 3, and the results in Sect. 4. Finally, conclusions are drawn in Sect. 5.

2   Robust Problem Formulation

2.1   Sets, Parameters and Decision Variables

The ALDP is defined on a graph $G = (V, E)$. V is given by $I \cup J$, where $I = \{v_1, \ldots, v_n\}$ and $J = \{v_{n+1}, \ldots, v_{n+m}\}$ represent all demand zones and potential waiting positions, respectively. $E = \{(v_i, v_j) : v_i, v_j \in V\}$ is the set of the edges connect-
ing the nodes in the graph. A demand zone $v_i \in I$ is characterized by the coordinates
of its centroid and a demand d_i, which is considered to be uncertain in our framework. A potential waiting position $v_j \in J$ is defined as a demand zone where p_j

vehicles ($p_j \geq 1$) can be located and from where they move to join the different calls. A travel time t_{ij} is associated to each edge $(v_i, v_j) \in E$.

Moreover, K denotes the set of all available vehicles, and Z_i the set of vehicles in the dispatch list of $v_i \in I$. Finally, we define the capacity W_i as the maximum number of demands an ambulance i can serve in a given period, and the busy fraction q_i as the time fraction an ambulance i is expected to be busy (and therefore unavailable to answer calls).

The ALDP considers two types of decisions: location decisions to select a waiting position for each vehicle, and dispatch decisions.

The latter are made according to the list Z_i of each demand zone i, choosing the first available ambulance starting from the beginning of the list. However, for life-threatening calls, the nearest available vehicle is sent to the scene of the incident. If no vehicle is available in the dispatch list of a zone, the nearest available vehicle is sent. Finally, if no vehicle is available at all, the call is placed on a waiting queue or redirected to another service; in the latter case, we consider an arbitrary value T for the response time.

The following assumptions are included (see [4, 5]):

1. Location and dispatch decisions are taken for a given planning horizon.
2. The number of vehicles available in the planning horizon is fixed and known.
3. All vehicles have the same workload capacity ($W_i = W \ \forall i$).
4. All vehicles have the same busy fraction ($q_i = q \ \forall i$); although this assumption needs to be empirically confirmed, it is broadly used in EMS design problems.
5. All of the dispatch lists have the same cardinality, without loss of generality.

Two groups of decision variables are defined to adequately consider location and dispatch decisions. They are summarized in Table 1, together with problem sets and parameters.

2.2 Objective Function and Constraints

A brief description of the deterministic model is given in the following; details can be found in [4] and [5].

The ALDP aims at minimizing the overall expected response time. Based on the dispatch lists, the expected response time for a demand zone is given by three contributions: *i*) the sum of the response times of the vehicles in the dispatch list, weighted by their probability to answer the call; *ii*) the time corresponding to the vehicles available to respond to emergency calls, but not in the dispatch list; *iii*) the time of the calls placed in queue or referred to another service because no vehicle is available. Only the first contribution is considered when formulating the ALDP; the others are calculated afterwards when the solution is executed, based on the decisions and the system characteristics.

Table 1 ALDP sets, parameters and decision variables

Sets	
I	demand zones
J	potential waiting positions for vehicles
K	available vehicles
Z_i	dispatch list of zone i (all with the same cardinality $|Z|$)
Parameters	
d_i	demand of zone i
t_{ji}	travel time from zone j to zone i
p_j	maximum number of vehicles in j
W	capacity, i.e. maximum number of demands a vehicle can serve in the time
	horizon (same for all vehicles)
q	busy fraction (same for all vehicles)
Decision variables	
w_i^{zk}	equal to 1 if vehicle k is in position z of the dispatch list of zone v_i,
	0 otherwise
y_{ij}^{zk}	equal to 1 if vehicle k, located in zone v_j, is in position z of the dispatch list
	of v_i, 0 otherwise

The formulation is completed by three sets of constraints. The first set ensures that each vehicle is located at a waiting position, guaranteeing that the maximum number of vehicles p_j is respected. The second set ensures that the demands assigned to a vehicle respect its capacity, considering for each vehicle the busy fraction and the presence in one or more dispatch lists. Finally, the third set imposes that a vehicle cannot occupy more than one position in the dispatch list of each zone, and that exactly one vehicle is at each position of each dispatch list.

2.3 Cardinality-Constrained Robust Formulation

To model the uncertain demands, we apply the cardinality-constrained approach to the parts of the model in [4, 5] where parameters d_i appear. First, we convert the objective function into a set of constraints by adding a new variable η_k. Thus, the new objective function is:

$$\min \sum_{k \in K} \eta_k \tag{1}$$

with the additional constraints:

$$\sum_{i \in I} \sum_{z \in Z} \sum_{j \in J} (1 - q)q^{z-1} d_i t_{ji} y_{ij}^{zk} \leq \eta_k \quad \forall k \tag{2}$$

Then, we consider the uncertain demands as independent random variables $\tilde{d}_i$ ($i \in I$). According to [8], each of them is characterized by a nominal value $\bar{d}_i$ and a maximum variation $\hat{d}_i$, i.e. $\tilde{d}_i \in [\bar{d}_i - \hat{d}_i, \bar{d}_i + \hat{d}_i]$.

Thus, demands can be expressed as $\tilde{d}_i = \bar{d}_i + \alpha_i \hat{d}_i$, where each $\alpha_i \in [-1, 1]$ represents the deviation of the demand of zone i from its nominal value $\bar{d}_i$, standardized by the half-length of the uncertainty interval $\hat{d}_i$. For example, $\alpha_i = 0$ corresponds to $\tilde{d}_i = \bar{d}_i$, $\alpha_i = 1$ to $\tilde{d}_i = \bar{d}_i + \hat{d}_i$, and $\alpha_i = -1$ to $\tilde{d}_i = \bar{d}_i - \hat{d}_i$.

Demands $\tilde{d}_i$ appear in the new constraint (2) and in the workload capacity limit for each vehicle (see [4, 5]) which are rewritten in the cardinality-constrained robust framework as:

$$\sum_{i \in I} \sum_{z \in Z} \sum_{j \in J} (1 - q)q^{z-1} \bar{d}_i t_{ji} y_{ij}^{zk} + \sum_{i \in I} \sum_{z \in Z} \sum_{j \in J} (1 - q)q^{z-1} (\alpha_i \hat{d}_i) t_{ji} y_{ij}^{zk} \leq \eta_k \quad \forall k \tag{3}$$

$$\sum_{z \in Z} \sum_{i \in i} (1 - q)q^{z-1} \bar{d}_i w_i^{zk} + \sum_{z \in Z} \sum_{i \in i} (1 - q)q^{z-1} (\alpha_i \hat{d}_i) w_i^{zk} \leq W \quad \forall k \tag{4}$$

Satisfying constraints (3) and (4) for all possible demand realizations, i.e. for all combinations of $\{\alpha_i, \ i \in I\}$, would lead to a too conservative (and unrealistic) solution. In fact, it is very unlikely that all of the demand coefficients assume their worst (highest) values simultaneously (i.e. $\alpha_i = 1 \ \forall i \in I$). Thus, in the cardinality-constrained approach, we limit the number of zones that ask for the highest demand, in each constraint, by means of the robustness parameters $\{\Gamma_k, \ k \in K\}$. Indeed, the robust cardinality-constrained solution guarantees that the solution remains feasible if up to Γ_k parameters α_i go to the maximum value equal to 1 in each constraint, while the others remain at the nominal value equal to 0.

In the following, we consider the same value Γ for all vehicles (i.e. $\Gamma_k = \Gamma \ \forall k$); however, the robust counterpart we derive can be easily extended to the case in which the robustness parameters $\{\Gamma_k, \ k \in K\}$ vary from vehicle to vehicle. Parameter Γ controls the level of robustness of the solution and can be set equal to $\{0, 1, \ldots, |I|\}$ (we consider only integer values). Fixing $\Gamma = 0$ guarantees feasibility only if all of the random variables assume their nominal value (deterministic solution), whereas setting $\Gamma = |I|$ means no restrictions (most conservative solution).

We underline that, in our formulation, the optimal values α_i can be different from constraint to constraint. This simply increases the level of robustness of the solution and has to be accounted while analyzing the impact of Γ.

Briefly, to derive the robust counterpart, (3) and (4) are rewritten as:

$$\sum_{i \in I} \sum_{z \in Z} \sum_{j \in J} (1-q) q^{z-1} \bar{d}_i t_{ji} y_{ij}^{zk} + \beta_k \leq \eta_k \qquad \forall k \qquad (5)$$

$$\sum_{z \in Z} \sum_{i \in i} (1-q) q^{z-1} \bar{d}_i w_i^{zk} + \gamma_k \leq W \qquad \forall k \qquad (6)$$

where β_k and γ_k are the optima of the two following knapsack problems (generated for each k):

$$\beta_k = \max \sum_{i \in I} \sum_{z \in Z} \sum_{j \in J} (1-q) q^{z-1} (\alpha_i \hat{d}_i) t_{ji} y_{ij}^{zk}$$

$$\sum_{i \in I} \alpha_i \leq \Gamma \qquad (7)$$

$$\alpha_i \in [0,1] \quad \forall i \in I$$

$$\gamma_k = \max \sum_{z \in Z} \sum_{i \in i} (1-q) q^{z-1} (\alpha_i \hat{d}_i) w_i^{zk}$$

$$\sum_{i \in I} \alpha_i \leq \Gamma \qquad (8)$$

$$\alpha_i \in [0,1] \quad \forall i \in I$$

Applying the Strong Duality Theorem [8], we obtain their dual problems, which can be substituted in (5) and (6) to obtain the following robust formulation:

$$\min \sum_{k \in K} \eta_k \qquad (9)$$

s.t.

$$\sum_{i \in I} \sum_{z \in Z} \sum_{j \in J} (1-q) q^{z-1} \bar{d}_i t_{ji} y_{ij}^{zk} + \Gamma a_k^{of} + \sum_{i \in I} b_{ki}^{of} \leq \eta_k \qquad \forall k \qquad (10)$$

$$\sum_{z \in Z} \sum_{i \in i} (1-q) q^{z-1} \bar{d}_i w_i^{zk} + \Gamma a_k^{con} + \sum_{i \in I} b_{ki}^{con} \leq W \qquad \forall k \qquad (11)$$

$$a_k^{of} + b_{ki}^{of} \geq (1-q) q^{z-1} \hat{d}_i t_{ji} y_{ij}^{zk} \qquad \forall k,i,z,j \qquad (12)$$

$$a_k^{con} + b_{ki}^{con} \geq (1-q) q^{z-1} \hat{d}_i w_i^{zk} \qquad \forall k,i,z \qquad (13)$$

$$\sum_{z \in Z} w_i^{zk} \leq 1 \qquad \forall k,i \qquad (14)$$

$$\sum_{k \in K} w_i^{zk} = 1 \qquad \forall z,i \qquad (15)$$

$$w_i^{zk} = \sum_{j \in J} y_{ij}^{zk} \qquad\qquad \forall z, i, k \qquad (16)$$

$$w_i^{zk} \in \{0, 1\} \qquad\qquad \forall z, i, k \qquad (17)$$

$$y_{ij}^{zk} \in \{0, 1\} \qquad\qquad \forall z, i, k, j \qquad (18)$$

$$\eta_k, a_k^{of}, a_k^{con} \geq 0 \qquad\qquad \forall k \qquad (19)$$

$$b_{ki}^{of}, b_{ki}^{con} \geq 0 \qquad\qquad \forall k, i \qquad (20)$$

Constraints (14)–(18) are the same as in the deterministic model, while (10)–(13) and (19)–(20) are those modified or added in the robust counterpart. The new variables $a_k^{of}, b_{ki}^{of}, a_k^{con}, b_{ki}^{con}$ are the dual of those appearing in the two knapsack problems.

3 Computational Experiments

Numerical tests have been run considering a set of instances based on the case of Montréal, QC, Canada. Instances (see [10]) have been generated using public annual reports published by *Urgences-santé* (2006) and *Statistics Canada* (2011). They include 30 demand zones, which represent the central part of the city of Montréal, whose nominal demands $\bar{d}_i$ range from 41 to 496. Moreover, vehicle capacity W has been set equal to 1500, and busy fraction q to 0.5.

The deterministic model has been solved considering the nominal demand $\bar{d}_i$ for each zone $i \in I$. Then, for the robust model, we have set the maximum variation $\hat{d}_i$ equal to $0.25\,\bar{d}_i$ for each demand zone $i \in I$.

We analyze the impact of Γ in terms of feasibility and price to pay for the improved feasibility (price of robustness). For this purpose, we consider 11 values of Γ, ranging from $\Gamma = 0$ (the deterministic model) to $\Gamma = |I|$ (the case in which each demand takes its maximum value) as follows:

$$\Gamma = \frac{k}{10}|I|, k \in \{0, 1, 2, 3, 4, 5, 6, 7, 8, 9, 10\}.$$

Being $|I| = 30$, the considered values of Γ are integer.

3.1 Execution and Solutions Evaluation

Three types of demand scenarios have been generated to evaluate the behavior of the obtained solutions, based on the values $\bar{d}_i$ and $\hat{d}_i$ in each zone i:

- *normal scenarios*: each d_i follows a Normal distribution centered in $\bar{d}_i$, and $\hat{d}_i$ represents about the 95% quantile, i.e. $d_i \sim N\left(\mu = \bar{d}_i, \sigma = \frac{\bar{d}_i + \hat{d}_i}{2}\right)$.
- *uniform scenarios*: each d_i is equal to $\bar{d}_i + \alpha_i \hat{d}_i$, where each α_i follows a uniform distribution in the interval $[-1, 1]$.
- *worst case scenarios*: each d_i is equal to $\bar{d}_i + \alpha_i \hat{d}_i$, where each α_i follows a uniform distribution in the interval $[0, 1]$. Unlike the previous cases, these demands are not centered in their respective $\bar{d}_i$. Therefore, these scenarios refer to a situation in which the demands have been underestimated, which represent an interesting case for health care managers.

For each alternative, 100 Monte Carlo samples have been drawn from each demand distribution, thus obtaining as many execution scenarios of each type.

To evaluate the feasibility of a solution in a scenario, we compute the maximum value of the workload W_{max} as:

$$W_{max} = \max_{k \in K} \sum_{z \in Z} \sum_{i \in i} (1 - q) q^{z-1} d_i w_i^{zk} \tag{21}$$

where d_i denotes here the demand in the scenario. A solution is considered to be *unfeasible* if the associated W_{max} is greater than the workload capacity W used to solve the problem. Similarly, to evaluate the *price of robustness* in a scenario, we first compute the value of the objective function OF as:

$$OF = \sum_{i \in I} \sum_{z \in Z} \sum_{k \in K} \sum_{j \in J} (1 - q) q^{z-1} d_i t_{ji} y_{ij}^{zk} \tag{22}$$

where d_i denotes, once again, the demand in the scenario. Then, we compute the price of the robust solution (i.e. its additional cost when executed) as the difference Δ_{OF} between the robust OF and the corresponding deterministic OF when $\Gamma = 0$:

$$\Delta_{OF} = OF_{robust} - OF_{deterministic} \tag{23}$$

4 Results

All instances have been solved to optimality within 1 hour, although the computational times increase from 108 up to 2514 seconds as the value of Γ increases.

Results are provided in Fig. 1. Solutions are always feasible for $\Gamma \geq 0.3|I|$ in the normal and uniform scenarios, while more than the 75% of the solutions are feasible for $\Gamma \geq 0.5|I|$ in the worst case scenario. As for the price of robustness, we observe that Δ_{OF} values are always below 6 s; thus, considering that OF values are around

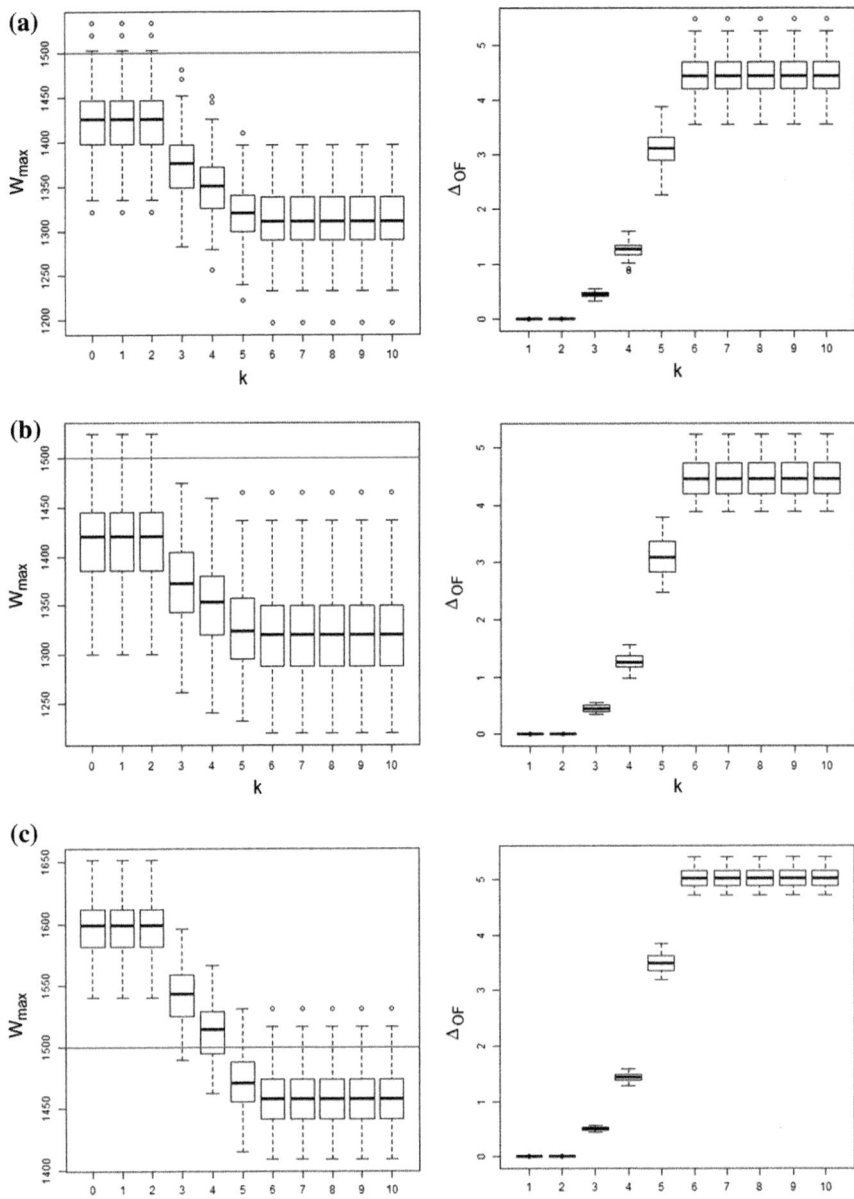

Fig. 1 Feasibility W_{max} (left column) and price of robustness Δ_{OF} (right column) for normal scenario (**a**), uniform scenario (**b**), and worst case scenario (**c**).

1500 s, they can be considered negligible. In other words, we can conclude that the price of robustness is highly affordable in order to guarantee feasible solutions.

5 Conclusions

In this paper, we propose and validate a robust counterpart of the ALDP in [4, 5] based on the cardinality-constrained approach. Results from the application to a realistic test case show that demand variations with respect to the expected values impair the feasibility of the deterministic solution, while its robust counterpart performs better for proper values of parameter Γ. In particular, in the considered test case, values of Γ between the 30 and the 50% of the demand zones (meaning that the $30-50\%$ of the demand zones assume the worst case value) allow the solution to remain feasible when tested against several demand scenarios. At the same, the observed increase of the objective function is negligible.

We may conclude that including the robustness in the ALDP problem is promising, at least in the tested case, because of the capability to increase the feasibility of the solutions while keeping limited the price to pay in terms of increased objective function value.

References

1. Addis, B., Carello, G., Grosso, A., Lanzarone, E., Mattia, S., Tànfani, E.: Handling uncertainty in health care management using the cardinality-constrained approach: advantages and remarks. Oper. Res. Health Care **4**, 1–4 (2015)
2. Aringhieri, R., Bruni, M.E., Khodaparasti, S., van Essen, J.T.: Emergency medical services and beyond: addressing new challenges through a wide literature review. Comp. Oper. Res. **78**, 349–368 (2017)
3. Bandara, D., Mayorga, M.E., McLay, L.A.: Priority dispatching strategies for EMS systems. J. Oper. Res. Soc. **65**, 572–587 (2014)
4. Bélanger, V., Lanzarone, E., Ruiz, A., Soriano, P.: The ambulance relocation and dispatching problem. Technical report CIRRELT 2015–59 (2015) https://www.cirrelt.ca/DocumentsTravail/CIRRELT-2015-59.pdf
5. Bélanger, V., Lanzarone, E., Ruiz, A., Soriano, P.: The ambulance location and dispatching problem. Paper under second review (2017)
6. Bélanger, V., Ruiz, A., Soriano, P.: Recent Advances in Emergency Medical Services Management. Technical report CIRRELT 2015–28 (2015) https://www.cirrelt.ca/DocumentsTravail/CIRRELT-2015-28.pdf
7. Beraldi, P., Bruni, M.E.: A probabilistic model applied to emergency service vehicle location. Eur. J. Oper. Res. **196**, 323–331 (2009)
8. Bertsimas, D., Sim, M.: The price of robustness. Oper. Res. **52**, 35–53 (2004)
9. Carello, G., Lanzarone, E.: A cardinality-constrained robust model for the assignment problem in home care services. Eur. J. Oper. Res. **236**, 748–762 (2014)
10. Kergosien, Y., Bélanger, V., Soriano, P., Gendreau, M., Ruiz, A.: A generic and flexible simulation-based analysis tool for EMS management. Int. J. Prod. Res. **53**, 7299–7316 (2015)

11. Schmid, V.: Solving the dynamic ambulance relocation problem and dispatching problem using approximate dynamic programming. Eur. J. Oper. Res. **219**, 611–621 (2012)
12. Zhang, Z., Jiang, H.: A robust counterpart approach to the bi-objective medical service design problem. Appl. Math. Model. **38**, 1033–1040 (2014)

A Practical Approach to Machine Learning for Clinical Decision Support

Projects at Lucile Packard Children's Hospital Stanford in Partnership with Stanford Engineering

Daniel Miller, David Scheinker and Nicholas Bambos

Abstract Machine learning has produced effective clinical decision support tools. The impact of such work is limited by the difficulty of implementing such tools outside the institution where they were designed. The recent wide-spread adoption of Electronic Medical Record systems (EMRs) makes possible the development and application of tools across institutions. We describe three machine learning projects to develop generalizable, EMR-based clinical decision support tools at the cardiac care units of Lucile Packard Children's Hospital Stanford: false alarm suppression, detection of critical events, and automated identification and detection of drug-drug interactions. These projects utilize flexible statistical and deep learning frameworks to enable automated, patient-specific care. We focus on the practical challenges of implementing such methodology and describe our progress on producing tools useful for our institution.

Keywords Healthcare · Machine learning · Artificial intelligence · Decision support · Electronic medical record · Neural networks · Deep learning

1 Introduction

Machine learning (ML) and artificial intelligence (AI) have been used to produce useful clinical decision support tools (CDSTs). These tools range from those based on a single, standard type of input data, e.g., tumor identification through imaging [4] or ultrasound [29]; to those that collect data from a variety of patient streams and

D. Miller (✉) · D. Scheinker (✉) · N. Bambos (✉)
Stanford University Department of Electrical Engineering, Stanford University
Department of Management Science and Engineering, and Lucile Packard
Children's Hospital Stanford, Palo Alto, CA, USA
e-mail: danielrm@stanford.edu

D. Scheinker
e-mail: dscheink@stanford.edu

N. Bambos
e-mail: bambos@stanford.edu

© Springer International Publishing AG 2017   111
P. Cappanera et al. (eds.), *Health Care Systems Engineering*, Springer Proceedings
in Mathematics & Statistics 210, https://doi.org/10.1007/978-3-319-66146-9_10

analyze these with a dedicated data infrastructure [9, 14]. The resources necessary for a hospital to implement such tools can range from a small team project, to a major organizational, multi-year partnership with a company such as IBM, Google, or Apple.

Our goal is to design tools that can be easily modified and implemented at any institution with an EMR. We illustrate this effor with our work on three ML-based CDSTs at the Lucile Packard Children's Hospital (LPCH) Children's Heart Center. Our specific contributions are:

- A flexible deep learning paradigm well suited to healthcare applications
- A detailed examination of a novel, highly acute pediatric population
- A unified approach to expanding the methods of this work beyond LPCH, to any institution with a comparable EMR.

2 False Alarm Suppression

In hospital intensive care units, high rates of false arrhythmia alarms result in the fatigue and desensitization of attending medical staff. This "false alarm fatigue" causes workers' response times to slow, and leads to detrimental decreases in the quality of patient care [1]. Excessive alarms can also lead to sleep deprivation and depressed immune systems among patients [3]. False alarm rates of up to 86% have been reported in pediatric intensive care units [16]. Such high rates suggest the potential for significant improvements by applying modern methods to identify and suppress false alarms in real time.

False alarm (FA) suppression is a well-examined problem in the machine learning community. Advances have been driven by the recent rise in public availability of well-curated and labeled data. One of the most prominent such databases is the Medical Information Mart for Intensive Care (MIMIC), a set developed by the MIT Lab for Computational Physiology which contains deidentified health data associated with roughly 40,000 critical care patients [13]. The MIMIC-II waveform dataset was filtered, and used for the PhysioNet Computing in Cardiology Challenge 2015: Reducing False Arrhythmia Alarms in the ICU [10]. Through its open nature, this competition led many researchers to implement a wide range of techniques for solving this problem.

While we worked in parallel to construct the database of bedside monitor waveforms for LPCH, we reviewed the approaches used by competition participants, and built a range of exploratory predictive models for this generalized dataset. We hypothesize that applying modern machine learning algorithms can significantly improve the specificity of the alarm system, while maintaining the true alarm sensitivity. Implementing these improvements in a modern, EMR-enabled hospital would lead to a more efficient ICU environment by minimizing the detrimental effects of alarm fatigue on both staff and patients.

We now describe some technical insights from our false alarm work, the opportunity this work reveals for further development at LPCH, and the general lessons of such an approach for any other EMR-enabled hospital.

We consider the five alarms triggered by the following critical conditions: asystole, extreme bradycardia, extreme tachycardia, ventricular tachycardia, and ventricular flutter/fibrillation. These are triggered by electrocardiogram (ECG) and pulsatile waveforms recorded by bedside monitors. The standard criteria are simple thresholds on the extracted vitals. For example, a ventricular tachycardia alarm triggers if the heartrate exceeds 140 beats per minute for 17 consecutive beats. This alarm threshold is a fixed value common to all patients, and disregards any potentially relevant patient-specific information like age, diagnosis, or established historic baselines on that patient's heartrate.

Our previous work reviewed traditional classification models, including logistic regression, support vector machines, decision trees, and random forests. These models balance simplicity and efficiency with raw predictive power. In general, all machine learning procedures conform to the following steps:

1. The data is prepared, and split into train and test sets.
2. Train data is used to develop the model, and estimate parameter values.
3. Model performance is evaluated on the distinct test set.
4. Predictions are generated, and inferences drawn from the results.

After comparing results, the most powerful model we built was a boosted classification tree trained on a range of heart rate, blood pressure, and signal quality features extracted from the ECG, photoplethysmogram (PPG), and arterial blood pressure (ABP) waveforms [27]. This feature selection was motivated by previous competitors in the Physionet competition, and their subsequent work [2, 7]. While the competitors focused on maximizing a specific weighted ratio of the true positive, false positive, true negative, and false negative rates, we examined the full sensitivity /specificity range for each arrhythmia type, providing a more holistic view of model performance.

Our results provided significant false alarm suppression, and key insights by examining error types and relative variable importances. For example, the specific model was far less important than the choice of variables–a common thread in machine learning. One factor limiting both traditional ML and basic statistical models is a reliance on very specific, intelligent feature engineering. This process usually requires a great deal of domain-specific knowledge. Often, what we call "machine learning" requires a great deal of human learning and design of which model coefficients to use. The machine simply performs the tedious numerical optimization to find the precise coefficient values.

In contrast, these methods and their results have the advantage of broad generality and adaptability. The preliminary waveform data we examined was randomly drawn from four different hospitals in the USA and Europe, and our analysis did not rely on any hospital specific information. Despite this, the resulting models are both general enough to be applied to a wide range of hospital types and environments, and adaptable enough to take advantage of the additional information and data available in any EMR-enabled hospital.

For example, as we consider implementing a CDS system for quality improvement in patient care at LPCH, our model can be extended to include patient-specific information like age, length of stay, primary diagnosis, or other protected health information (PHI) not usable in a publicly available, de-identified dataset. Our current work focuses on extracting these significant data from the LPCH EMR, cleaning and filtering the data, and providing them as inputs to these types of models via an accessible framework.

We are motivated by the opportunity to re-use these data. Often, datasets are collected for a single, limited purpose, such as administrative record-keeping. Despite the intended scope, the data often contain information usable for many different problems. In each problem, a feature might be more relevant than in others. For example, to determine a patient's immediate vitals, we might consider the specific attributes of recent QRS complexes in the ECG waveform. Alternatively, if primarily concerned with the patient's past treatment responses, we may examine long-term trends or deviations from that patient's established baseline. In each case we would design a feature set uniquely tuned to the problem, and based on a clinical expert's knowledge.

Recent advancements in deep learning focus on automating this feature selection process, emphasizing the *artificial* in artificial intelligence [25]. This automation often comes at the cost of model interpretability and simplicity, but has been validated by raw performance improvements demonstrated in fields from language translation to playing Go. These powerful, yet adaptable models allow us to reuse large collections of data like the LPCH physiological waveforms to solve many previously intractable and difficult problems.

3 Adverse Event Prediction with Deep Learning

Congestive Heart Disease (CHD) affects nearly 1% of all births in the US, at roughly 40,000 per year [12]. Of those born with CHD, roughly one in four do not survive their first year [22]. Hospitalized patients often exhibit predictive indicators 24–48 h prior to experiencing a life-threatening event [6, 11]. Detecting and filtering these predictors can enable early identification of patient decline, allowing preventative measures to be taken. Current algorithms utilize select clinical variables for predictive modeling, but have generally excluded congenital heart disease due to the heterogeneity of both patient and disease.

Standard methods for quantifying patient condition rely heavily on manually collected data, such as vital signs, and other specifically selected discrete data points [8]. As with false alarm suppression, most systems for detecting patient decline rely on simple methods like thresholding to a fixed range. In some cases this range may be personalized to a degree, usually by allowing for age or weight input during the initial bedside monitor configuration. However, most patient-specific metrics available within the EMR often go unused. One reason is that the wealth of information automatically collected by bedside monitors and subsequently consolidated in EMR databases is often difficult to extract and parse into a usable form. That these data are not currently fully utilized for early warning systems suggests a significant opportunity.

We hypothesize that the information necessary to predict patient decline is embedded within these continuous data streams. Recent advances in AI and deep learning allow for the autonomous selection and extraction of predictive features [25], often making manual feature selection unnecessary. By leveraging patterns deep within our data, we can hope to achieve significant improvements in early warning of critical patient events as compared to standard methods.

We next describe technical insights from our designs in adverse event prediction, how this work may be adapted to address other problems at LPCH, and lessons in applying this framework to a similar hospital.

Our data comes from two distinct datasets. The first is the EMR records for 2628 inpatients admitted to the LPCH ICU. These contain 33403 sparse clinical variables. The second is a historical warehouse of the bedside monitor data for all LPCH patients. This contains waveform data for roughly 38,000 patients, with each waveform sampled at 125 Hz, and pre-processed vital signs sampled once per minute. Additionally, all alarms–and most notably all arrhythmia alarms–generated by the bedside monitor are recorded here. The immensity of these data severely limits the options for standard statistical methods, and requires the development of new techniques to take advantage.

We are currently examining the application of convolutional neural networks (CNNs) to these physiological waveform signals. CNNs are a deep learning framework that have experienced a recent surge in popularity due to their groundbreaking success in image classification in 2012 [15]. The CNN was originally inspired by the visual cortex and its ability to focus on local regions to extract information from a larger picture. Since then, these networks have been adapted to a variety of data and applications ranging from speech recognition using auditory waveforms, to epileptic diagnosis using EEG waveforms [21]. Most relevantly, CNNs have shown success in analyzing multi-channel time series data like our physiological waveforms [30]. Motivated by these examples, we have designed and are refining an adaptable multi-channel deep CNN for processing our physiological data. We describe here a specific use example, and present a simplified version of the CNN structure. This basic network may be easily extended to increase the model's predictive power, or adapted to incorporate additional sources of data, including both general publicly available data, and patient-specific clinical or demographic information.

As a motivational example, consider the basic problem to predict the risk of a patient experiencing a cardiopulmonary arrest (CPA) within the next 24 h. We assume this prediction is performed at intervals of 1 hour, and that we have the patient's ECG, PPG, and respiratory waveforms for the full duration. We therefore define our sample input as a $(3 \times L)$ matrix, with L our sampling frequency times the sample duration. We define labels as *True/False* if the patient experienced a CPA within the 24 h following the sample.

To solve this problem, we construct the following CNN from basic neural network layers: We first construct a fixed-width filter to capture local waveform information. This width controls whether the filter focuses on short-term patterns like those within individual QRS complexes, or extracts longer-term trends, patterns, or interactions like those between sequences of heartbeats. The filter will be applied to all 3 of our waveforms concurrently to capture all available information in a specific time window. The filter is *convolved*, or slid across our waveform to extract a filtered output. We next apply a non-linear function, allowing us to model a wider range of complex, non-linear functions. This is a key step distinguishing neural networks from traditional linear classifiers. Finally, we use a fully connected layer and a *softmax* to generate an estimated probability for the CPA event, and predict the most probable class. This network is then trained on a large historical dataset, empirically determining the values for all internal filters and parameters. Training a deep neural network requires a large amount of both data, and computation. However, once the network is trained, it can generate predictions extremely efficiently.

While an actual CNN used for this type of problem is more complex than the above example, it shares many similarities, and the same basic layers. Full details of the model we use to predict adverse events in CHD patients at LPCH are forthcoming [20]. In general, we build this type of network empirically by providing a computer with a large training and development set of data, and allowing it to semi-autonomously decide the exact number of network layers, the ordering of these layers, the number of filters, and the exact sizes of each filter and weight. Finally, we compare results with a simple statistical baseline model, and present the relative performances. This type of hyper-parameter tuning is parallel by design, and can be performed quickly on a modern GPU or similar parallel system. Furthermore, CNNs were chosen specifically for their efficiency and capacity for parallelization. Deep learning thereby allows us to automate the feature selection process, enabling agile implementation on a variety of problems by sharing a flexible modeling framework.

For example, as we develop a CNN framework to predict adverse events from the raw LPCH patients' waveforms, we can also search the waveform data for critical medication administrations, identify resting states, or perform diagnostics by simply changing the labels used to train the CNN. Alternatively, we can improve prediction results and allow for a wider range of applications by adding input channels for a drug infusion pump or aligning discrete inputs for lab results. Where a traditional machine learning model might require rigorous feature selection and re-design, we can simply re-train the deep learning model in the new paradigm to learn the optimal feature structure. Furthermore, by designing our data streams in a modular and re-usable manner, we can build models that share resources between even fundamentally different problems.

4 Drug-Drug Interaction Identification

The CHD patients who we have been discussing require a broad drug regimen to maintain a delicate system balance, particularly between their heart and kidneys. These drugs include ACE inhibitors and Beta Blockers to control blood pressure, anticoagulants to prevent blood clots, and diuretics to counter fluid overload. Many of these drugs may interact, and the effects of these interactions must be weighed against their benefits. These drugs may also share toxicity, even if they do not otherwise interact.

For each drug, a broad range of detailed information is available in public databases. These databases are generally assembled manually by specialists reviewing the assorted research, revising on a regular basis. For example, DrugBank contains chemical, pharmacological and pharmaceutical drug data that was assembled over several years by a team of archivists and annotators including two accredited pharmacists, a physician, and three bioinformaticians with dual training in computing science and molecular biology/chemistry [28]. Based on these data, many common CHD medications are known to have specific types of pairwise interactions. One issue forestalling us from directly using such data is its focus on general drug information, with no specific relevance to the patient population, and no direct account for patient-specific information. Furthermore, it relies on a static corpora that cannot quickly adapt to the continuously published research on drug interactions and applications.

We hypothesize that much of the CHD drug interaction information is not only documented in the published literature, but can be extracted by a natural language processing (NLP) system based on deep learning. Such a system would incorporate information from a broad range of sources, and could continuously update and re-evaluate previous results as new research is released. Though often less powerful than statistical models that use a specifically designed feature set, NLP is able to incorporate otherwise unusable free-text data, and may be ensembled with traditional models to boost performance.

Extracting general drug-drug interactions from natural language corpora has been well-studied in recent years, and has been the focus of several computing and machine learning challenges [26]. As is often the case with such challenges, the ready availability of clean and labeled data led wide range of research and modeling techniques to be applied to the problem [5]. Most of the best-performing methods have incorporated data from multiple sources, and several notable results utilized deep learning techniques [17].

We now describe technical insights from our DDI extraction work, how this work may be implemented as a CDS tool at LPCH, and general considerations for applying these methods to new patient populations at other hospitals.

Our work draws inspiration from previous general DDI extraction successes to address the problem specific to pediatric CHD. The following supervised learning problem examines how well NLP could identify known CHD drug interactions: We begin by considering a set of pediatric cardiology drugs. For each cardiac drug, we extract from the DrugBank database all pairwise interactions between that cardiac drug and any other drug. This forms the super-set of all drugs considered in this analysis. As an input data source to this problem, we parsed the PubMed Central Open Access Subset [24], a publicly available corpus of research publications which currently includes a total of 663,597 papers. To reduce the size of this problem, we consider only the paper abstracts. Any abstract that contains mention of at least one of the relevant drugs is defined as a "Cardiac Abstract" and included in the analysis. This filtering process left 69,713 remaining abstracts, roughly 10% of the initial total.

For each cardiac drug, we create a pairwise sample for every other drug, and assign a indicator label to the pair as (*True, False*) if the two drugs interact. This label is used for the binary classification supervised learning problem, in which we predict whether two drugs interact. We also consider the type of interaction, as provided by DrugBank. By removing the drug names in each interaction, we obtain 53 distinct interaction types. For example: *"The serum concentration of (drug 1) can be increased when it is combined with (drug 2)."* These interaction types label a supervised multi-class classification problem to predict the actual type of interaction between two drugs, if any exists.

Thus far, we have examined a range of ML classification techniques using simple word count statistics extracted from the relevant abstracts as input features. This approach demonstrated limited success using both support vector machines and Lasso classifiers to handle the high dimensionality of the word counts [18]. We also examined the application of traditional classification techniques to lower-dimensional representations using a bag-of-words model in conjunction with recently developed advances in semantic word embedding to vectorize our abstracts [23]. Our current work focuses on using recurrent neural networks structures to extract semantic embeddings from variable-length word sequence inputs, like the PubMed research abstracts. This model allows us to maintain the syntactic relationships present in word order. While computationally intensive, these recurrent networks have demonstrated remarkable success in similar NLP tasks, and are central to nearly all current dominant NLP models. More details on this approach are forthcoming [19].

For this project, we considered a set of 44 drugs identified as specifically relevant for treating CHD by pediatric cardiologists at Lucile Packard Children's Hospital. By examining the medical administration records for a cohort of 60 CHD patients, we automatically identify when a patient receives two interacting medications within a given time frame. Based on the level of associated risk predicted by our model, we could use this type of event to trigger an automated alert for the primary care physician to review the administered medications. Such a decision support tool provides significant value by automating the tedious task of literature review, therefore reducing the workload on the physician and allowing them to more efficiently make decisions.

5 Discussion

Machine learning is an powerful tool to improve patient care, extract useful information from dense data sources and provide clinical decision support. While not applicable to all CDS tasks–requiring sufficient data and infrastructure–these methods present significant benefits where applicable. Our work builds a central framework to bring together disparate and scattered sources of data, enabling discrete yet cooperative projects to improve patient care. Sharing methodologies between projects enables us to quickly address new problems.

By designing flexible models and using data sources standard to most EMR systems, we provide an adaptable framework that may be applied and tuned to specific hospital environments, patient populations, and opportunities.

Acknowledgements Our work on these projects would not have been possible without the motivation, and feedback received from LPCH physicians and staff. In particular, we thank Christopher Almond, David Rosenthal, Shannon Feehan, Andrew Shin, and John Dykes.

References

1. Aboukhalil, A., Nielsen, L., Saeed, M., Mark, R.G., Clifford, G.D.: Reducing false alarm rates for critical arrhythmias using the arterial blood pressure waveform. J. Biomed. Informat. **41**(3), 442–451 (2008)
2. Behar, J., Oster, J., Li, Q., Clifford, G.D.: Ecg signal quality during arrhythmia and its application to false alarm reduction. Biomed. Eng. IEEE Trans. **60**(6), 1660–1666 (2013)
3. Chambrin, M.C., et al.: Alarms in the intensive care unit: how can the number of false alarms be reduced? Critic. Care London **5**(4), 184–188 (2001)
4. Chang, R.F., Wu, W.J., Moon, W.K., Chou, Y.H., Chen, D.R.: Support vector machines for diagnosis of breast tumors on us images. Acad. Radiol. **10**(2), 189–197 (2003)
5. Chowdhury, M.F.M., Lavelli, A.: Fbk-irst: A multi-phase kernel based approach for drug-drug interaction detection and classification that exploits linguistic information. Atlanta, Georgia, USA **351**, 53 (2013)
6. Churpek, M.M., Yuen, T.C., Huber, M.T., Park, S.Y., Hall, J.B., Edelson, D.P.: Predicting cardiac arrest on the wards: a nested case-control study. CHEST J. **141**(5), 1170–1176 (2012)
7. Couto, P., Ramalho, R., Rodrigues, R.: Suppression of false arrhythmia alarms using ECG and pulsatile waveforms. In: Computing in Cardiology Conference (CinC), IEEE, pp. 749–752 (2015)
8. Cretikos, M., Chen, J., Hillman, K., Bellomo, R., Finfer, S., Flabouris, A., Investigators, M.S., et al.: The objective medical emergency team activation criteria: a case-control study. Resuscitation **73**(1), 62–72 (2007)
9. Eom, J.H., Kim, S.C., Zhang, B.T.: Aptacdss-e: a classifier ensemble-based clinical decision support system for cardiovascular disease level prediction. Exp. Syst. Appl. **34**(4), 2465–2479 (2008)
10. Goldberger, A.L., Amaral, L.A.N., Glass, L., Hausdorff, J.M., Ivanov, P.C., Mark, R.G., Mietus, J.E., Moody, G.B., Peng, C.K., Stanley, H.E.: (June 13)) PhysioBank, physioToolkit, and physioNet: components of a new research resource for complex physiologic signals. Circulation **101**(23), e215–e220 (2000)
11. Hillman, K., Bristow, P., Chey, T., Daffurn, K., Jacques, T., Norman, S., Bishop, G., Simmons, G.: Antecedents to hospital deaths. Intern. Med. J. **31**(6), 343–348 (2001)

12. Hoffman, J.I., Kaplan, S.: The incidence of congenital heart disease. J. Am. College Cardiol. **39**(12), 1890–1900 (2002)
13. Johnson, A.E., Pollard, T.J., Shen, L., Lehman, LwH, Feng, M., Ghassemi, M., Moody, B., Szolovits, P., Celi, L.A., Mark, R.G.: Mimic-iii, a freely accessible critical care database. Scientific Data 3 (2016)
14. Kohn, M., Sun, J., Knoop, S., Shabo, A., Carmeli, B., Sow, D., Syed-Mahmood, T., Rapp, W., et al., Ibms Health Analytics and Clinical Decision Support. IMIA Yearbook pp. 154–162 (2014)
15. Krizhevsky, A., Sutskever, I., Hinton, G.E.: Imagenet classification with deep convolutional neural networks. In: Advances in Neural Information Processing Systems, pp. 1097–1105 (2012)
16. Lawless, S.T.: Crying wolf: false alarms in a pediatric intensive care unit. Critic. Care Med. **22**(6), 981–985 (1994)
17. Liu, S., Tang, B., Chen, Q., Wang, X.: Drug-drug interaction extraction via convolutional neural networks.In: Computational and Mathematical Methods in Medicine (2016)
18. Miller, D.: Automated Identification of Drug-Drug Interactions in Pediatric Congestive Heart Failure Patients. Unpublished. arXiv: 1702.04615v1 (2016)
19. Miller, D.: Nlp and Deep Learning for DDI Identification in Pediatric Congestive Heart Failure, Manuscript in preparation (2017)
20. Miller, D., Scheinker, D., Shin, A.: Application of Deep Learning to Predict Crisis Events in Hospitalized Infants and Children with Congestive Heart Disease, Manuscript in preparation (2017)
21. Mirowski, P.W., LeCun, Y., Madhavan, D., Kuzniecky, R.: Comparing svm and convolutional networks for epileptic seizure prediction from intracranial eeg. In: Machine Learning for Signal Processing, 2008. MLSP 2008. IEEE Workshop on, IEEE, pp. 244–249 (2008)
22. Oster, M.E., Lee, K.A., Honein, M.A., Riehle-Colarusso, T., Shin, M., Correa, A.: Temporal trends in survival among infants with critical congenital heart defects. Pediatrics. peds–2012 (2013)
23. Pennington, J., Socher, R., Manning, C.D.: Glove: global vectors for word representation. In: Empirical Methods in Natural Language Processing (EMNLP), pp. 1532–1543 (2014)
24. Sayers, E.W., Barrett, T., Benson, D.A., Bolton, E., Bryant, S.H., Canese, K., Chetvernin, V., Church, D.M., DiCuccio, M., Federhen, S., et al.: Database resources of the national center for biotechnology information. Nucl. Acids Res. **39**(suppl 1), D38–D51 (2011)
25. Schmidhuber, J.: Deep learning in neural networks: An overview. Neural Netw. **61**, 85–117 (2015)
26. Segura-Bedmar, I., Martínez, P., Herrero-Zazo, M.: Lessons learnt from the ddiextraction-2013 shared task. J. Biomed. Informat. **51**, 152–164 (2014)
27. Ward, A., Miller, D., Miller, K.: Reducing False Arrhythmia Alarms in the Intensive Care Unit. Unpublished report: https://cs229.stanford.edu/proj2015/288_report.pdf (2015)
28. Wishart, D.S., Knox, C., Guo, A.C., Shrivastava, S., Hassanali, M., Stothard, P., Chang, Z., Woolsey, J.: Drugbank: a comprehensive resource for in silico drug discovery and exploration. Nucl. Acids Res. **34**(suppl 1), D668–D672 (2006)
29. Gm, Xian: An identification method of malignant and benign liver tumors from ultrasonography based on glcm texture features and fuzzy svm. Exp. Syst. Appl. **37**(10), 6737–6741 (2010)
30. Zheng, Y., Liu, Q., Chen, E., Ge, Y., Zhao, J.L.: Time series classification using multi-channels deep convolutional neural networks. In: International Conference on Web-Age Information Management, Springer, pp. 298–310 (2014)

User-Centered Development of an Information System in Patient's Motor Capacity Evaluation

Justine Coton, D. Vincent-Genod, Guillaume Thomann, Carole Vuillerot and François Villeneuve

Abstract Many medical devices are created and rejected because of their lack of adequacy to the clinician needs and situation. The implication of a clinician in the design process may prevents the creation of solution than seems pushed to the users but on the contrary create a solution calibrated to their usages. In this paper a UCD cycle was applied to the development of an information system for patients' motor evaluation using the motion analysis sensor Kinect. The system should support the therapist in its evaluations and provide a way to improve the evaluation. The context exploration, requirements definition, solution proposition and adequacy evaluation was applied. The model allowed to emphasize the important design aspects and those who were correctly answered but also the revisions needed for a second UCD cycle to generate an acceptable device.

Keywords User-centered design · Medical device · Information system
Medical population investigation

1 Introduction

Technologies are invading all aspects of our life. They help manage our task, assist the person by moderating, providing information or support etc. They can have different aims: improve the quality, generate cost economy, or provide support by given access to new information or by making a process more precise and effective.

J. Coton (✉) · G. Thomann · F. Villeneuve
Grenoble INP (Institute of Engineering), G-SCOP, University Grenoble Alpes,
CNRS, 38000 Grenoble, France
e-mail: justine.coton@grenoble-inp.fr

D. Vincent-Genod · C. Vuillerot
Escale PMR department, Hospices Civils de Lyon,
University of Lyon, 69000 Lyon, France

© Springer International Publishing AG 2017
P. Cappanera et al. (eds.), *Health Care Systems Engineering*, Springer Proceedings
in Mathematics & Statistics 210, https://doi.org/10.1007/978-3-319-66146-9_11

121

Medical Devices in Physiotherapy

The medical personals are led to use different types of medical devices or services to help or support their activities. In the case of physiotherapy, a category of tools exists to help them understand the movements and capacities of their patients. Those devices are from the simplest, such as an assessment manual, to the more complex such as for the gait analysis who use 3 different types of technologies: the VICON, a motion analysis tools using infrared cameras and markers, a walk structure equipped with Force plates, and EMG (electromyography) to monitor the muscle activities. Lots of technologies are developed but a lot are also rejected or abandoned for several reasons: they are too complex to use, take too much time, are too expensive or need a specific place that is not available. Most of those aspects could be prevented by having a better understanding of the factors impacting the device. The inclusion of the users and the consideration of all stakeholders in the design process may prevent the device rejection. In those case, the User-Centered Design (UCD) process, a design method putting the users in the center of the process, may be applied.

Design Process and User Centered Design

Different design processes exist to develop a product. Each process have its pro and cons: some are adapted to short-terms projects, some support prototyping, some are efficient for well-defined problem etc. In the case of a medical product there is a clear gap of knowledge between the users (clinicians) and the designer team. In our case it will be the first introduction of technologies into this type of practices which means the impacts will be difficult to define clearly in advance. Those 2 facts mean the process should emphasis the user research to define requirements and evaluate their responses and comprehension with prototypes, it needs to accommodate to an evolution of requirements. Classic systematic and linear models which used few to no prototype such as the Pahl and Beitz approach [1] or even V-model risk to deliver a product that is not in good accordance with the user's needs. An iterative process using prototypes will be needed, as with agile processes but the needs for a clear study of the users' needs make the User-Centered model more adapted even if those two approaches are not necessarily exclusive [2].

The User-Centered Design (UCD) approach is to involve the final users into the design process and was normalized by the ISO 9241-210 [3]. This process is composed of 6 steps that should be iterated until the system matches correctly the user's needs (see Fig. 1). The aim is to calibrate the product to the users. The functionalities should not be "pushed" to the users. The UCD has already be applied into a wide variety of industries. Its main advantages is its capacities to improve the qualities, effectiveness and usability of the solution. A major drawback is the difficulty to maintain the users involved during the whole process [4]. Those features are also valid for the medical device industry, as was investigated by Shah and Robinson [5], notably the difficulty to have access to the users both in terms of time and cost but the extracted knowledge about the users increase the possible success of the product [6, 7].

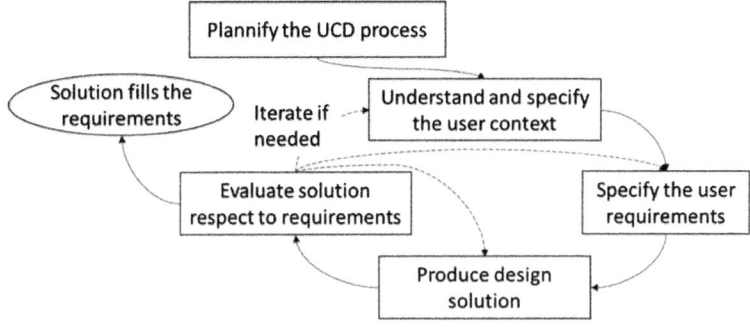

Fig. 1 UCD process [3]

Context of This Study

In physiotherapy, technologies for motion analysis is mostly used in specific and punctual analysis such as the gait analysis. Many technologies are also developed for more regular needs such as in rehabilitation [8]. The apparition of affordable motion analysis sensors, as the Kinect from Microsoft, has generated a renewed interest for assessment and rehabilitation tools and/or e-health in general [9–11]. The development of such tools should be made in close relation with clinicians to lower the risk of rejection.

This paper will present an application of the UCD model in collaboration with physiotherapists to produce a motion analysis information system using the Kinect sensor. This device has to help with the capture and analysis of the patients' movements during an assessment used to define the motor capacities of patients with progressive diseases: the Motor Function Measure (MFM) scale[1] [12, 13]. This scale, administrated once a year per patient, permits to quantify the patient's motor functions. It is composed of 32 exercises rated from 0 (fails to executes the activities) to 3 (do the activities "normally" with controlled motion, regular speed etc.).

2 Case Study: Use of the Kinect Sensor in the MFM Scale

A first UCD cycle was applied (see Fig. 2) which included one therapist to explore and manage the whole process and the punctual implication other therapists for the requirements definition and the evaluation of the solution. Three mains steps will be presented:

1. The context and requirements definition: In this step an analysis of the current context was made and an investigation with 6 therapists was realized to produce a lists of functions and criteria for the tool. The aim was to define the needs and

---

[1]MFM website: http://www.motor-function-measure.org/ [17/02/2017].

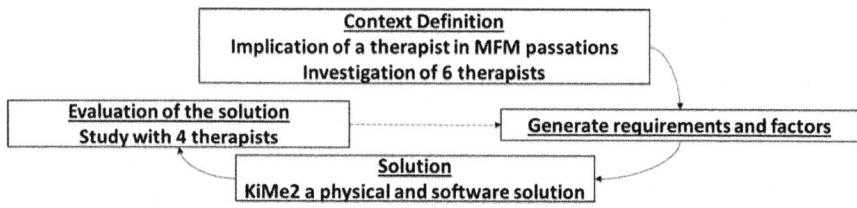

Fig. 2 Application of the UCD process in the Kinect case study

improvements needed for this tool but also to explore the changes and impacts its introduction will have on the process, and its participants (therapists, patients, hospital…).

2. A solution (in terms of physical installation, service and use) was proposed based on those functions and criteria.
3. Evaluation: The tool was then evaluated in regards to the requirements. The efficiency of some key functions were evaluated during a study including 4 therapists.

2.1 Context and Requirements Definition

The needs and opinions on this new tool were explored on one hand with the inclusion of a physiotherapist throughout the whole process to evaluate the current state of the process and to conduct the design evolution. And on the other hand with an investigation based on the "Unified theory of acceptance and use of technology" model [14] to define the impact this new technology may have on the process.

2.1.1 The MFM Assessment

An MFM assessment is realized in a space equipped with chairs and tables (adaptable to the patient morphology) and a physiotherapist table. In addition to those utilities the MFM include a manual as a reference for scoring and some commonly found objects (a tennis ball, a pencil, sheet of paper and 10 coins). The MFM is composed of 32 exercises. An exercise takes between less than a minute to 2 min and the whole scale take on average 30 min.

Only one physiotherapist is needed. For each exercise the therapist gives the instruction to the patient, he may ask if the patient think he can do it and adapt the instruction to the patient's capacities, and should check every time on the manual the score levels. In case of doubt, the lesser score is chosen. The patient can do the exercise twice to give the best score, doing anymore trials may be detrimental because this type of patient is easily fatigable. During the whole evaluation the therapist stays near the patient to administrate care. If the patient is a child then the parents may also stay during the evaluation.

Finally all scores and commentaries are noted on a paper scoring sheet which is kept in the patient folder, the therapist should then compute the results in an excel file to generate a graphic of the successives MFM and send the result to the MFM database.

Several points can be extracted from this description:

- The MFM is a tool made to be easily accessible for clinicians: the objects used are easily found in a "common" physiotherapist or even ergo-therapist room.
- The MFM is quick to administrate: up to 35 mn.
- There is a strong relationship between the therapist and its patients: the therapist has to stay near the patient for his care but also to maintain him motivated and to discuss with him, to adapt the exercise to its capacities and mood.
- The MFM manual is a reference that should always be at hand.
- A training is necessary.

But other points leave room for improvement:

- The clinician may miss a point if he is tired or concentrated on another point.
- Some movements can be difficult to explain or represent into commentaries.
- The storage of the result in the MFM database has to be manually done. This tends to lead some clinicians to not update it.

2.1.2 The Impact of the Introduction of Technology in the MFM

The impacts of the introduction of this technology was investigated with semi-structured interviews and questionnaires. For this investigation 2 groups of therapists were mobilized (for a total of 6 therapists):

- 1 group of non-users: 3 therapists with no prior knowledge of the Kinect.
- 1 group of users: 3 therapists with prior knowledge of the Kinect as a tool for evaluation.

The system acceptance seems to be correlated with 3 factors:

1. The system performance. It's the condition sine qua none for its usability: the data provided have to be reliable and the new information to be interesting enough for the disease comprehension and score's attribution.
2. Distributed cognition: the cognitive attention of the therapist, previously centered on the patient, will now be distributed between the patient and the Kinect-system.
3. The social influence: this new system will impact the profession and the view on the work done

The resulting functions and factors are listed below:

System Performance

- The system should provide validated data (accurate, objective, reproducible, and validated by the community etc.).
- The system should provide additional information useful for the analysis of the disease and the assessment and facilitating the follow-up.
- The system should not stretch the duration of the assessment (limit the increase of workload).
- The system should conserve the "simple and accessible" feature of the classical MFM by remaining into a moderate pricing range and requiring technologies easily accessible.
- The system is an information system and the therapist has to keep the control.
- The system has to be safe and hygienic.

Distributed Cognition

- The capture of the patient movements should be done without

  - Hindering the evaluation capacity of the therapist.
  - Taking too much additional time in installation and in use.

- The analysis of the movements and the data representations should allow the therapist to understand and trust the system.

  - The system should avoid the "black-box" effect.
  - The system need to provide information calibrated to the clinician needs.
  - The analysis should be quick and the access to desired data easy.

- The system should not hinder the relation between the therapist and the patient but try to improve it. It should not distract the patient or the therapist.

Social Influence

- The system should not be seen as an intrusion or a surveillance tool.
- The therapists should not have the impression to lose the control.
- The system should not lead the therapists to lose their expertise.

2.2 The Application

The following installation, named KiMe2 (Kinect Medical Measurement), is proposed. The patient realizes its activities in front of a Kinect sensor with the therapist being at the patient side and able to touch him and interact with him. The therapist is equipped with a tablet that contains its MFM manual and a wireless connection with the computer linked to the Kinect sensor and the KiMe2 software (see Fig. 3).

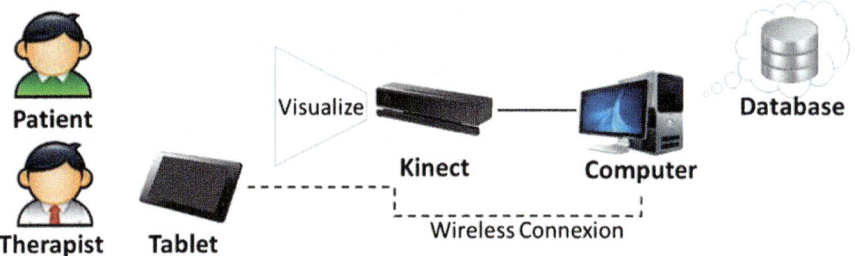

Fig. 3 Physical installation

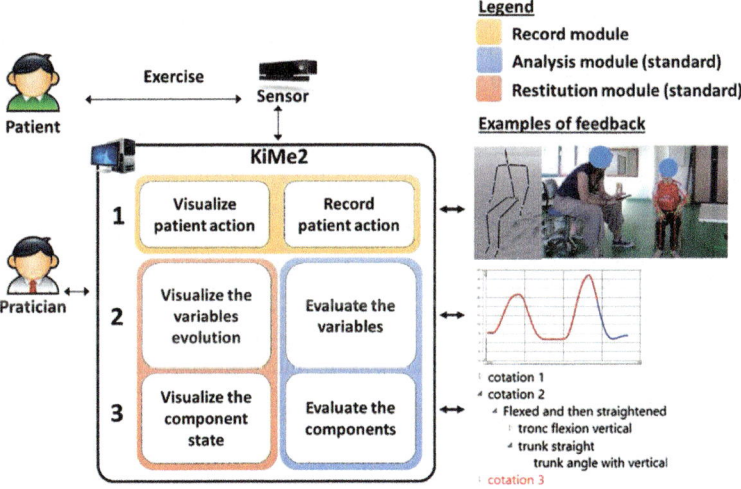

Fig. 4 Software architecture

This tablet and software enable the therapists to see what the Kinect see and if the system function correctly (if the patient is correctly positioned for the sensor and for the activities etc.). The therapist can then ask the patient to do its exercises and record them and then provide its decisions on the tablet which saves them and send them into the database with the record of the movements. The results can be analyzed by the software whom is described on the Fig. 4.

The therapist can use the tablet and thus software to analyze the exercises and can revise or validate its decisions depending on the results. For this, the system provides a score and its justifications. The justification indicates which components (scoring criteria) were validated such as if "a support was used", if a posture were attained, etc. If the therapist has a doubt on a score or more specifically on a component (scoring criteria) of the score, the system provides for each score the list of the components (scoring criteria) mobilized and their states and the numerical variables used for their calculations (see line 3 in Fig. 4). If the therapist has a doubt on a value in particular he can generate the plot of the variables (numerical variable, see line 2 in Fig. 4). Those graphs are associated to the component to facilitate their access. The exercise can also be reconstructed to re-visualize the exercise (see line 1 Fig. 4). This structure makes the scoring system more transparent to the therapist.

2.3   Adequacy of the Proposition to the Requirements

System's Performance Adequacy

The system answered several requirements: (1) The system Tablet—Kinect—Computer can easily be found and remains on a pricing range way lower than other movement analysis tools on the market, but it should be noted that it is substantially more complex and costly than the current MFM. (2) The system does not bring any physically harm or any hygienic difficulties but the flow of data has to be secured. (3) The therapist control over the final score was kept and comforting: the aim is to provide reassurance and additional information. (4) The record of the exercise takes more time than a classic MFM but stays in an admissible range. The analyzing process takes a significant time that need to be improved. (5) The treatment process (create the MFM graph of the disease and updating the result into the database) could be facilitated and help provide more systematically data on rare disease. (6) On the other hand it should be noted that the Kinect is not quite performant enough for now. The measurement have to be made more accurate and robust.

Adequacy to the Distributed Cognitive Factors

The cognition distribution was evaluated with 4 therapists: they had to use the system to see if they were able to record and analyze correctly the MFM exercises with the KiMe2 tool.

In terms of capture: the therapists were able to easily position themselves in a way that does not hinder the recording of the activities. However the system presence itself made the therapists less incline, even if possible, to be near or in contact with the patient which can be dangerous for the patient care. Globally the system did not hinder the evaluation but the fact the therapist has to stay at the patient's side changes the patient's stare orientation which can be bad for some exercises. During the evaluation of a posture the fact to not be able to go in front of the patient to check the posture is also an inconvenient and can hinder the therapist evaluation.

In terms of analysis: The system enables the therapist to understand the software results proposition. The therapists can know if they have to revise their judgments or maintain it. They understood what were reasons for a score and if the software were missing or misinterpreting a knowledge. They were able to see if this system can monitor components that can be difficult to analyze or be overlooked. The mobilized knowledge were easy to understand and the parallel between the mobilized knowledge and the manual description was made even if the terms needed to be more medical oriented. The data were easily accessible: the graphs were easy to find and generate. The therapists were able to find the sought information (an angle, what was the knowledge mobilized etc.). The possibility to represent the amplitudes and displacements of movements into graphs were an interesting improvement of the system.

In general the relationship between the therapist and the patient remains the same. The system may distract a bit the therapist from the patient, but it can also be a tool to generate motivation and discussion with the patient. But therapists can have difficulty in managing the patient, the activities and the system (too much dispersion).

Several points were highlighted and will have to be corrected into the next UCD cycle. This system make notable change on the therapists practices. The therapists and the patient, not positioned as before, change their stares orientation which is not straight anymore, the system makes the displacement around the patient more difficult (since the therapist should not pass in front of the sensor) and naturally tends to stay far from the view of the sensor, this makes the evaluation of posture more difficult and may endanger the patient care. The introduction of the system distracted a bit the therapists which did not behave as usual and thus made mistake during their evaluation. Finally the installation and system increase the assessment duration and the workload but the simplicity of the result and of the installation in regards to other system as the VICON for the gait analysis was noted and the capacity to understand the system analyze was well appreciated.

Adequacy to the Social Influence Factors

A formation will be needed to prevent the negative factors in the social influence sphere. The formation should emphasis the fact that this tool does not monitor the therapist's activities and has to be seen as a support system for the therapists and not a replacement for any role of the therapist (the therapist stays the main relation with the patient, he keeps the control on the control and result, etc.). It should also emphasis the fact that the system is not a perfect but that it may provide more consistency on specific values or data but should in no way replace the therapist or its expertise which will always be essential.

3   Conclusion

In this paper a UCD process, involving a physiotherapist, was applied to develop a medical device for motion analysis during assessments. The process allows the description of the needs, the creation of a solution and its evaluations. The evaluation of this systems provided 4 main good points to maintain:

1. The systems allows an easy capture of the exercises with a good visual feedback to follow the assessment.

2. The analysis allows a new standardized point of view.
3. The software analysis can be understood and interpreted by therapist. The representations allowed a good transcription of the exercise.
4. The system may brought new information with new representation of the movement such graphs of displacement and amplitude not available until now.

But the systems still need improvements on 4 mains points, and will ask for an iteration of the process

5. The systems increases the assessment duration and the workload of the analysis. The used of the representation tools taking times.
6. The system is still heavy for the cognitive: the therapist has to manage the assessment, the patient and the system.
7. The system changes the practice and notably the position of the therapist and of the patient that can hinder the assessment of some exercise.
8. The Kinect system needs to be perfected to provide more reliable score.

The UCD cycle allows to anticipate positive factors and prevents the negative factors that will be corrected on the next cycle. This type of process should facilitate the introduction of new medical device that may be helpful in common medical practice.

References

1. Pahl, G., Beitz, W., Feldhusen, J., Grote, K.-H.: Engineering Design, vol. 11. Springer, London (2007)
2. Fox, D., Sillito, J., Maurer, F.: Agile methods and user-centered design: how these two methodologies are being successfully integrated in industry. In: Agile 2008 Conference, pp. 63–72 (2008)
3. ISO 9241-210: International Organization for Standardization. Ergonomics of Human-System Interaction—Part 210: Human-Centred Design for interactive Systems, p. 36 (2010)
4. Vredenburg, K., Mao, J.-Y., Smith, P.W., Carey, T.: A survey of user-centered design practice. In: Proceedings of the SIGCHI Conference on Human Factors in Computer Systems: Changing Our World, Changing Ourselves—CHI '02, no. 1, p. 471 (2002)
5. Shah, S.G.S., Robinson, I.: Benefits of and barriers to involving users in medical device technology development and evaluation. Int. J. Technol. Assess. Health Care. 23(1), 131–137 (2007)
6. Teixeira, L., Ferreira, C., Santos, B.S.: User-centered requirements engineering in health information systems: a study in the hemophilia field. Comput. Methods Programs Biomed. 106(3), 160–174 (2012)
7. Rinkus, S., Walji, M., Johnson-Throop, K.A., Malin, J.T., Turley, J.P., Smith, J.W., Zhang, J.: Human-centered design of a distributed knowledge management system. J. Biomed. Inform. 38(1), 4–17 (2005)
8. Zhou, H., Hu, H.: Human motion tracking for rehabilitation—a survey. Biomed. Signal Process. Control 3(1), 1–18 (2008)
9. Hondori, H.M., Khademi, M.: A review on technical and clinical impact of Microsoft Kinect on physical therapy and rehabilitation. J. Med. Eng. (Hindawi Publishing Corporation) 2014, 16 (2014)

10. Webster, D., Celik, O.: Systematic review of Kinect applications in elderly care and stroke rehabilitation. J. Neuroeng. Rehabil. **11**, 108 (2014)
11. Da Gama, A., Fallavollita, P.: Motor rehabilitation using Kinect: a systematic review. GAMES Heal. J. Res. Dev. Clin. Appl. **4**(2), 13 (2015)
12. Bérard, C., Payan, C., Hodgkinson, I., Fermanian, J.: A motor function measure for neuromuscular diseases. Construction and validation study. Neuromuscul. Disord. NMD **15** (7), 463–470 (2005)
13. Vuillerot, C., Payan, C., Iwaz, J., Ecochard, R., Bérard, C.: Responsiveness of the motor function measure in patients with spinal muscular atrophy. Arch. Phys. Med. Rehabil. **94**(8), 1555–1561 (2013)
14. Venkatesh, V., Morris, M.G., Davis, G.B., Davis, F.D.: User acceptance of information technology: toward a unified view. MIS Q. **27**(3), 425–478 (2003)

A Hybrid Simulation Approach to Analyse Patient Boarding in Emergency Departments

Paolo Landa, Michele Sonnessa, Marina Resta, Elena Tànfani
and Angela Testi

Abstract This paper deals with the problem of patient boarding in Emergency Department (ED) due to the lack of availability of stay beds in inpatient hospital wards. The boarding of emergent patients is a major reason of ED overcrowding, which in turn creates long waiting times in ED, as well as patient and staff dissatisfaction. In this paper a hybrid simulation framework is proposed to describe how emergent and elective patients flows interact each other inside the hospital. The framework combines System Dynamics (SD) and Discrete Event Simulation (DES). Flows are generated at macro level by the SD model, while in DES model they are disentangled into single entities following a detailed process-oriented pathway. The hybrid model has been then applied and validated to data from a medium-size public hospital located in Genova (Italy). Furthermore, some metrics (such as waiting times to be admitted in hospital, number of trolleys in ED, inpatient bed occupancy rates and elective patients delayed) are proposed as key indicators to assess the system performance.

Keywords Discrete event simulation · System dynamics · Patient flows management · Bed management · Patient boarding · Emergency department

---

P. Landa
Medical School, University of Exeter, Exeter, UK
e-mail: P.Landa@exeter.ac.uk

M. Sonnessa (✉) · M. Resta · E. Tànfani · A. Testi
Department of Economics and Business Studies, University of Genova, Genoa, Italy
e-mail: michele.sonnessa@edu.unige.it

M. Resta
e-mail: resta@economia.unige.it

E. Tànfani
e-mail: etanfani@economia.unige.it

A. Testi
e-mail: testi@economia.unige.it

© Springer International Publishing AG 2017
P. Cappanera et al. (eds.), *Health Care Systems Engineering*, Springer Proceedings
in Mathematics & Statistics 210, https://doi.org/10.1007/978-3-319-66146-9_12

133

1 Introduction and Problem Addressed

Overcrowding of Emergency Department (ED) together with long waiting lists for elective surgery are two of the main topical issues in many healthcare systems. In this paper, we focus on the problem of patient boarding in ED. This problem deals with emergent patients that, after the first intervention and diagnosis into an ED, should be admitted in a hospital inpatient ward to have further investigations and treatments. If a bed is not available patients have to wait in ED for a free bed: this status is called *boarding*. The literature stresses that the ED boarding is a major reason of ED overcrowding and elective admission postponements [5]. Two main issues therefore arise: first, how to guarantee the completion of care pathway timely and properly for emergent patients that were already diagnosed and in charge of an ED; second, how to avoid the delay of care delivery for elective patients waiting to be admitted to receive their timely and proper care.

At a first glance, the ED boarding may be originated by the limited number of inpatient ward beds. In recent years, the situation has worsened worldwide, due to the reduction of inpatient beds owing to public budget restrictions. Indeed, the optimal levels of beds should take into consideration the holding cost of unoccupied resources together with the shortage cost of delaying care delivery [22]. However, in the examined case, the issue is particularly tricky, because it is necessary to assume a dynamic point of view. Supply and demand cannot match only on average, but also dynamically, taking the patient flow variability into account. Moreover, the same bed capacity engenders different costs and benefits depending on how activity is organized.

In this paper, we do not focus on strategic issues (provided that the total number of beds cannot be changed, as, currently, this is the more realistic assumption), but rather on tactical and operational issues. This implies that the improvement of performance and stakeholders' (hospital, patients, society) utility is not only a problem of insufficient resources, but it can be achieved also by organizational changes. Thus, the objective of this paper is to develop a Simulation Framework for modelling the overall ED and hospital system and predict the impact of organizational changes without increasing the number of beds.

Simulation has been deeply explored as the technique for modelling complex health care systems [16]. Many studies had focused on simulation models for the analysis and improvement of ED performance [2, 3, 13, 21].

Discrete Event Simulation (DES) models generally capture details about the supply side and service level of the system under study, and are therefore well suited for modelling EDs and capturing their performance [11, 14, 18].

System Dynamics (SD) models, on the other hand, focus on feedback and non-linear relationships so that they are a natural choice when the dynamic complexity of the patient flow modelling into EDs and its natural variability needs to be considered [20, 23, 25].

Recently the attention has grown on the adoption of hybrid simulation models, where the combination of DES and SD methodologies helps in having a wider and holistic point of view of the system complexity, thus improving the analysis and the achievement of objectives [6, 8]. Chahal and Eldabi [10] argued that the combination of SD and DES allows decision makers to face decisions from both the microscopic (capturing detail up to individual level) and macroscopic (holistic system wide interactions) viewpoints. In their work, they explored different modes of governance in UK healthcare and the applicability of hybrid simulation with each of them, proposing three possible ways of using hybrid simulation: Hierarchical, Process-environment and Integrated. Brailsford et al. [9] analysed in detail the approach of combining DES and SD modelling, describing weakness and strength of the approaches and they apply them to two practical healthcare problems. Viana et al. [24] have developed a composite model with the combination of DES and SD, used to address the sexually transmitted infection Chlamydia. The authors analyzed the results of both two techniques. The results show that the composite model better captures longer term effects of untreated infections linked to the operational decisions and clinic performance than the "SD only" model.

Other hybrid approaches deal with the adoption of DES and Agent-Based (AB) simulation for system modelling of different services such as Radiology service access [1] or Emergency Medical Services management [4, 17] where AB models human behaviour in decision making and DES can capture the different patient flows.

According Heath et al. [15] the hybrid simulation methodology is not precisely defined yet, but in most cases, we can think of the combination of continuous and discrete paradigms as a hybrid. Heath et al. [15] stressed how different simulation techniques can represent different world views. While DES approach gives a process-oriented world view, SD are able to model qualitative aspects through tools like *causal loop* or *influence* diagrams.

In this paper, we start by a previous study examining the emergent and elective patient flows using a SD model [19], and develop a Process-environment Hybrid Simulation Framework. The framework is able to get more knowledge on the ED/hospital system and to evaluate the performance of the system and the impact of improvement organization strategies. Moreover behavioural aspects are introduced to explain the intrinsic system adaptability and to reproduce coping mechanisms.

The paper is organized as follows. In Sect. 2 the Hybrid Simulation Framework is analysed and the main elements and components explained, while in Sect. 3 the real case study and data collection analysis are introduced. In Sect. 4 the validation and some preliminary results, on the application of the framework to the real case under study, are reported. Finally, conclusions and future research direction end the paper.

2   The Hybrid Simulation Framework

The modelling approach herein proposed combines SD and DES in a hybrid simulation environment to model the overall ED/hospital system over time-based fluctuations of incoming patient flows.

In order to design the roles of the sub-models it is very important to highlight the feedbacks underneath the system going to be reproduced. As in [7], in SD modelling, the concepts of feedback and causal effects are important and they should be a priori deeply analysed by combining qualitative and quantitative aspects. They seem to be very effective in catching qualitative phenomena, behaviours and complex adaptive mechanisms difficult to be detailed as rules in a DES environment.

In Fig. 1 the causal loop diagram of the whole system is reported.

Indeed, the evidence resulting from observable data is that the system can generate ED overcrowding but it never blows up.

A possible explanation relies in the existence of some feedback maintaining the system in equilibrium when peculiar events cause critical conditions.

Looking at Fig. 1, we assume three behavioural mechanisms described by the red arrows. The first behavioural mechanisms is represented by the increase in the

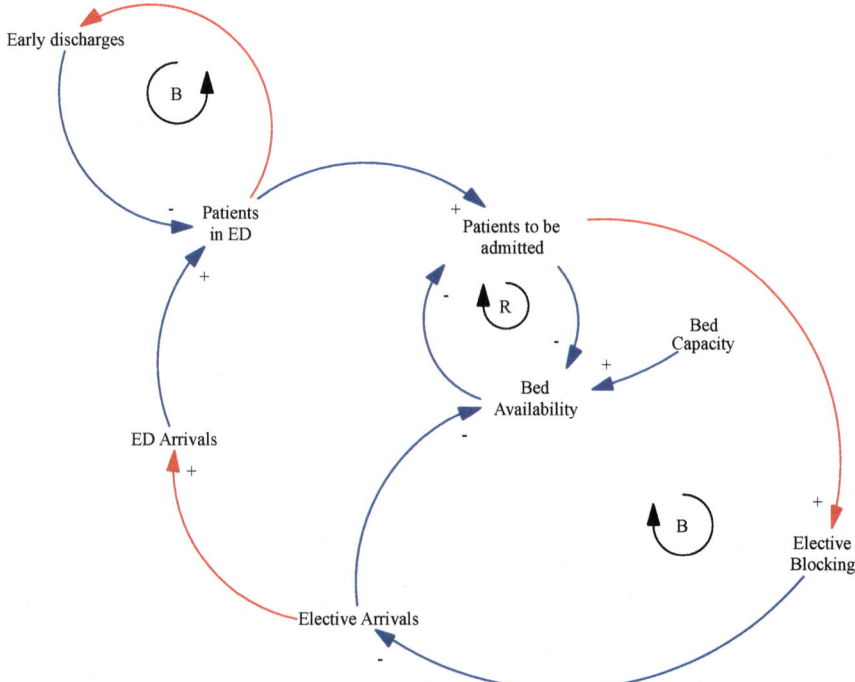

Fig. 1 Causal loop of the system under study

ED productivity, here named Exit Rate Modifier (ERM). This behaviour is captured by the balancing loop represented on the top left side of the diagram. The arrow "Early discharges" refers to higher and faster discharges from ED, transferrals of patients to other hospitals or other care facilities, trigged by increase in arrivals.

This increase, however, entails increasing the number of patient who require admissions in inpatient (see the feedback loop labelled by 'R' centred in the Fig. 1). At this point the second behavioural mechanism Elective Blocking", is used, as in the current practice this is the usual tool to allow emergent admission by reducing the planned arrivals in inpatient wards.

Blocking elective patients, however, triggers the third behavioural mechanism: when the waiting time to be admitted becomes too long for elective patients, some of them in more critical conditions bypass the waiting list and address the ED directly, thus further increasing the "ED arrivals". Such effect is here called Arrival Rate Modifier (ARM). This is due, partly, as a direct result of the worsening of their health status and dissatisfaction and, partly, as a behavioural response of doctors wishing to get their patients admitted 'by the back door'.

The SD component of the hybrid model is intended to assess the mutual interaction of the three behavioural mechanism that are able to maintain equilibrium.

In conclusion, the SD component assumes a holistic view of the system at an aggregate high level of patient flows. Thus, it is necessary to implement also the DES component able to develop the productivity aspects, at a detailed individual level [12]. In particular, it is possible to find out bottlenecks and excessive waiting. The combination of SD and DES techniques can, therefore, offer significant benefits to the ED-hospital modelling and assess the impact of organizational changes.

The hybrid model, represented in Fig. 2, considers the above described environment.

The first component is the SD sub-model which generates the patient arrival flows to ED using arrivals rates empirically derived by way of seasonal time series. The SD sub-model applies an exit rate to patients' stocks with a time delay. Such exit flows are split into admitted and discharged patients. The patients-to-be-admitted flow generated by the SD sub-model is then disentangled into single entities, following process-oriented pathways described by the DES sub-model. The SD model calibrates the stock of patients in the system by regulating productivity rate (the ERM function). In this way it is then able to provide a more realistic flow of patients to the DES model, in presence of sudden demand variation. In addition, SD component is able to receive signals from the DES model to modify the demand. In fact, a small part of the elective patients rejected by blocked elective list are supposed to address the ED (the ARM function). The two models are so forth interconnected: the output of the SD model gives the input to the DES model, while the output of the DES component may modify the arrival flows (for both emergent and elective patients) generated by the SD model. In particular, the DES sub-model reproduces the pathways of elective and emergent patients into inpatient wards. Each admitted patient is associated with a stochastic LOS generated from empirical distributions. The DES sub-model implements the so-called Bed Manager (BM) rule, which models the decision of the BM function

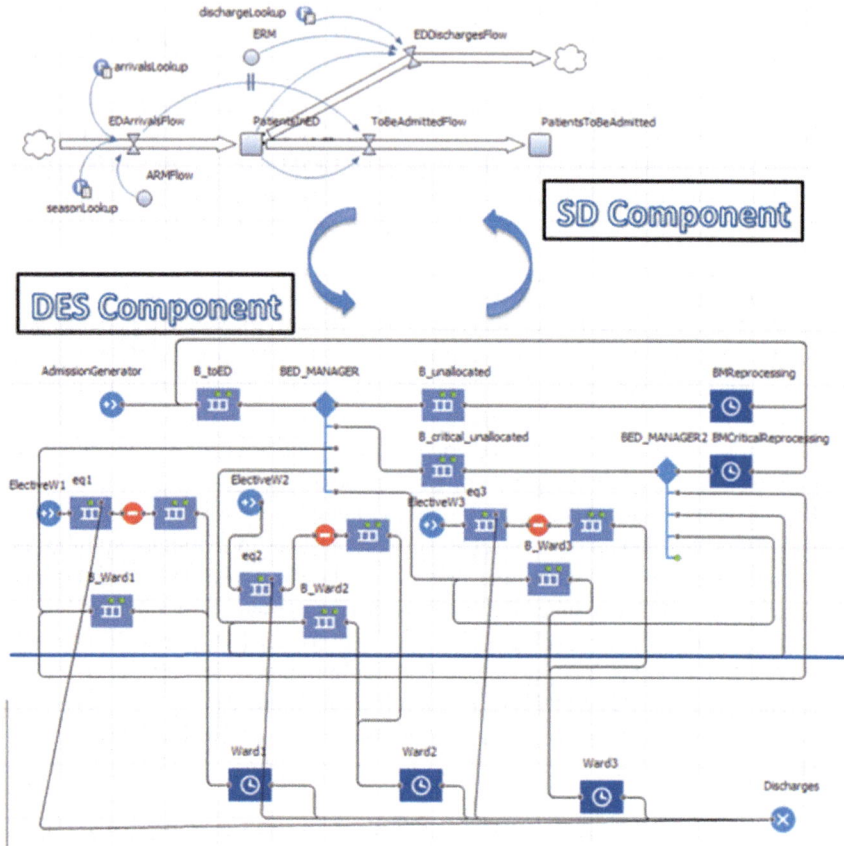

Fig. 2 Hybrid simulation model representation

inside the hospital system that may decide to activate the *elective admissions blocking* when the number of patients boarding in ED reach a critical threshold parameter a priori fixed.

3   Case Study

Thanks to the collaboration with the Local Health Authority of the Liguria region, an observational analysis was conducted based on data collected over a 1-year period (from January to December 2015) in a public hospital of a large city's health district. Inpatient hospital wards were classified as "Medicine" or "Surgery" main specialty area. During the 1 year period under examination the total number of ED accesses was equal to 44194; among them 6278 patients (the 14% of the total

Table 1 Number and percentage of hospital admission for each group of specialties

Ward	Emergent admissions		Elective admissions		Total	
	# of patients	%	# of patients	%	# of patients	%
Medicine	4227	67	1276	29	5503	52
Surgery	2051	33	3113	71	5164	48
Total	6278	100	4389	100	10667	100

number of ED accesses) needed to be admitted in a hospital wards for further treatments, while 37916 (86%) had been discharged after the ED diagnosis and treatment. During the same period, the total number of elective admissions amounted to 4389 patients. The total number of admissions in hospitals was 10667, the greatest percentage of admitted patients (59%) came from the ED, while the remaining 41% was made up by programmed admissions. This created a great level of complexity and uncertainty in the inpatient wards. In Table 1 the number and percentage of total, emergent and elective admissions for each macro group of specialties are reported.

A first look to the data highlights that of the 6278 emergent admissions, 4227 (67%) were directed to medical wards, while about a half of this number, i.e. 2051, needed further treatments in surgical inpatient wards.

In deepest detail, looking at Table 1, row by row, one can then see that the number of emergent and elective admissions were 4227 and 1276 respectively, in the Medicine group; 2051 and 3118, respectively in the Surgery group. The total number of admissions arc almost thc samc for thc two groups: 5503 and 5164, respectively, for Medicine and Surgery. Thus, the proportion between emergent and elective admissions is opposite in the two groups.

Moving to the supply side, Table 2 provides the number of available beds and the average length of stay (LOS) for Medicine and Surgery. The number of beds equals to 210 and 135, respectively. The average LOS for the Medicine group is almost twice than in the Surgery.

In order to estimate the distributions of inter-arrivals for emergent and elective patients, data have been analysed and ad hoc empirical distributions derived.

In Figs. 3 and 4 the emergent and elective patient admissions into medicine and surgery hospital wards and the ED flows are reported. Different behaviours are observed with respect to months, weekends and working days. In order to define regularity, we developed 72 time slots.

Table 2 Number of beds and average LOS for emergent and elective admissions and for each group of specialties

Ward	Number of beds	Average LOS (in days)	
		Emergent	Elective
Medicine	210	13.2	10.5
Surgery	135	9.3	4.6

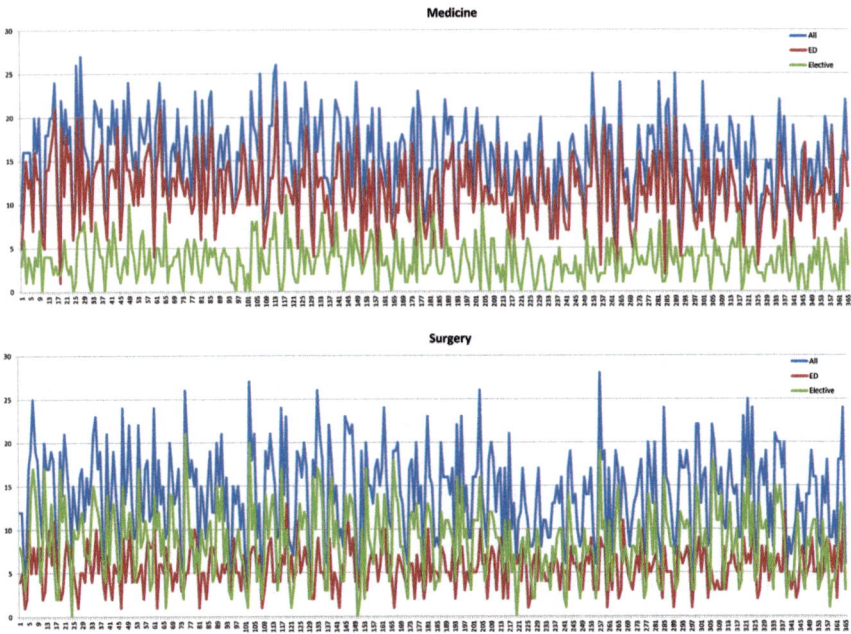

Fig. 3 Emergent and elective patient admission into medicine and surgery wards

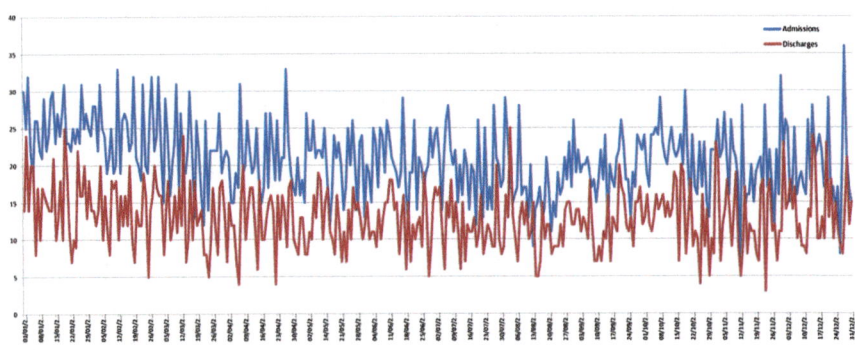

Fig. 4 ED admissions and discharges patient flows

The patient admissions in medicine and surgery ward present an interesting seasonality. Medicine has a smoother variability than surgery, and the number of elective admission is a small part of the overall number of admissions, while surgery present a higher variability in terms of peaks of activities. This difference is due to the shorter patient LOS in surgery than in medicine patients, and to the different impact of emergent patient in surgery admissions. Also, ED admissions and discharges present a high variability that influences the ward activities and the overall bed availability.

4 Validation and Preliminary Results

A baseline scenario has been defined with the following parameters and model settings. The Bed Manager computes the ward beds available each hour as the sum of the available beds to those that will be released in the next 2 h minus the number of elective patients that will be admitted in the next 36 h. The BM activate a 72 h' elective admissions blocking when more than 10 emergent patients are waiting to be admitted in ED, i.e. boarded.

The hybrid model has been implemented using the simulation software Anylogic. Special attention was devoted to validate the model to ensure that the simulation outputs were adequately represented in the system under investigation. One-year steady state results, based on 20 IID replications (with different random seeds) and a warm up period of 3 months, are reported. Validation is based on t-Student test comparing the empirical and average simulated outputs of the monthly ward occupancies. The null hypothesis (H_0) was tested under a probability of rejecting the model at the $\alpha = 0.01$ level. The corresponding critical value of the test was $t_{19,0.99} = 2.539$. Table 3 reports the statistical test for each month and ward group. The results show the ability of the model to represent the real system and its seasonality patterns.

The model well fits the empirical bed occupancy, which is the result of concurring parameters such as ED and elective arrivals, emergent and elective patient's length of stay, the coping behaviour of clinicians and nurses in day-by-day hospital management, which changes over time.

Table 3 T-student test for ward occupancy

Months	Medicine				Surgery			
	Real measures	Avg. Simulated output	Δ_{opt}	T-test (t_{19})	Real measures	Avg. Simulated output	Δ_{opt}	T-test (t_{19})
1	190,52	184,41	0,03	1,20	89,52	90,99	−0,02	1,82
2	194,87	186,51	0,04	2,53	92,45	89,39	0,03	1,00
3	194,68	186,76	0,04	3,30	86,94	91,20	−0,05	1,58
4	191,61	195,06	−0,02	12,24	86,77	90,46	−0,04	2,44
5	186,55	191,86	−0,03	6,62	92,45	87,78	0,05	−0,12
6	174,19	184,96	−0,06	1,75	91,63	89,95	0,02	1,26
7	180,26	181,47	−0,01	−0,68	83,06	88,69	−0,07	−0,05
8	163.71	182,87	−0,12	0,24	81,45	90,43	−0,11	1,20
9	177,19	180,50	−0,02	−1,29	88,06	90,38	−0,03	1,27
10	183,57	192,12	−0,05	6,30	89,93	90,51	−0,01	2,04
11	177,03	187,95	−0,06	3,87	92,03	93,51	−0,02	4,22
12	174,65	181,48	−0,04	−0,62	87,70	91,52	−0,04	0,22
Total	2188,83	2235,94	−0,02		1062,00	1084,81	−0,02	
90% confidence				2,539				2,539

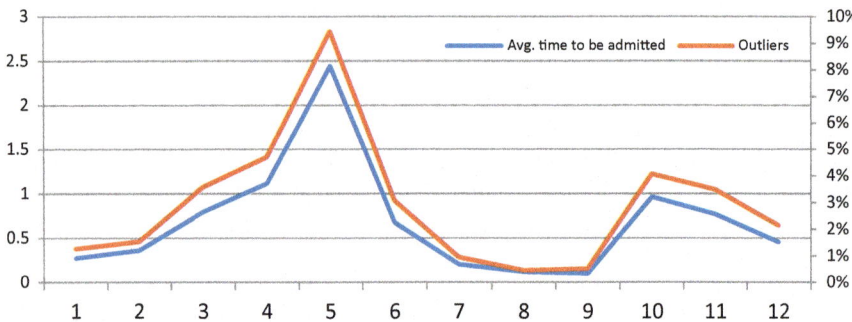

Fig. 5 Average time to be admitted and outliers

We then focused on the average time length to admit emergent diagnosed patients, which is one of the most relevant measures to evaluate the about 1 h waiting is needed to obtain a bed in the impatient wards. Drilling down this datum into monthly dynamics, Fig. 5 shows how this time length varies from 0.11 h in September to 2.4 h in May. Indeed, the regional NHS imposes the limit of 8 h for patient waiting time for admissions as a constraint for every hospital. In the critical month of May, the percentage of patients overtaking 8 h (outliers) waiting for obtaining a bed reaches the 10% of all admitted patients.

Table 4 briefly compares the global outcomes of simulation model with observable statistics. To such aim is noteworthy to observe that certain measures such as, average time to be admitted, more than 8 h waiting and Blocked elective measures cannot be evaluated in observable systems, because they are not included in the hospital's dataset.

The model is able to well mimic the ward occupancies. Due to the effect of the ERM and ARM functions the balancing between arrivals and discharged patients is not precise. Nevertheless the model is able to produces an input flow that reproduce very accurately hospital inpatient ward dynamic. The validation on the model on both the ED and ward flows can help to estimate very important measures likewise

Table 4 Simulation results

Measure	Simulated output	Real measures
Ed admitted	6.235	6.278
Elective admitted	4.142	4.389
ED discharges	35.512	37.916
Average time to be admitted (h)	0.68	–
Average patients in ED	29.56	25.53
Medicine wards utilization rate (%)	86	85
Surgery wards utilization rate (%)	66	66
More than 8 h waiting	186.45	–
Blocked elective	226.00	–

the time spent by emergent patients to obtain a free bed and so forth the number of patients overtaking the 8 h limit imposed by Local Health policy. This measure is currently not recorded by hospital information system despite its relevance in terms of efficiency measurement. The model can be used to provide an estimation of this parameter and it can be used to evaluate alternative what-if scenario analysis using the hybrid model.

5 Conclusions and Further Works

The model herein developed well mimic the reality of the complex dynamics of patient flows and interactions between Emergency department and hospital wards.

Provided the difficulty of adapting hospital systems, in short and medium periods, in terms of resources and bed availability, adopting this hybrid model might be useful for the hospital management to understand the system itself and the dynamics that lead to the main problem related to ED boarding and bed management. The introduction of behavioral aspects and auto-adaptive mechanisms allow to better reproduce real life phenomena. The results from seasonality analysis and validation show the model robustness and its ability to mimic internal dynamics and to adapt the system response to external demand variations.

On the other hand, the introduction of adaptive behaviors as well as hypothesis about capacity of the system to react to exogenous conditions introduces additional challenges to perform an extensive model validation and results interpretation.

Future works will be oriented to use the model for a scenario analysis aimed at comparing and evaluating the impact of alternative reorganization scenarios on system performances (i.e. waiting times to be admitted in hospital, number of trolleys in ED, inpatient bed occupancy rates and elective patients delayed).

References

1. Abdelghany, M., Eltawil, A.B., Abdou, S.F.: A discrete-event and agent-based hybrid simulation approach for healthcare systems modelling and analysis. In: IEOM Society International (eds.) Proceedings of the 2016 International Conference on Industrial Engineering and Operations Management, pp. 1921–1928 (2016)
2. Abo-Hamad, W., Arisha, A.: Simulation-based framework to improve patient experience in an emergency department. Eur. J. Oper. Res. **224**(1), 154–166 (2013)
3. Aboueljinane, L., Sahin, E., Jemai, Z.: A review on simulation models applied to emergency medical service operations. Comput. Ind. Eng. **66**, 734–750 (2013)
4. Anagnostou, A., Nouman, A., Taylor, S.J.E.: Distributed hybrid agent based discrete event emergency medical services simulation. In: Proceedings of the 2013 Winter Simulation Conference: Simulation: Making Decisions in a Complex World, pp. 1625–1636 (2013)
5. Bagust, A., Place, M., Posnett, J.: Dynamics of bed use in accommodating emergency admissions: stochastic simulation model. BMJ **310**(7203), 155–158 (1999)

6. Brailsford, S., Churilov, L., Dangerfield, B.: Discrete-Event Simulation and System Dynamics for Management Decision Making. Wiley (2014)
7. Brailsford, S.C.: System dynamics: what in it for healthcare simulation modelers. In Mason, S.J., Hill, R.R., Mönch, L., Rose, O., Jefferson, T., Fowler, J.W. (eds.) Proceedings of the 2008 Winter Simulation Conference, Piscataway, New Jersey, USA (2008)
8. Brailsford, S.C.: Hybrid simulation in healthcare: new concepts and new tools. In: Yilmaz, L., Chan, W.K.V., Moon, I., Roeder, T.M.K., Macal, C., Rossetti, M.D. (eds.) Proceedings of the 2015 Winter Simulation Conference, pp. 1645–1653, Piscataway, New Jersey, USA (2015)
9. Brailsford, S.C., Desai, S.M., Viana, J.: Towards the Holy Grail: combining system dynamics and discrete-event simulation in healthcare. In: Johansson, B., Jain, S., Montoya-Torres, J., Hugan, J., Yucesan, E. (eds.) Proceedings of the 2010 Winter Simulation Conference, pp. 2293–2303, Baltimore, Maryland, USA (2010)
10. Chahal, K., Eldabi, T.: Applicability of hybrid simulation to different models of governance in UK healthcare. In: Mason, S.J., Hill, R.R., Monch, L., Rose, O., Jefferson, T., Fowler, J.W. (eds.) Proceedings of the 2008 Winter Simulation Conference, pp. 1469–1477, Miami, Florida, USA (2008)
11. Duguay, C., Chetouane, F.: Modeling and improving emergency department systems using discrete event simulation. Simulation 83(4), 311–320 (2007)
12. Fakhimi, M., Mustafee, N.: An investigation of hybrid simulation for modelling sustainability in healthcare. In: Proceedings of the 2015 Winter Simulation Conference (2015)
13. Gül, M., Guneri, A.F.: A comprehensive review of emergency department simulation applications for normal and disaster conditions. Comput. Ind. Eng. 83(5), 327–344 (2015)
14. Gunal, M.M., Pidd, M.: Discrete event simulation for performance modelling in health care: a review of the literature. J. Simul. 4, 42–51 (2010)
15. Heath, S.K., Brailsford, S.C., Buss, A., Macal, C.M.: Cross-paradigm simulation modeling: challenges and successes. In: Proceedings of the 2011 Winter Simulation Conference, pp. 2788–2802 (2011)
16. Katsaliaki, K., Mustafee, N.: Applications of simulation within the healthcare context. J. Oper. Res. Soc. 62(8), 1431–1451 (2010)
17. Kausshal, A., Zhao, Y., Peng, Q., Strome, T., Weldon, E., Zhang, M., Chochinov, A.: Evaluation of fast track strategies using agent-based simulation modeling to reduce waiting time in a hospital emergency department. Socioecon. Plann. Sci. 50, 18–31 (2015)
18. Landa, P., Sonnessa, M., Tànfani, E., Testi, A.: Managing emergent patient flow to inpatient wards: a discrete event simulation approach. In: Obaidat, M.S., Ören, T., Kacprzyk, J., Filipe, J. (eds.)Advances in Intelligent Systems and Computing, Chapter 17, pp. 333–350. Springer (2016a)
19. Landa, P., Sonnessa, M., Tànfani, E., Testi, A.: System dynamics modelling of emergent and elective patient flows. In: Matta, A., Li, J., Vandaele, N.J., Guinet, A., Sahin, E. (eds.) Springer Proceedings in Mathematics and Statistics. Springer, New York (2016)
20. Lane, D.C., Monefeldt, C., Rosenhead, J.V.: Looking in the wrong place for healthcare improvements: a system dynamics study of an accident and emergency department. J. Oper. Res. Soc. 51, 518–531 (2000)
21. Paul, J.A., Lin, L.: Models for improving patient throughput and waiting at hospital emergency departments. J. Emerg. Med. 43(6), 1119–1126 (2012)
22. Proudlove, N.C., Black, S., Fletcher, A.: OR and the challenge to improve the NHS: modelling for insight and improvement in in-patient flows. J. Oper. Res. Soc. 58, 145–158 (2007)
23. Vanderby, S., Carter, M.W.: An evaluation of the applicability of system dynamics to patient flow modelling. J. Oper. Res. Soc. 61, 1572–1581 (2010)
24. Viana, J., Brailsford, S.C., Harindra, V., Harper, P.R.: Combining discrete-event simulation and system dynamics in healthcare setting: a composite model for Chlamydia infection. Eur. J. Oper. Res. 237(1), 196–206 (2014)
25. Wong, H.J., Morra, D., Wu, R.C., Caesar, M., Abrams, H.: Using system dynamics principles for conceptual modelling of publicly funded hospitals. J. Oper. Res. Soc. 63(1), 79–88 (2012)

Estimation of Case Numbers at Pandemics and Testing of Hospital Resource's Sufficiency with Simulation Modeling

Pınar Miç and Melik Koyuncu

Abstract Influenza pandemics have occured throughout the past two centuries, killed millions of people worldwide. Although it is impossible to predict when/where the next pandemic will occur, proper planning is still needed to maximize efficient use of hospital resources and to minimize loss of life and productivity. Thus, it is highly important to estimate case numbers and to test hospital resources' sufficiency to take actions about this area. One of the most common tools used to estimate case numbers in an influenza pandemic is Basic Reproduction Number (R_0). In this study, we estimated case numbers using different R_0 values for a possible influenza pandemic. The developed simulation model used the estimated case numbers as input parameters, for testing important healthcare resources' sufficiency, which are non-ICU (non-intensive care unit) hospital beds, ICU (intensive care unit) hospital beds and ventilators.

Keywords Pandemic influenza · Basic reproduction number (R_0) · Simulation modelling · Health services

1 Introduction

Commonly known as flu, influenza is one of the long-lasting major health issues throughout the world. This single disease alone causes hundreds of thousands of deaths annually. The most severe known pandemic influenza is "The Spanish Flu"

P. Miç (✉) · M. Koyuncu
Engineering-Architecture Faculty, Industrial Engineering Department,
Cukurova University, Adana, Turkey
e-mail: pmic@cu.edu.tr; pinarmic@gmail.com

M. Koyuncu
e-mail: mkoyuncu@cu.edu.tr

© Springer International Publishing AG 2017   145
P. Cappanera et al. (eds.), *Health Care Systems Engineering*, Springer Proceedings
in Mathematics & Statistics 210, https://doi.org/10.1007/978-3-319-66146-9_13

(A/H1N1) pandemic of 1918–1919, estimated to be resulted in 20–50 million deaths worldwide, with unusually high mortality among young adults [1]. "The Asian Flu" (A/H2N2) at 1957 and "The Hong Kong Flu" (A/H3N2) at 1968 pandemics were less severe, with the highest mortality in the elderly [2]. Lots of influenza pandemics emerged many times around the world until late history such as at Hong Kong in 1997, at Thailand, Vietnam, Indonesia, Cambodia and China in 2003 and in 2006 at Turkey, Russia, Europe and some African nations resulting 225 cases [3].

This global threat requires advanced planning, early preparedness and rapid action. Not only may it increase demand upon health services, but it may also result in substantial socioeconomic problems and loss of life [5]. Because of these important reasons, policymakers have begun to realize the severity of this threat and they have started to develop preparedness plans across many layers of government [4]. These plans are based on World Health Organization (WHO) guidelines (published at 1999). But, with the emergence of H5N1 avian influenza virus in Asia and the outbreaks in the Europe and elsewhere, concerns about human influenza have grown up more [6]. Thus, in April 2005, WHO represented its concern about the general lack of global preparedness for influenza pandemics and updated its 1999 global influenza preparedness plan to outline the components that each country's plan should involve [7].

In the Risk Assessment section of WHO checklist (updated at 2005); possible case numbers (numbers of an instance of disease with its attendant circumstances [9]), hospital admissions and deaths are asked for with conduct modelling [8]. In addition, in the "Health Service Facilities" section of WHO checklist, it is requested to determine potential alternative sites for medical care. Therefore, some of the most important health care resources in an influenza pandemic are non-ICU hospital beds, ICU hospital beds and ventilators [10]. It is required to estimate the need of these resources at the recommended alternative sites as WHO checklist offered and it should be suitable making these estimates with models.

By examining related literature, we realized that there is not any study of case numbers in Turkey using Basic Reproduction Number (R_0). Furthermore, there is not common study testing hospital resource's sufficiency based on this estimation in a pandemic case. This paper aims to fill this gap at this area. In this way, hospitals can rationalize their inadequate resources and can be prepared to the possible future pandemic.

Thereby, we can divide our methodology into two sections; (1) Estimation of case numbers by using different R_0 and average duration of infectiousness (D) (which is the duration of infection's stay within individual's body) values, (2) Testing the most critical healthcare resources by a simulation model, which uses estimated case numbers as input parameters.

2 Materials and Methods

2.1 Estimation of Case Numbers

Estimation of case numbers' methods can be divided into two main categories in literature:

(i) Clinical attack rate: Clinical attack rate is defined as a biostatistical measure of frequency of morbidity, or speed of spread, in an 'at risk population' [11]. An 'at risk population' is defined as one that has no immunity to the attacking pathogen which can be either a novel or established pathogen [12].
(ii) Estimating case numbers by using Basic Reproduction Number [9]: The second and more common method is Basic Reproduction Number and usually denoted by "R_0". It is the most frequently used approach in the estimation of case numbers to be occured in a pandemic and time-dependent case numbers can be estimated using R_0 [13].

2.1.1 Estimation of Case Numbers with Basic Reproduction Number

The basic reproduction number, R_0, defined as the average number of secondary cases generated by a primary case introduced into a completely susceptible population, is a crucial quantity for identifying the intensity of interventions required to control a pandemic [10].

At this point, the model structure should reflect the natural history of the infection and therefore important disease categories and transitions need to be described as well as important categories in the population itself. The Susceptible-Pre-infectious-Infectious-Recovered (SEIR) model which shows fitting spreading of the infection to the influenza is used in this study. Many researchers used the SEIR model such as Chowell et al. [9], Flahault et al. [13], Dukic et al. [14], Verikios et al. [15], Pollicott et al. [16] and Farah et al. [5] in their studies for pandemic influenza.

Tables 1 and 2 are the summary of commonly used symbols and basic formulas used in the study to estimate case numbers via R_0 [11].

At Table 2, there are some notations need-to-know:

S_{t+1} Number of susceptible individuals at time $t + 1$,
E_{t+1} Number of pre-infectious individuals at time $t + 1$,
I_{t+1} Number of infectious individuals at time $t + 1$,
R_{t+1} Number of immune individuals at time $t + 1$,
$R_{t=0}$ Number of immune individuals at time $t = 0$ (at the start),
$I_{m(0)}$ Rate of infectious individuals at the start,

Table 1 Summary of the commonly used symbols [14]

Symbol	Definition
β	Rate at which two specific individuals come into effective contact per unit time (equivalent to the per capita rate at which two specific individuals come into contact). This notation is used when we assume that individuals mix randomly
D	Duration of infectiousness
E_t	Number of individuals who are infected but not yet infectious at time t
f	Rate of onset of infectiousness
I_t	Number of individuals who are infectious at time t
λ_t	Force of infection at time t (rate at which individuals are infected per unit time)
N	Total population size
r	Rate at which individuals recover from being infectious
R_t	Number of individuals who are immune ('recovered') at time t
R_0	Basic reproduction number (average number of secondary infectious persons resulting from the introduction of an infectious person into a totally susceptible population)
s_t	Number of susceptible individuals at time t
t_e	Average time to the infectiousness

Table 2 Summary of basic formulas used in the study [14]

$\beta = \frac{R_0}{N \cdot D}$	$\lambda_t = \beta \cdot I_t$	$f = 1/t_e$	$r = 1/D$
$S_{t+1} = S_t - \lambda_t \cdot St$	$E_{t+1} = Et + \lambda t \cdot St - f \cdot Et$	$I_{t+1} = I_t + f \cdot E_t - r \cdot I_t$	$R_{t+1} = Rt + r \cdot It$
$R_{t=0} = N \cdot I_{m(0)}$	$NC_{t+1} = E_t \cdot f \cdot CI$	$I_{new(t+1)} = E_t \cdot f$	

$I_{new(t+1)}$ Total number of new infectious individuals per unit time,

NC_{t+1} Total number of new reported case numbers per unit time,

CI   Proportion of those infectious who are reported as cases.

In this study, case numbers are estimated with formulas explained above, according to the method given in "An Introduction to Infectious Disease Modelling" by Vynnycky and White [14].

2.2 Analyzing the Healthcare Resource's Sufficiency with Simulation Model

For different values of R_0, case numbers can be estimated as explained at Sect. 2.1.1. A major part of these cases will get over the disease with mild symptoms, but some patients will need critical medical interventions. The patients will generally need non-ICU hospital beds, ICU hospital beds and ventilators for these interventions [10].

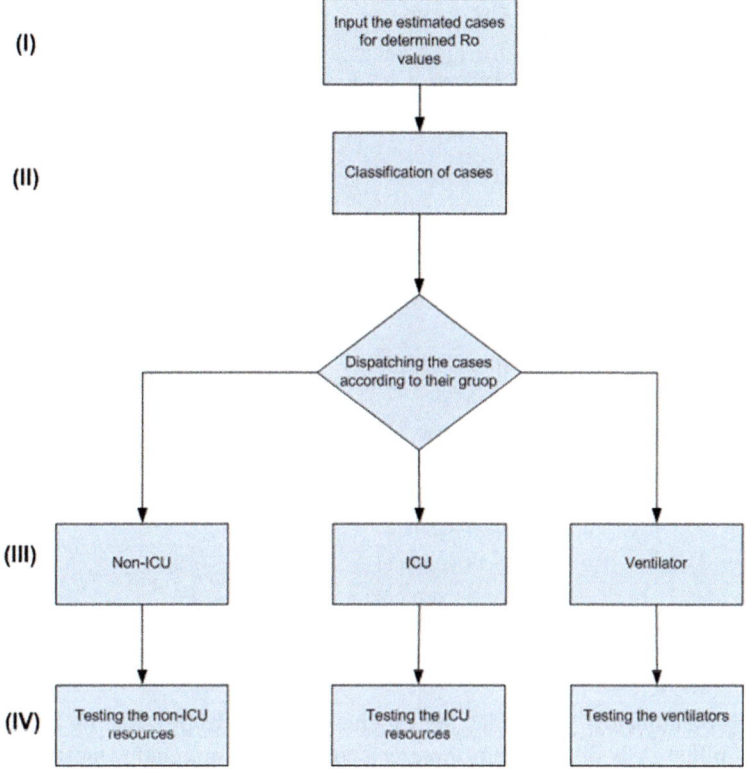

Fig. 1 The flow chart of developed simulation model

Developed simulation model enables to test the sufficiency of critical and important resources under different scenarios. Figure 1 shows the flow chart of the developed simulation model.

I. stage of model consists of input parameters which are the estimates of case numbers calculated by the method detailed in Sect. 2.1.1 [14]

When a pandemic influenza emerged in a population, because of the illnesses in the past and chronic diseases, all cases will not affected by the influenza at the same severity. Within the population, a definite percentage of community may have more medical risk. For this reason, it should be better to break down the population into risk groups. For this reason, at II. stage of simulation model, the population is divided into six risk groups according to age and medical risk. Table 3 shows the notation of the risk groups and their assumed proportions in the total population. Table 4 shows the hospitalization rates of these groups. Both Tables 3 and 4 are the notations and values of some input parameters based on the references given. Considering the past influenza pandemic data, its reasonable to consider that the

150   P. Miç and M. Koyuncu

Table 3 The rates of high and non-high risk groups [10]

Risk groups	Age groups		
	0–19	20–64	65+
High risk	%2.24	%8.32	%2.83
Non-high risk	%32.88	%49.45	%4.25

Table 4 Population-based rates (per 1000 persons) of pandemic influenza hospitalizations [10]

Influenza outcome	Rate per 1000 persons		
	Minimum	Most likely	Maximum
Hospitalizations			
High risk			
0–19 years old	2,1	2,9	9
20–64 years old	0,83		5,14
65+ years old	4		13
Non-high risk			
0–19 years old	0,2	0,5	2,9
20–64 years old	0,18		2,75
65+ years old	1,5		3

severity of influenza pandemics are different. Therefore, the severity of the pandemic influenza is divided into three categories as of minimum, most likely and maximum as at Table 4.

The patients grouped in the II. stage are sent to related hospital resources (non-ICU hospital beds, ICU hospital beds and ventilators) according to the rates determined by the user at the III. stage of the model. For the patients who will need related hospital resources at a case of pandemic influenza, the sufficiency (whether they are enough for all patients or not) of these resources are tested by developed simulation model with the parameters identified by the user in the last and fourth stage of the model.

3   Case Study and Results

3.1   Case Study

Table 5 shows the population and hospital resources in a county, Adana, Turkey, which are used to test the developed simulation model.

For the I. stage of developed simulation model; case numbers are estimated starting with $R_0 = 2$ and it is increased by 0.5 range until $R_0 = 4$ and these case numbers are entered to the model as input parameters.

Table 5 Inputs and assumptions (assumptions are based on [10])

	Default
Inputs	
Population of locale by age groups	
0–19	753,159
20–64	1.241,61
65+	130,863
Total non-ICU beds	6236
Assumed occupancy rate (%)	75
Total ICU beds	556
Assumed occupancy rate (%)	75
Total number of ventilators	452
Assumed occupancy rate (%)	75
Influenza pandemic duration (weeks)	28
Basic reproduction number (interval)	2,0–4
Assumptions	
1. Average length of non-ICU hospital stay for influenza-related illness (days)	5
2. Average length of ICU hospital stay for influenza-related illness (days)	10
3. Average length of ventilator usage for influenza-related illness (days)	10
4. Average proportion of admitted influenza patients will need ICU care (%)	15
5. Average proportion of admitted influenza patients will need ventilators (%)	7,5

Table 6 The constant input parameters' notation, definition and supposed values determined by user

Notation	Definition	Supposed value
N	Total population size	2.125.635 person
$I_{t=0}$	Number of infectious individuals at the start	0,8 person
CI	Proportion of those infectious who are reported as cases	0.77
$PI_{t=0}$	Proportion immune at the start	0.3
API	Average pre-infectious period (days)	2 days
R_0	Basic reproduction number	[2–4]
D	Average duration of infectiousness	[1.5–2.5]

The constant input parameters, which are used to estimate the case numbers in this study, notations, definitions and assumed values are shown at Table 6.

Figure 2 shows the comparison of daily case number estimations for $R_0 = 2, 5$ and D = 1, 5; 2 and 2, 5 values. For example, for D = 2 value, case numbers have reached to the maximum level at about 45th day (6th week) and case numbers on the peak of pandemic for D = 2 is approximately 65,000 cases.

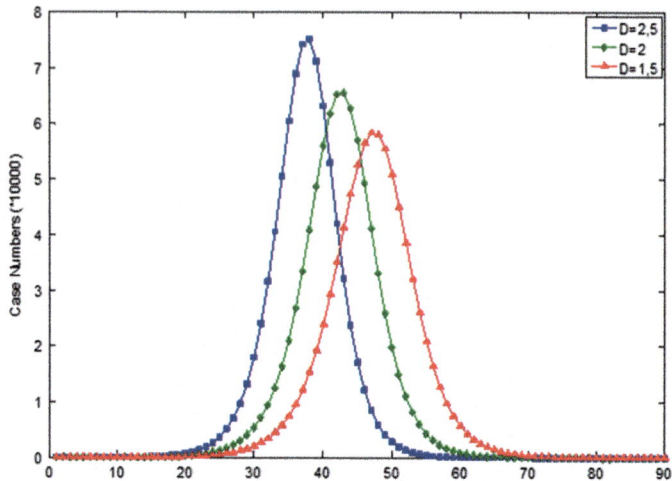

Fig. 2 The comparison of daily estimated case numbers for D = 1,5; 2 and 2,5 and $R_0 = 2,5$

3.2 Results

As a result of the analyses for all scenarios (maximum, most likely and minimum), its concluded that all resources are insufficient for maximum scenario and its necessary to increase them due to the possible needs. Due to the page limit, here we present only some of maximum scenario analyses.

Figure 3.a shows the rejected patients for non-ICU hospital beds, ICU hospital beds and ventilators. We estimated that increasing the number of non-ICU hospital beds to 2270 instead of 1559, the number of ICU hospital beds to 570 instead of 139 and the number of ventilators to 280 instead of 113 will be sufficient even the worst scenario.

4 Suggestions for Further Studies

Case numbers can be estimated with different parameter values. In addition, grouping the population to non-high risk and high risk groups and the rates of hospitalizations can be altered and revised in the simulation model; thereby this process might bring a different approach to the analyses. Furthermore, the simulation model can be developed and analyses can be carried out for larger districts and countries.

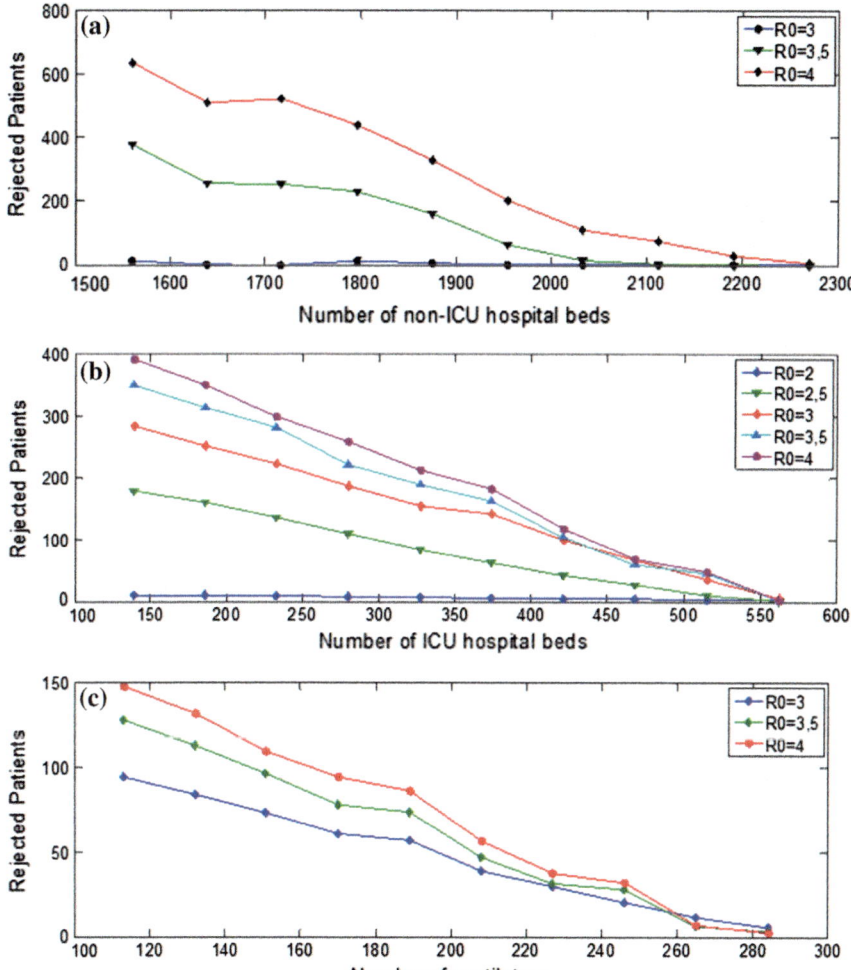

Fig. 3 Rejected patients due to insufficient capacity **a** for non-ICU hospital beds, **b** for ICU hospital beds, **c** for ventilators

References

1. World Health Organization (2010). http://www.wpro.who.int/en/
2. Cox, N.J., Tamblyn, S.E., Tam, T.: Influenza pandemic planning. Elsevier Vaccin. **21**, 1801–1803 (2003)
3. Rahman, S.M.A., Zou, X.: Flu epidemics: a two-strain flu model with a single vaccination. J. Biol. Dyn. **5**(5), 376–390 (2010)
4. Nigmatulina, K.R., Larson, R.C.: Living with influenza: impacts of government imposed and voluntarily selected interventions. Eur. J. Oper. Res. **195**, 613–627 (2009)

5. Farah, M., Birrell, P., Conti, S., De Angelis, D.: Bayesian emulation and calibration of a dynamic epidemic model for A/H1N1 influenza. J. Am. Statist. Assoc. **109**, 1398–1411 (2014)
6. Fraser, C., Donnelly, C.A., Cauchemez, S., Hanage, W.P., Van Kerkhove, M.D., Hollingsworth, T.D., Griffin, J., Baggaley, R.F., Jenkins, H.E., Lyons, E.J., Jombart, T., Hinslry, W.R., Grassly, N.C., Balloux, F., Ghani, A.C., Ferguson, N.M.: Pandemic potential of a strain of influenza A (H1N1): early findings. Science **234**, 1557–1561 (2009)
7. Mounier-Jack, S., Jas, R., Coker, R.: Progress and shortcomings in European national strategic plans for pandemic influenza. Bull. World Health Organ. **85**, 923–929 (2007)
8. Bull. World Health Organ. **90**(11), 793–868 (2012). http://www.who.int/bulletin/volumes/90/11/12-021112.pdf?ua=1
9. Chowell, G., Nishiura, H.: Quantifying the transmission potential of pandemic influenza. Elsevier Phys. Life Rev. **5**, 50–77 (2008)
10. Zhang, X., Meltzer, M.I., Wortley, P.M.: FluSurge—a tool to estimate demand for hospital services during the next pandemic influenza. Med. Decis. Making **26**, 617–623 (2010)
11. Gordon, A., Saborío, S., Videa, E., Lôpez, R., Kuan, G., Balmaseda, A., Harris, E.: Clinical attack rate and presentation of pandemic H1N1 influenza versus seasonal influenza A and B in a pediatric cohort in nicaragua. Clin. Infect. Dis. **50**(11), 1462–1467 (2010)
12. Juckett, G.: Avian influenza: preparing for a pandemic. Am. Fam. Phys. **74–5**, 783–790 (2006)
13. Hens, N., Van Ranst, M., Aerts, M., Robesyn, E., Van Damme, P., Beutels, P.: Estimating the effective reproduction number for pandemic influenza from notification data made publicly available in real time: a multi-country analysis for influenza A/H1N1v 2009. Elsevier Vaccin. **29**, 896–904 (2011)
14. Vynnycky, E., White, R.: An Introduction to Infectious Disease Modelling. Oxford University Press, New York (2010)
15. Flahault, A., Vergu, E., Coudeville, L., Grais, R.F.: Strategies for containing a global influenza pandemic. Elsevier **24**, 6751–6755 (2006)
16. Dukic, V., Lopes, H.F., Polson, N.G.: Tracking epidemics with Google flu trends data and a state-space SEIR model. J. Am. Statist. Assoc. **107**(500), 1410–1426 (2012)

Empirical Data Driven Intensive Care Unit Drugs Inventory Policies

Paola Cappanera, Maddalena Nonato and Roberta Rossi

Abstract This paper proposes a drugs inventory policy at point-of-use level, tailored for the Intensive Care Unit (ICU) case study and aimed at relieving nurses of the time-wasting task of drugs ordering and refilling. The policy aims at jointly reducing order occurrences and imposing service regularity, while keeping stock value as low as possible. An optimization model is proposed and solved on a one-month period real instance and on a set of realistic ones derived from drugs consumption data collection at the ward. The potentially conflicting priorities of three stakeholders (nurses, administration and clinicians) have been successfully incorporated and their impact on order occurrences and stock value has been discussed. Computational results suggest that it is possible to optimize the time-consuming order process currently adopted at the ICU case study. This study is part of a more comprehensive project in which the optimization block will be integrated with a demand forecasting tool and deployed in a rolling horizon framework.

Keywords Supply chain management · Drugs inventory policies · Replenishment ICU · Antibiotics

1 Introduction

Drugs logistics optimization is crucial in containing steadily increasing healthcare costs occurring in industrialized countries [9]. This study addresses drugs inventory policies in an Intensive Care Unit (ICU) at a renowned hospital in Italy. Worldwide, ICUs represent highly peculiar settings [2] due to the following key-features:

P. Cappanera · R. Rossi (✉)
Dipartimento di Ingegneria dell'Informazione, University of Florence, Florence, Italy
e-mail: roberta.rossi@unifi.it

P. Cappanera
e-mail: paola.cappanera@unifi.it

M. Nonato
Dipartimento di Ingegneria, University of Ferrara, Ferrara, Italy
e-mail: maddalena.nonato@unife.it

© Springer International Publishing AG 2017   155
P. Cappanera et al. (eds.), *Health Care Systems Engineering*, Springer Proceedings
in Mathematics & Statistics 210, https://doi.org/10.1007/978-3-319-66146-9_14

(i) severe clinical conditions of patients, (ii) extreme variability in patients' Length of Stay (LoS), (iii) clinical conditions may worse very quickly (great dynamicity), (iv) few ward beds, (v) no stock-out is allowed. We propose an optimization model aiming to: (i) reduce the nurse burden due to drug orders management (lower number of orders in the planning period); (ii) allow service regularity in drug orders management (for each drug, the same quantity is ordered every time an order is triggered for that drug); (iii) control the value of stock, i.e., the amount of money corresponding to the cost of the drugs daily stored at the ward.

Since stock value at ward is not such a direct financial cost as it is in other logistics settings, inventory cost containment is not usually pursued at ICUs. On the contrary, over stocking is often used as a means to prevent undesirable drug shortages which would deeply affect critical patients. However, uncontrolled over stocking not only gives rise to a suboptimal use of budget but, when it is due to frequent drug orders, distracts nurses from patient care. Conversely, lowering the number of orders and imposing regularity in the order process would increase the time nurses can spend at bedside. We propose an Integer Linear Programming (ILP) model which reduces order occurrences, imposes orders regularity, guarantees demand satisfaction and incorporates capacity constraints.

Besides a basic version of the model, we propose three variants that introduce the perspective of three possibly conflicting stakeholders (nurses, hospital's General Direction and clinicians) and we evaluate the resulting trade-offs on a realistic data set.

Material logistics and inventory management have been largely studied in manufacturing and service industries setting. If applied to different settings such policies require to be adapted and verified on the field. Indeed, recent attempts have been made to customize classical inventory policies to account for healthcare settings [9] and they have mainly been applied at pharmacy point-of-use and for large-scale systems [5]. The contribution of this paper is to adapt classical lot-sizing problem and reorder quantity policies to a dynamic and highly peculiar setting such as ICU [10]. Differences mainly concern demand, which is extremely variable. The objective is indeed significantly different: the attention is focused on order events and on the number of different drugs ordered, since they reflect the amount of time spent to manage drug reordering and refilling. Moreover there are no storage costs and stock is, conversely, a value to prevent shortages (we don't have perishable drugs and we can reasonably assume that every ordered drug will be used before the expiry date). With respect to the healthcare extant literature [9], our model reflects an inventory policy based on a continuous review cycle rather than on a periodic one [8], on a constant reorder quantity rather than on an order-up-to-level quantity [1], it is focused on point-of-use inventory type rather than on a multi-echelon one [7], and on strict demand coverage rather than on minimization of stock-out costs [6].

This study is the first step of an ongoing, more comprehensive, project, due to develop and integrate the missing elements required to make it self sufficient as an operative tool. In particular, the optimization module presented in this work will act as the building block of a decision support system where it will be integrated with a demand forecasting tool.

2 Drugs Inventory Policy for ICU

2.1 Problem Description

This study focuses on antibiotics, which are crucial at ICUs since [3]: (i) they are largely used, (ii) some are quite expensive, (iii) inappropriate prescriptions may occur (iv) which might give rise to antibiotic resistance. Nurses are in charge of managing drug orders, a very frequent and time-consuming task which distracts them from patients care. In our case study, lead time is one day but on Saturday: drugs ordered on Saturday will be available on the next Monday. Urgent orders have few hours lead time but are restricted to emergencies. Orders are based on the patients population at ward. There is no demand forecasting and ICU tends to order more than required to avoid disruptions due to a sudden therapy switching. Indeed, when clinicians suspect the presence of microorganisms, an empirical medical treatment begins based on broad-spectrum antibiotics and a clinical test request is issued to the microbiology laboratory. If test results confirm infection, a targeted medical treatment starts consisting in antibiotics targeted to the identified microorganisms [4]. A therapy switching may also occur due to patients antibiotic resistance. Drugs are stored in a medicine cabinet located in the inpatient room for prompt use and in other shelving and cabinets for storage. Each drug has both its own dedicated storage space and a shared space. Some drugs may have specific requirements, such as low temperature. On these premises drugs are partitioned in groups where drugs in the same group share a dedicated storage unit. When a storage unit is devoted to a single drug, its capacity is expressed in terms of number of boxes of that drug; conversely, when a storage unit can host different drugs, possibly with different boxes, its capacity is expressed in terms of units of volume each drug box occupies.

2.2 Optimization Model

Tables 1 and 2 introduce the mathematical notation used in the problem formulation. Then, the description of the model in terms of objective function and constraints is given in (1)–(19). Observe that constraints (2)–(4) mirror the weak lot-sizing formulation [10] but our model has a distinctive objective function with set up costs, no time-dependent production costs and with time-independent storage costs.

A hierarchical objective function (1) characterizes the model that first minimizes the number of order events (by means of Big-M) and at a lower level, keeps stock down by minimizing the sum of daily stock value. Constraints (2)–(4) are classical flow conservation constraints regulating the stock level respectively on the first day of the planning horizon (2), on week days when an order occurred the previous day increases the stock level (3), and on Sunday when no order is received (4). Without loss of generality, we assume that the first day of the planning horizon ($d = 0, w = 1$) is Sunday and that the initial stock of each drug (l^f) meets first day demand (q_{01}^f).

Table 1 Sets and parameters

F	Set of drugs (indexed by f)
G	Set of drug groups (indexed by g)—drugs in a group share the same storage unit
$F_g \subseteq F$	Set of drugs in group $g \in G$
$D = \{0, \ldots, 6\}$	Ordered set of days (indexed by d, 0 corresponds to Sunday, 1 to Monday etc.)
W	Set of weeks (indexed by w, with $w \geq 1$)
q_{dw}^f	Demand of drug f on day d of week w, expressed in number of doses
U^f	Number of doses in each box of drug f
c^f	Cost of each dose of drug f
B_{dw}	Maximum allowed stock monetary value on day d of week w
l^f	Doses of drug f in ward at time 0
C^f	Capacity of storage unit dedicated to drug f, expressed in number of boxes
$\overline{C}^f$	Maximum number of drug f boxes in shared storage unit
V^f	Volume of a drug f box in number of units in shared storage unit
$\overline{V}_g$	Capacity of shared storage unit for group g, expressed in number of units
$\Gamma^f = C^f + \overline{C}^f$	Upper bound on the maximum number of boxes of drug f in stock

Table 2 Variables

s_{dw}^f	Stock level of drug f at the end of day d of week w, expressed in number of doses
y_{dw}^f	Order quantity of drug f on day d of week w, expressed in number of boxes
v_{dw}^f	Equal to 1 when an order of drug f occurs on day d of week w; 0 otherwise
υ_{dw}	Equal to 1 when an order occurs on day d of week w; 0 otherwise
Δ^f	Order quantity of drug f, expressed in number of boxes
x_{dw}^f	Number of boxes of drug f in the shared storage unit on day d of week w

$$\min \quad M \sum_{d \in D, w \in W} v_{dw} + \sum_{f \in F} \sum_{d \in D, w \in W} c^f s_{dw}^f \tag{1}$$

$$s_{01}^f = l^f - q_{01}^f \qquad\qquad\qquad \forall f \in F \tag{2}$$

$$s_{dw}^f = s_{d-1,w}^f - q_{dw}^f + y_{d-1,w}^f U^f \qquad \forall f, \forall d \in D, d \neq 0, \forall w \geq 1 \tag{3}$$

$$s_{dw}^f = s_{6,w-1}^f - q_{dw}^f \qquad\qquad \forall f \in F, \ d = 0, \forall w \geq 2 \tag{4}$$

$$s_{dw}^f \leq U^f \left(C^f + x_{dw}^f \right) \qquad\qquad \forall f \in F, \forall d \in D, \forall w \in W \tag{5}$$

$$x_{dw}^f \leq \overline{C}^f \qquad\qquad\qquad \forall f \in F, \forall d \in D, \forall w \in W \tag{6}$$

$$\sum_{f \in F_g} x_{dw}^f V^f \leq \overline{V}_g \qquad\qquad \forall g \in G, \forall d \in D, \forall w \in W \tag{7}$$

$$\sum_{f \in F} c^f s_{dw}^f \leq B_{dw} \qquad\qquad \forall d \in D, \forall w \in W \tag{8}$$

$$v_{dw} = 0 \qquad\qquad\qquad\qquad d = 6, \forall w \in W \tag{9}$$

$$\Delta^f \leq \Gamma^f \qquad\qquad\qquad\qquad \forall f \in F \tag{10}$$

$$y_{dw}^f \leq v_{dw}^f \Gamma^f \qquad\qquad\qquad \forall f \in F, \forall d \in D, \forall w \in W \tag{11}$$

$$y_{dw}^f \leq \Delta^f \qquad\qquad\qquad \forall f \in F, \forall d \in D, \forall w \in W \tag{12}$$

$$y_{dw}^f \geq \Delta^f - (1 - v_{dw}^f)\Gamma^f \qquad \forall f \in F, \forall d \in D, \forall w \in W \tag{13}$$

$$v_{dw}^f \leq v_{dw} \qquad\qquad\qquad \forall f \in F, \forall d \in D, \forall w \in W \tag{14}$$

$$y_{dw}^f \in Z^+ \qquad\qquad\qquad \forall f \in F, \forall d \in D, \forall w \in W \tag{15}$$

$$x_{dw}^f \in Z^+ \qquad\qquad\qquad \forall f \in F, \forall d \in D, \forall w \in W \tag{16}$$

$$v_{dw} \in \{0, 1\} \qquad\qquad\qquad \forall d \in D, \forall w \in W \tag{17}$$

$$v_{dw}^f \in \{0, 1\} \qquad\qquad\qquad \forall f \in F, \forall d \in D, \forall w \in W \tag{18}$$

$$s_{dw}^f \geq 0 \qquad\qquad\qquad\qquad \forall f \in F, \forall d \in D, \forall w \in W \tag{19}$$

Stock level is computed at the end of the day and the quantity y_{dw}^f of drug f ordered in day d can be used promptly to meet the demand of day $d + 1$. Constraints (5)–(7) concern capacity. Specifically, (5) impose an upper bound on the maximum stock level that a given drug may assume on a given day of the planning horizon, by taking into account the quantity of that drug both in the dedicated and in the shared space and by properly converting the number of boxes in number of doses. Constraints (6) impose, for each drug and each day, an upper bound on the maximum number of boxes present in the common storage space. Constraints (7) are knapsack-like constraints imposing that the overall volume of all the boxes sharing a common storage unit g does not exceed its volume $\overline{V}_g$. Constraints (8) guarantee that the cost of drugs stocked at the end of a day does not exceed the daily budget B_{dw}; in addition to storage constraints, they prevent over stocking. Constraints (9) impose that no order takes place on Saturday. Constraints (10) impose an upper bound on the lot size of drug f which is the same during all the planning horizon. The same upper bound Γ^f is used in (11) that fix variable y_{dw}^f to zero when drug f is not ordered on that day ($v_{dw}^f = 0$). Constraints (12) and (13) guarantee that if drug f is ordered on a specific day, then the lot size is Δ^f. Constraints (14) assure that the variable related to an order event on a given day is equal to one if at least one drug is ordered on that day. Finally, (15)–(19) define variable domain.

The ILP mathematical model described above aims at optimizing the time-consuming management of drug orders related tasks that mainly depends on the number of times orders take place during the planning horizon, what we call *order events*. Further investigating in this direction, we observe that the drug supply chain management involves at least the three following stakeholders: (i) head nurse, (ii) hospital's General Direction, and (iii) clinicians. Each of them has its own perspective in evaluating a solution, possibly conflicting with the others. On one side, nurses in charge of orders management, besides minimizing the number of order events, ask for homogeneous orders in terms of number of drugs in each order. This would allow them to allot in advance a constant portion of their shift to accomplish drugs orders related tasks and to better plan their activities. Indeed, the time required to store drugs into cabinets and update the system data (so far on paper) mainly depends on the number of different drugs involved in the order rather than on the number of boxes, so they would like to keep this number steady. On the other side, the hospital's General Direction aims at minimizing the total quantity of drugs ordered exceeding demand, possibly weighted by coefficients that reflect the importance of drugs or their cost. Finally, clinicians are interested in maximizing the daily variety of drugs involved in broad-spectrum therapy so as to reduce the latency of therapy switching due to patient's resistance to the currently prescribed antibiotic.

We are interested in evaluating the trade-offs arising when model (1)–(19) is enriched with constraints that take into account stakeholders' perspectives. To this aim, we assume that a set I of scenarios is available, where I is a meaningful sample of representative drug demand realizations in a monthly period. Then, scenario by scenario, we solve (1)–(19) and record the three values representing the three perspectives above described. The average value of each criterion over the whole set of scenarios can be computed. Then, perspective by perspective, model (1)–(19) is tightened by a constraint imposing that the solution value according to the considered criterion does not deviate too much from the average value of the criterion computed over the set of scenarios. Specifically, we can manage this latter constraint with three increasing levels of flexibility with respect to average value deviation: low, medium, high. In the following, we denote by the superscript (i) the optimal value a variable assumes in the optimal solution of instance $i \in I$. For example, $v_{dw}^{f(i)}$ represents the value variable v_{dw}^f assumes in the optimal solution obtained when model (1)–(19) is run on instance i.

To take into account nurses' perspective, we define, for each instance $i \in I$:

$$v_{\max}^{(i)} = \max_{d \in D, w \in W} \sum_{f \in F} v_{dw}^{f(i)} \quad \text{as well as} \quad v_{\min}^{(i)} = \min_{\substack{d \in D, w \in W, \\ \text{s.t. } v_{dw}^{(i)} = 1}} \sum_{f \in F} v_{dw}^{f(i)}$$

where $v_{\max}^{(i)}$ and $v_{\min}^{(i)}$ denote respectively the minimum and maximum number of drugs in an order over the planning period when model is run on instance i. Since nurses consider orders as homogeneous when the difference between $v_{\max}^{(i)}$ and $v_{\min}^{(i)}$ is low, the average difference over the set of instances I is computed as

$$\bar{v} = \left\lceil \sum_{i \in I} (v_{\max}^{(i)} - v_{\min}^{(i)}) / |I| \right\rceil$$

and model (1)–(19) is run again with the following additional set of constraints:

$$\sum_{f \in F} v_{dw}^{f} \leq v_{\max} \qquad\qquad \forall d \in D, \forall w \in W$$

$$v_{\min} \leq \sum_{f \in F} v_{dw}^{f} + (1 - v_{dw})|F| \qquad\qquad \forall d \in D, \forall w \in W$$

$$v_{\max} - v_{\min} \leq \bar{v}.$$

Concerning General Direction's perspective, denote $b^{f} = \lceil \sum_{d \in D, w \in W} q_{dw}^{f} / U^{f} \rceil$ as a lower bound on drug f boxes required to satisfy demand and w^{f} as the weight associated with drug f. Then, the average weighted number of boxes ordered in excess with respect to the demand over all instances in I is:

$$\bar{\gamma} = \frac{1}{|I|} \sum_{i \in I} \sum_{f \in F} w_f \left(\sum_{d \in D, w \in W} y_{dw}^{f(i)} - b^{f} \right)$$

and model (1)–(19) is run again with the following additional constraint:

$$\sum_{f \in F} w_f \left(\sum_{d \in D, w \in W} y_{dw}^{f} - b^{f} \right) \leq \bar{\gamma}.$$

In regards to clinicians' perspective, we introduce the additional definitions: $F^{e} \subseteq F$ as the set of equivalent drugs in broad-spectrum therapy; n^{f} as drug f daily dosage; $T = \{T^{j}, T^{j} \subseteq F^{e}\}$ as a set of therapies where each T^{j} is a subset of drugs in F^{e}; K as the daily target number of available therapies. We compute how many days in the planning period, averaged over I, stock exhibits the desired variety in terms of equivalent therapies, namely $\bar{\delta}$ and run model (1)–(19) with the following additional set of constraints:

$$\alpha_{dw}^f \le s_{dw}^f / n^f \qquad\qquad \forall f \in F^e, \; \forall d \in D, \; \forall w \in W$$

$$\beta_{dw}^{T^j} = L(\alpha_{dw}^f \text{ with } f \in T^j) \qquad \forall T^j \subseteq T, \; \forall d \in D, \; \forall w \in W$$

$$\lambda_{dw} \le \sum_{j:T^j \subseteq T} \beta_{dw}^{T^j} / K \qquad\qquad \forall d \in D, \; \forall w \in W$$

$$\sum_{d \in D, w \in W} \lambda_{dw} \ge \overline{\delta}$$

where α_{dw}^f is a binary variable equal to one when a surplus of at least n^f doses of drug $f \in F^e$ is present on day d week w and zero otherwise; $\beta_{dw}^{T^j}$ is a binary variable equal to one when therapy T^j is available on day d week w while λ_{dw} is a binary variable equal to one when at least K therapies in T are available in surplus on day d week w. Variables $\beta_{dw}^{T^j}$ are the results of a logic function ($\vee$ and $\wedge$ predicates) of the drugs involved in therapy T^j that can be linearized in standard way.

3 Computational Results

We present and discuss the results obtained with several variants of the optimization model on a realistic data set made of (i) a real instance related to drug consumption in June 2016 and (ii) a set of realistic scenarios generated out of microbiology laboratories data: for each patient for whom a request has been issued to the laboratory, they contain the treatment plan of the drug which is most likely to stop infection. All instances involve a monthly planning period. The ILP model was coded in Mosel and solved by FICO ®XPress [11] on a Dell Latitude E6530, with Intel i7-3540M cpu. The first set of experiments concerns the real instance and aims at investigating the impact of the number of order events (OEs for short) on the financial records of the process: these are stock value SV, i.e., the sum on each day of the value of stocked drugs, and the excess cost EC, i.e., the amount spent on ordered drugs minus the value of demand, respectively. To this aim, first we compute $\bar{Y}$, i.e., the minimum number of OEs necessary to fulfill demand complying with capacity and budget constraints. Then, we solve a set of parametrized instances in which the number of OEs ranges from $\bar{Y}$ to almost an order per day, and minimize stock value on each such instance, thus obtaining the Pareto front for the two criteria, i.e., OEs and SV, as depicted in Fig. 1a. SV, reported in euro on the left vertical axis, is highest at $\bar{Y}$ but soon decreases and stabilizes around 8 OEs. This is quite reasonable since reducing OEs requires to cluster different orders on the same day, which can only be accomplished by placing an order before usage much earlier than what lead time would require. It follows that drugs stay longer in storage, thus increasing total SV. This is not necessarily a minus, since at ICU SV is an *opportunity cost* rather than a real cost, because it provides more flexibility if a quick change in therapy is needed. Regarding the second financial indicator EC, we compute and plot its value for each instance (see its scale on the right vertical axis of Fig. 1a). While EC is highest at $\bar{Y}$ and decreases as the number of OEs increases, it still is a small fraction of demand value (15122 €), namely a mere 3%, which decreases to about 0.3% when OE doubles (10 OEs). This comes at no surprise due to an important difference with production logistics. In our case there are no time dependent production costs nor real inventory costs but only the money spent on drugs, so that what is not consumed in the current period will be the inventory at the next one, without additional costs. Indeed, the SV on the last day yielded by the minimum OE policy is about twice the one yielded by a double order frequency one (the "10 order events policy"), which in turns is about 1/3 of initial SV. This is quite a trivial finding: the more order frequency increases the closer the associated order policy gets to *just in time*. This would be a plus in supply chain management; on the contrary, in healthcare and at ICU in particular, less tangible values must be considered such as being able to quickly react to sudden therapy changes or to critical patients admission, which means that in defining drug inventory policies a trade-off must be searched between cost and robustness. Indeed, we speak of stock value rather than inventory cost. This analysis leads us to approve the policy based on (1)–(19), namely the minimum OE policy with constant lot size, despite of the higher stock value since it allows to considerably reduce nurses effort in drug inventory management in a financial sustainable manner, while complying with storage and budget constraints. It is worth mentioning that in the real case 13 OEs take place to satisfy the same demand on the same set of drugs.

Now that the minimum OEs policy has been experimentally approved, we broaden the discussion to take into account additional stakeholders and analyze how safeguarding their priorities may affect the number of OEs and SV. We present the results of the method discussed above. We generated 15–30-days-demand scenarios involving 18 drugs (antibiotics) and set three thresholds for the right hand side of the additional constraints modeling the stakeholder priority. We consider one stakeholder at a time (nurses, General Direction and clinicians) and present one boxplot for each threshold regarding two criteria: the net increase of OEs and the percentage variation of SV.

As mentioned, nurses value service regularity, such as dealing with about the same number of drugs at each OE. Let us call as *drug variance in order* (DViO) the difference between the maximum and the minimum number of different drugs involved in a single order during the planning period. On the 15 instances, DViO values range from 3 to 9 and the three integer rounded thresholds used as the right hand side ($\bar{v}$) of the additional constraint modeling nurses perspective, namely 6, 4, and 7, represent respectively the average, the average minus standard deviation, and average plus standard deviation. For each instance, once constrained we obtain the same number of OEs placed on the same days and with the same overall ordered quantities, but some drug orders have been anticipated (with respect to the results of the unconstrained version). For example, drug 14 in instance 1 is ordered 4 times and lot size is 8 in both cases, but orders that took place on days 2, 9, 17, 26 now occur on days 2, 9, 17, 22. As a result, the only variations are observed in SV as drugs tend to stay longer in storage, the highest variation occurring for the lowest threshold. In Fig. 1b the boxplots of SV for the three thresholds are depicted. Regarding General Direction perspective, the runs on the 15 instances yielded $\bar{\gamma} = 588$ and the thresh-

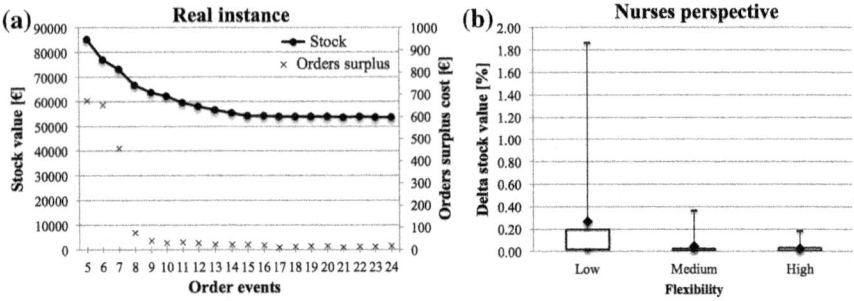

Fig. 1 **a** Pareto front of OEs and SV (left axis) and OEs and extra cost (right axis) in €. **b** Boxplot of SV for Nurses perspective

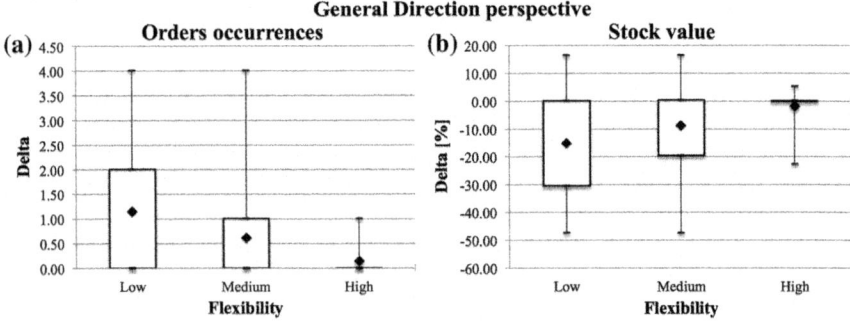

Fig. 2 Hospital General Direction perspective. Boxplots of OEs (**a**) and SV (**b**) on the 15 instances when the cost of extra drugs is constrained below a threshold (function of $\bar{\gamma}$)

olds for extra cost are 187, 588 and 990 €. This constraint impacts on OEs (Fig. 2a) as well as on SV (Fig. 2b). If extra cost has to be reduced, then orders must further adapt to demand irregular trends and then more chances for ordering are needed (a request of 9 boxes is either met by 2 OEs of 5 boxes each at some extra cost, or by 3 OEs of 3 boxes each at no extra cost). As shown for the real case, SV decreases when OEs increase. Finally, the request of drugs surplus not related to actual demand acts as a *shadow demand* and affects OEs and SV only when the drugs involved in the addressed therapies are not routinely used. Therefore, the instances in set I react quite differently to the inclusion of clinician priorities in the drugs inventory policy. Again, we set three thresholds related to the number of days over the 30 days planning horizon when surplus should be present, namely 20, 25 and 30. Boxplots are depicted in Fig. 3. Note that since the binding constraints are of the $\geq$ kind, the highest threshold is the most affecting one.

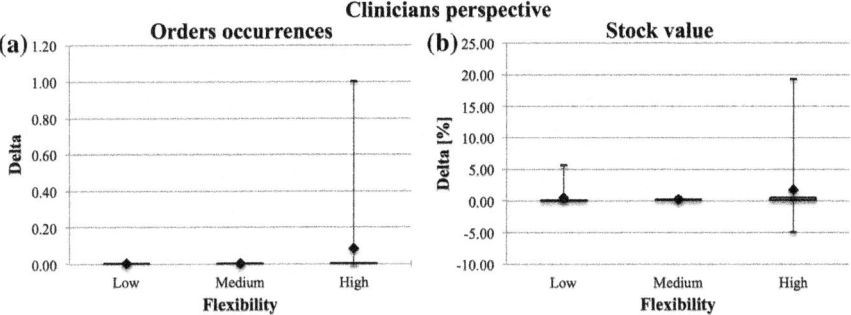

Fig. 3 Clinicians perspective. Boxplots of OEs (**a**) and SV (**b**) on the 15 instances when the number of days stock exhibits the desired variety is above a threshold (function of $\bar{\delta}$)

4 Conclusions

This paper proposes drugs inventory policies at point-of-use level which are specifically customized to take into account the features of the ICU case study, with its own specific targets and constraints. The policy aims at partially relieving nurses from the burden of drugs ordering and refilling by jointly reducing the number of order events and imposing regularity on the orders in terms of drug order size, while keeping stock value as low as possible. As expected, order regularity obtained by means of constant lot size may increase drugs stay in storage. On the one hand, high stock levels keep the value of stocked material high in the overall period. For this reason our model includes the minimization of stock value as a secondary objective. On the other hand, high stock levels seldom pose a challenge in terms of storage space in our instances where storage space is not a tight resource. Computational results suggest that the current time-consuming order process can be rather improved. In addition, the perspectives of three, possibly conflicting, stakeholders have been successfully incorporated in our inventory policies and their impact on orders occurrences and stock value have been discussed.

In conclusion, this study simultaneously addresses: (i) continuous review, (ii) constant order quantity, (iii) different stakeholders' perspectives, and (iv) a peculiar setting. The introduction of scenarios and of stakeholders' perspectives is just one first step in addressing demand variability that characterizes such a setting. The comprehensive project this study is part of, aims to integrate the optimization module, which requires complete demand knowledge in the planning period, with a demand forecasting tool and to deploy it in a rolling horizon framework. Further research also concerns: including urgent orders into the inventory policies, tightening the ILP model formulation, and investigating the mutual relations between stakeholders perspectives.

Acknowledgements The authors are indebted to the ICU medical staff and nurses, who answered to all questions with endless patience and made this fruitful collaboration possible. This work has been partially supported by LINFA (Logistica INtelligente del FArmaco) project, funded by Regione Toscana under the call PAR FAS 2007–2013, Linea dAzione 1.1—Bando FAR FAS 2014.

References

1. Kelle, P., Woosley, J., Schneider, H.: Pharmaceutical supply chain specifics and inventory solutions for a hospital case. Oper. Res. Health Care. **1**, 54–63 (2012)
2. Knaus, W.A., Wagner, D.P., Zimmerman, J.E., Draper, E.A.: Variations in mortality and length of stay in intensive care units. Ann. Intern. Med. **118**, 753–761 (1993)
3. Luyt, C.-E., Bréchot, N., Trouillet, J.-L., Chastre, J.: Antibiotic stewardship in the intensive care unit. Crit. Care. **18**, 480 (2014)
4. Rhodes, A., Evans, L.E., Alhazzani, W., Levy, M.M., Antonelli, M., Ferrer, R., Kumar, A., et al.: Surviving sepsis campaign: international guidelines for management of sepsis and septic shock: 2016. Intensive Care Med. **43**, 304–377 (2017)
5. Rosales, C.R., Magazine, M., Rao, U.: Point-of-use hybrid inventory policy for hospitals. Decis. Sci. **45**, 913–937 (2014)
6. Rosales, C.R., Magazine, M., Rao, U.: The 2Bin system for controlling medical supplies at point-of-use. Eur. J. Oper. Res. **243**, 271–280 (2015)
7. Uthayakumar, R., Priyan, S.: Pharmaceutical supply chain and inventory management strategies: optimization for a pharmaceutical company and a hospital. Oper. Res. Health Care. **2**, 52–64 (2013)
8. Vila-Parrish, A.R., Ivy, J.S., King, R.E., Abel, S.R.: Patient-based pharmaceutical inventory management: a two-stage inventory and production model for perishable products with Markovian demand. Health Syst. **1**, 69–83 (2012)
9. Volland, J., Fügener, A., Schoenfelder, J., Brunner, J.O.: Material logistics in hospitals: a literature review. Omega **69**, 82–101 (2017)
10. Wolsey, L.A.: Integer Programming. Wiley (1998)
11. XPress Optimization Suite. http://www.fico.com/xpress

A Decision-Making Tool for the Calculation of a Robust Planning for Home Service Employees

Maria Di Mascolo, Marie-Laure Espinouse, Pierre Gruau and Jérôme Radureau

Abstract This paper deals with the scheduling and routing problem in a Home Service organization, and we are more particularly interested in a real case problem, taking place in Bourgoin Jallieu (France). The aim of this work is to create a decision support tool to help the planners improving an existing planning. Many constraints must be taken into account, as legislation, competences or preferences. The planners face two key challenges: minimizing the travel time of the employees, and replacing an absence. For the first one, we propose a local search heuristic, and for the second one, we develop a dedicated heuristic. A tool with these two proposals has been implemented and has been tested with real data.

Keywords Home Service · Scheduling · Routing · Robust planning Decision-making tool

1 Introduction

Home Service consists in offering assistance in the daily living tasks, or direct assistance to fragile persons (called beneficiaries), at their home. These beneficiaries are dependent persons for whom certain activities of everyday life are impossible.

M. Di Mascolo (✉) · M. -L. Espinouse
Grenoble INP (Institute of Engineering), G-SCOP, Univ. Grenoble Alpes,
CNRS, 38000 Grenoble, France
e-mail: Maria.Di-Mascolo@g-scop.grenoble-inp.fr

M. -L. Espinouse
e-mail: marie-laure.espinouse@g-scop.grenoble-inp.fr

P. Gruau
CEA Grenoble, Grenoble, France
e-mail: pierre.gruau@grenoble-inp.org

J. Radureau
Adomni-Adhap Services, Lyon and Bourgoin Jallieu, France
e-mail: j.radureau@adhapservices.eu

© Springer International Publishing AG 2017
P. Cappanera et al. (eds.), *Health Care Systems Engineering*, Springer Proceedings
in Mathematics & Statistics 210, https://doi.org/10.1007/978-3-319-66146-9_15

This includes people with disabilities, temporary immobilization and, for the vast majority, dependencies related to age. The offered services can be meal preparation, medication reminders, laundry, light housekeeping, or transportation, among other services. These home service solutions are increasingly popular because, for many seniors, they are an alternative to retirement homes.

Home Service is provided by life assistants (LA), who visit beneficiaries at home. One of the difficulties of this sector is the management of the turnover, for beneficiaries (death, hospitalization), as well as for the life assistants. Indeed, the emotional difficulty of the activity, as well as the very low social recognition of the profession, does not help the stability of the staff. It is therefore important to try to facilitate as much as possible the work of the LA to maintain a constant staff.

An issue, for good working conditions, is the building of a good planning for the LA. Indeed, travel times between tasks can be long and difficult to bear for the workers. In addition, high daily time range, as well as repeated breaks during the day do not facilitate the activity of life assistants. The creation of these schedules is a difficult and time-consuming task for planners. Many constraints and many objectives have to be taken into account. The planning work is therefore complex and requires taking into account the human factor.

Our goal is to propose a decision-making tool for planning, which, from an existing planning, proposes alternative solutions to reduce the travel time of the LA and replace the absences, and which is usable by planners, who are not specialists in optimization.

This paper is organized as follows: we define our problem in Sect. 2, and discuss related work in Sect. 3. Section 4 proposes some algorithms for solving our problem. Then, in Sect. 5, we describe the proposed tool. Some experiments are presented in Sect. 6, before a conclusion and further issues given in Sect. 7.

2 Problem Description

In our partner organization, the planners build, on Friday, the plannings for each employee for the next week. These plannings are build from previous plannings, with the aim of minimizing the travel times and smoothing the employees' workload. They have to take into account the following constraints:

Fixed tasks: when a new beneficiary is admitted, the times for the tasks to be performed at his/her home are fixed, with a maximum flexibility of +/− 1 h.

Legal constraints, strongly linked to French legislation and collective agreements: No more than 10 h of work (i.e. time spent by the employees to the beneficiaries' homes and travelling) per day—Less than 13 h for the time range of the day (i.e. time between the start and the end of the working day, including breaks and other interruptions)—A break of 20 min every 6 h of work—A maximum of 35 h per week—One day off per week.

Availability constraints for LA, which are strict constraints, expressed in terms of availability time windows (in hours) during the week.

LA's Skills: Each LA has a skill level (I to IV) and can therefore only perform the services compatible with his/her level of competence.

Continuity of care: It is crucial to maintain stability in the care and try to send the same LA (or known LA) to each care, for the same beneficiary. Unlike availabilities, continuity is a soft-constraint.

In addition to the building of these plannings, the planners also have to face last minute modifications, due to employee absences.

We aim at proposing a tool to help planners to improve existing plannings, and to manage the absence of employees. To be more accurate, our tool starts from an existing planning, and proposes alternative solutions allowing to:

- reduce the travel time of LA: currently, the planners try to allocate the tasks in order to create coherent tours, but without optimizing the travel times
- replace absences: absenteeism is common among staff, and planners spend a lot of time re-allocating tasks, which should in no case be suppressed

We propose algorithms for recalculation of tours, based on local search methods [10], and include them in a tool that is easy to use for non-specialists in optimization.

3 Problem Positioning Regarding the Literature

Home service is a relatively recent sector, but it is now in full-expansion in France [14]. However, we found no article dealing specifically with this topic. We can note that this sector is very close to Home Care, which consists in delivering medical and paramedical services to patients at their home. Many papers in the literature consider Home Care, and some of them also include service activities (see the works by Duque et al. [3], Fikar and Hirsch [4], Heching and Hooker [7], Redjem and Marcon [11], and Rest and Hirsch [12], for recent examples of studies considering both, Home care and Home service, and dealing with real cases).

Several issues are considered in the literature when dealing with resource planning in a Home Care organization, such as resource dimensioning, partitioning patients and resources into districts, admission of patients and their assignment to one or more caregivers, and, of course, the scheduling and routing of human and material resources. The first paper dealing with the problem of assignment and routing of nurses to patients home is Cheng and Rich [1]. They propose a mixed integer mathematical model and a heuristic approach, and give results for just 4 staff members and 10 patients. Since then, there has been many papers on the subject, and we can note an increasing interest in solving real-size instances, with up to 100 patients. The recent literature review by Fikar and Hirsch [5] shows the large variety of problems that are addressed when considering planning in Home Care organizations. Most works present some problems closely related to the VRP problem. The most common considered objective is the travel cost, and the most common considered constraints are time windows, legislative rules, and skill requirement. Nevertheless, there are also more and more papers dealing with constraints, which are closer to real-life cases, like time-dependency constraints, continuity of care, or preferences of patients.

Most of the articles are dealing with the static problems, whereas in real life, several uncertainties can affect the solutions, as traffic congestion, or absence of a staff member, or cancellation of a visit by the patient (see the work by Lanzarone and Matta [9] and Yuan et al. [13] for examples of robust assignment and scheduling of daily operations).

The specificities of our Home Service planning problem are the following:

- staff members do not require to start and finish their tours at the organization location, as it is the case when dealing with home care
- task duration can be very long, for example in case of housekeeping service.
- task dates are fixed, with very low flexibility
- a planning already exists and can not be entirely modified

To conclude this section, we can say that many of the characteristics encountered in the literature are present in our problem: the minimization of travel times, the consideration of skills, or the satisfaction of beneficiaries, are classical elements of home care. A notable difference with most encountered works concerns tasks. In many articles, the tasks are short (15 min duration) and must be done within a time window (e.g. between 2 and 4 pm), whereas in our problem the tasks can be longer (1–2 h) and must be done at fixed instants, contractualized between the beneficiaries and the Home Service organization. Moreover, we are dealing with a dynamic problem, since we aim at proposing a planning able to reallocate in an existing planning the tasks that were assigned to an employee, who is finally absent. The last characteristic of our work is that it provides a tool, usable in real situations (with 200 beneficiaries, and 55 LA) and whose functioning is understandable and convincing for planners who are not IT specialists.

4 Proposed Algorithms

4.1 Reduction of Travel Time

From an existing planning (which already takes into account the constraints concerning time windows, skills, continuity of care...), we aim at reducing the travel time of the LA by means of a descent with two operations, and without modifying too much the existing planning. This solution has the advantage of being easy to set up and relatively efficient. Local research is done here in an informed manner. First, we sort the tasks according to the travel time of their associated employee (decreasing). The aim is to seek to make changes prioritarily for employees traveling a lot. Two types of descents are carried out:

Insertion. The idea is to change the employee assigned to a task if this improves the travel time. Several constraints are taken into account during this operation: skill of the employee, legal constraints, respect of the hourly volumes of the contracts. The algorithm scans the tasks and, for each of them, tries to insert it to another employee, known by the beneficiary (in order to ensure the continuity of care), if this improves the solution. Tasks are sorted by the decreasing workload of employees (one wants to remove tasks from the employees who are the most loaded). Conversely, known employees are sorted by increasing load, in order to favor the addition of tasks to employees who need more working hours. If we call N the number of tasks on a day, and M the number of employees, we can calculate the complexity of this algorithm. In the worst case, there are M employees known by the beneficiary of the task. Moreover, the assignment of a task to an employee is realized by a sort, and has thus a complexity of $N * log(N)$. Finally, the complexity of this algorithm is: $\theta(M * N^2 * log(N))$. Since the number of tasks is greater than the number of employees, this complexity is no larger than: $\theta(N^3 * log(N))$.

In fact, eventhough this complexity seems important, the instances are not large enough to not being solved in reasonable time. For one day, there are at most 150 tasks, and the number of employees known by a beneficiary never exceeds 10.

The 2-OPT. It is a local search algorithm proposed by Croes et al. [2] to solve the problem of the commercial traveler by improving an initial solution. We adapted it to our problem by keeping the same principle: take two tasks from two different employees and exchange the employees if this improves the travel time. Tasks and employees are sorted in order to give priority to exchanges involving employees who make many travels. The constraints related to the skills of the employee and the legal constraints are taken into account. Only LA known by the two beneficiaries are considred. As for insertion, computation of complexity has been carried out and by the same reasoning one obtains: $\theta(N^3 * log(N))$.

Insertion and 2-OPT are performed successively and repeated as long as the objective is improved.

4.2   Replacement of an Employee

Our objective here is to help the planners to modify the existing planning in order to re-allocate some tasks in case of last minute absences. This is a very common situation (2–3 times per week). By giving the name of the employee who is absent, the start date and the end date of the absence, we seek to reallocate the tasks to the other employees, by respecting as much as possible the hourly volumes of the contracts and by seeking to minimize additional travel times. For this, heuristics to manage absences are inspired by the current functioning of the planners. The idea is to consider all the possible modifications, and propose the "best" ones to planners.

Pseudo code to find modifications to be done. For each task to be replaced, the algorithm will propose an enumeration of the possible modifications enabling to reschedule it. These changes are saved in a list that will be processed later. The services are first sorted according to LPT (Longest Processing Time) to insert first the longest tasks (therefore the most difficult to insert). All the changes are based on the same principle: try to insert the task to a known employee, and if there is a task in conflict with the one to be inserted, the task in conflict is shifted. Contracts between the Home Service Organization and the beneficiaries indicate that the tasks, which are fixed, can be shifted by more or less 1 h, if necessary. The most complex modification moves two tasks, and thus affects the planning of three employees. Moving more tasks would have been possible, from a complexity point of view, but would have represented too many changes in reality. The complexity of this search is $\theta(N^5 * log(N))$, and the algorithm takes less than one second to reschedule each task, which is definitely acceptable.

Choice of the "best" modifications. The aim of this tool is not to find THE best solution, but to offer options to the planners, by leaving them the choice. Thus, it was agreed with the planners to present two alternatives for each task. The difficulty was to choose these two modifications, taking into account the following criteria:

- Additional travel time: the reassignment of a service involves an additional detour which can be significant
- Workload of the chosen employee: it is necessary to give the preference to the employees whose workload is under the maximum value set in the contract
- Complexity of the modification: it is easier to change the schedule of an employee rather than that of three employees

Algorithm 1. Replacement of an Absence

```
Sort the known employees by increasing workload
Sort the tasks according to LPT

For each task t to be replaced
  For each employee e known by the beneficiary of task t
    If (e can do t)
        ListOfPossibleModificationsFort ← Modification (e does t)
    End If
    Let L be the list of the tasks done by e which are not compatible with t
    If (size of L=1)
      Let t' be the task which is not compatible with t
      For each employee e' known by beneficiary for task t'
        If (e' can do t')
            ListOfPossibleModificationsFort ← Modification (e does t and e' does t')
        End If
        Let L' be the list of tasks done by e' which are not compatible with t'
        If (size of L'=1)
          Let t'' be the task which is not compatible with t'
          For each employee e'' known by beneficiary for task t''
            If (e'' can do t'')
                ListOfPossibleModificationsFort ← Modification (e does t and e'' does t'')
            End If
          End For
        End If
      End For
    End If
    If (size of L=2)
      Let t' and t'' be the tasks which are not compatible with t
      For each employee e' known by beneficiary for task t'
        For each employee e'' known by beneficiary for task t''
          If (e' can do t' and e'' can do t'')
              ListOfPossibleModificationsFort ← Modification (e does t, e' does t', and e'' does t'')
          End If
        End For
      End For
    End If
  End For
  Choose and display the best two modifications
End For
```

This multi-criteria objective makes it very difficult to choose the two modifications. We tried numerous configurations with the planners to get closer to their needs, and the procedure is as follows: the possible modifications are first sorted according to the additional travel time, and, the two smallest are taken. Then, if other modifications have similar travel times (no more than 10 more minutes), and the chosen employee has a smaller workload (10% less), they are chosen. Finally, if several modifications have the same travel times and the same workload, the simplest is chosen.

5 Description of the Proposed Tool

The software chosen to build the decision-making tool is Eclipse. Figure 1 illustrates the environment of the tool. First, the data is exported in Excel format in the form of 3 files: a beneficiary file, an employee file, and a task file. In addition, a matrix of travel times between municipalities is also needed. An interface has been developed to allow planners to clearly express their demand in order to obtain the best answers. Finally, the results are displayed in the Eclipse console. The proposed modifications are presented on a daily basis. For each modification, the additional travel time (or the gain) is given in minutes. In addition, the employee's workload before assignment is shown (as a percentage of his/her contract).

Calculation of travel times between the different beneficiaries. To do this, the Quantum GIS mapping software was used. The address of the beneficiaries has been approximated to their municipality of residence. Thus, calculating the travel time between two beneficiaries is done by calculating the travel time between their two municipalities.

To obtain these times, a cartography of the 350 municipalities of the Nord Isère region was carried out. The centroids of these 350 municipalities were calculated in order to obtain Euclidean distances. To make the connection between these Euclidean distances and the distances by road, the formula of Gallez and Hivert [6] was used. It connects the Euclidean distances (D_e) with the distances by the road (D) in the rural area according to the expression: $D = D_e(1.1 + 0,3 * e^{De/20})$. This formula is used in particular by ADEME, a public organization carrying out numerous transport actions. We calculated the travel times considering an average of 50 km/h.

A comparison of the obtained travel times was made with Google Maps (indicating the exact addresses). The results of this comparison showed that the averages are similar and that the standard deviation is 13%. We can therefore conclude without loss of generality that this approximation is credible and in line with reality.

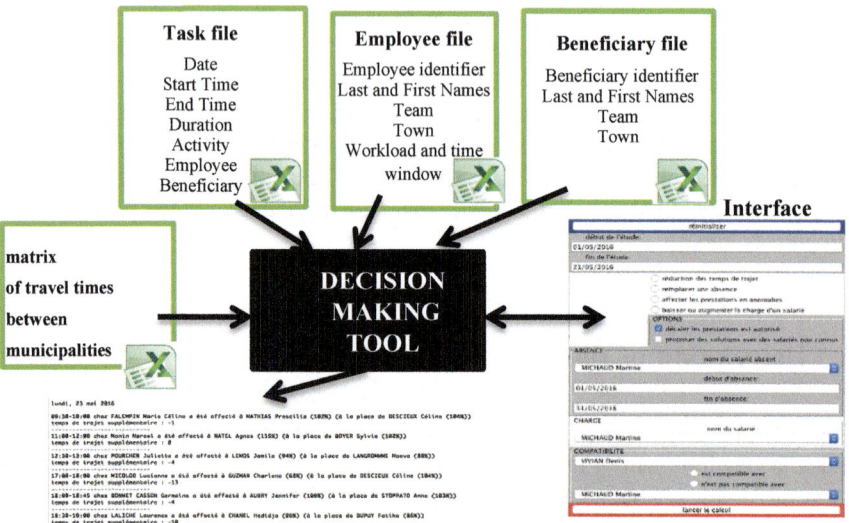

Fig. 1 Environment of the proposed tool

6 Tests and Implementation

In order to validate the decision-making tool, it was important to carry out tests on real data, both to evaluate the performance of the tool and to check its conformity with reality. Thus, we tested the tool on real instances (55 employees, and 200 beneficiaries). The tool was implemented on a computer of the partner organization, so that planners can use it easily. After each use, solutions are analyzed, any errors corrected, and certain aspects (displays, options …) possibly modified.

We made a series of tests to evaluate and quantify the contribution of the decision-making tool, using instances of April, May and June 2016. The evaluation of the improvement in travel times was made over 12 weeks. The aim was to minimize the sum of the travel times for employees. The results are given in Table 1.

These results show that the tool provides a significant improvement for travel times (around 8%). In addition, the calculation time is fast enough (10 s) for the planners to use the tool.

We tested the replacement of an absence on real situations. The absence of an employee during a week was simulated on 10 different employees during week 23. The objective was to find one or more solutions to replace each task. The results in Table 2 show that at least two solutions were found in more than 60% of cases and only one in 15% of the cases (no other alternatives to replace it). The decision-making tool therefore proposes at least one solution in 75% of the cases. Among the cases where no solution is found, the majority are those where it is impossible to find them (all the employees known by the beneficiary are on holiday) and where the decision

Table 1 Tests for the improvements of the travel times

Week	Total initial travel time (min)	Gain (min)	Gain (%)	Calculation time (s)	Number of operations done
14	5361	390	7	10.2	32
15	5230	483	9	11.6	34
16	5474	426	8	12.1	34
17	6223	368	6	12.8	25
18	4978	311	6	11.3	23
19	4543	309	7	8.9	24
20	4268	244	6	8.6	19
21	4635	410	9	6.4	32
22	5367	452	8	6.1	27
23	4914	471	10	5.1	33
24	4901	539	11	6.9	43
25	4644	405	9	9.1	38

can only belong to the planners. Finally, the quasi-instantaneous calculation time allows planners to use the tool as often as needed.

Experiments carried out on real-life situations made it possible to estimate the time saving for planners at more than one hour for the replacement of an employee absent during one week.

7 Conclusions and Future Issues

The central element of this work has been the development of a decision-support tool for planners of a home service organization. The challenge was to find a balance between advanced optimization and a tool usable in a real situation. Different approaches were studied before leading to the methods presented here. The main difficulty has been to make this tool usable in a real situation and by non-specialist users of optimization.

Implemented in Bourgoin Jallieu (France), the tool was greatly welcomed by the teams who found a way to save time in their work and improve the working conditions of planners. It is already usable on a daily basis, however, it could be improved, for example by allowing the LA to be assigned to tasks requiring a lower skill level (with a penalty), and considering real travel times between beneficiaries.

One future issue of this work could be to compare the results obtained by the proposed heuristic to optimal results, which could be obtained when we know the absence in advance and we allow modifying completely the original planning, or if we allow modifying it while trying to minimize the modifications.

Another issue would be to consider public transportation modes rather than private ones, as proposed by Fikar and Hirsch [4] and Rest and Hirsch [12] or combining these last two, such as Hiermann et al. [8] and Rest and Hirsch [12].

Table 2 Tests for the replacement of an absence

	nb of tasks to be replaced	Task replaced with at least 2 alternative solutions		Task replaced with only one alternative solution		Task not replaced – no solution was found		Task not replaced – no possible solution		Calculation time (s)
Employee 1	11	6	55%	4	36%	0	0%	1	9%	0.88
Employee 2	14	9	64%	2	14%	0	0%	3	21%	0.57
Employee 3	12	8	67%	0	0%	1	8%	3	25%	0.28
Employee 4	14	11	79%	1	7%	1	7%	1	7%	0.52
Employee 5	16	9	56%	6	38%	0	0%	1	6%	0.19
Employee 6	13	6	46%	2	15%	0	0%	5	38%	0.3
Employee 7	29	16	55%	5	17%	2	7%	6	21%	0.74
Employee 8	23	17	74%	2	9%	0	0%	4	17%	0.4
Employee 9	18	11	61%	3	17%	0	0%	4	22%	0.6
Employee 10	19	13	68%	1	5%	1	5%	3	16%	0.5

In this chapter, it is proven that combining transportation modes is highly advantageous to improve services and to decrease costs. That is why, considering this feature will be appreciated to improve real-life applications.

References

1. Cheng, E., Rich, J.L.: A home health care routing and scheduling problem. Technical report. https://www.researchgate.net/profile/Eddie_Cheng/publication/2754197_A_Home_Health_Care_Routing_and_Scheduling_Problem/links/02e7e5254ba01aaab2000000.pdf (1998)
2. Croes, G.A.: A method for solving traveling-salesman problems. Oper. Res. **6**(6), 791–795 (1958)
3. Duque, P.M., Castro, M., Sorensen, K., Goos, P.: Home care service planning the case of landelijke thuiszorg. Eur. J. Oper. Res. **243**(1), 292–301 (2015)
4. Fikar, C., Hirsch, P.: A matheuristic for routing real-world home service transport systems facilitating walking. J. Clean. Prod. **105**, 300–310 (2015)
5. Fikar, C., Hirsch, P.: Home health care routing and scheduling: a review. Comput. Oper. Res. **77**, 86–95 (2017)
6. Héran, F.: Des distances à vol d'oiseau aux distances réelles ou de l'origine des détours. Flux **2**(76–77), 110–121 (2009)
7. Heching, A., Hooker, J.: Scheduling home hospice care with logic-based benders decomposition. Lect. Notes Comput. Sci. **9676**, 187–197 (2016)
8. Hiermann, G., Prandtstetter, M., Rendl, A., Puchinger, J., Raidl, G.R.: Metaheuristics for solving a multimodal home-healthcare scheduling problem. CEJOR **23**(1), 89–113 (2015)
9. Lanzarone, E., Matta, A.: Robust nurse-to-patient assignment in home care services to minimize overtimes under continuity of care. Oper. Res. Health Care **3**(2), 48–58 (2014)
10. Pirlot, M.: General local search methods. Eur. J. Oper. Res. **92**(3), 493–511 (1996)
11. Redjem, R., Marcon, E.: Operations management in the home care services: a heuristic for the caregivers routing problem. Flex. Serv. Manuf. J. **28**(1–2), 280–303 (2016)

12. Rest, K.-D., Hirsch, P.: Daily scheduling of home health care services using time-dependent public transport. Flex. Serv. Manuf. J. **28**(3), 495–525 (2016)
13. Yuan, B., Liu, R., Jiang, Z.: A branch-and-price algorithm for the home health care scheduling and routing problem with stochastic service times and skill requirements. Int. J. Prod. Res. **53**(24), 7450–7464 (2015)
14. http://www.amelis-franchise.com/marche

Service Reconfiguration in Healthcare Systems: The Case of a New Focused Hospital Unit

Alessandro Stefanini, Davide Aloini, Riccardo Dulmin and Valeria Mininno

Abstract In the last years, hospitals have been pushed to change their services in the final attempt to maximize both care effectiveness and efficiency. In particular, emergent trends are prompting hospitals to reorganize current activities around patients and their diagnoses rather than in discipline focused departments. This research aims to support service reconfiguration by proposing a methodology exploiting the benefits of process mining techniques in the healthcare systems. In order to support decision makers during this process, the method mainly identifies/analyzes the patient flow and estimates the resources necessary for specific classes of patients. A case study also shows evidence deriving from its application to a new Patient-Focused Care Unit.

Keywords Process mining · Service configuration · Resource planning
Patient-Focused Care (PFC) · Healthcare · Service Planning

1 Introduction

In the last 25 years, hospitals have taken sundry strategic improvement initiatives in the organization of care activities.

A. Stefanini (✉)
Department of Enterprise Engineering, University of Rome Tor Vergata, Rome, Italy
e-mail: a.stefanini@ing.unipi.it

A. Stefanini · D. Aloini · R. Dulmin · V. Mininno
Department of Energy, Systems, Territory and Construction Engineering, University of Pisa, Pisa, Italy
e-mail: davide.aloini@dsea.unipi.it

R. Dulmin
e-mail: riccardo.dulmin@dsea.unipi.it

V. Mininno
e-mail: valeria.mininno@dsea.unipi.it

© Springer International Publishing AG 2017
P. Cappanera et al. (eds.), *Health Care Systems Engineering*, Springer Proceedings
in Mathematics & Statistics 210, https://doi.org/10.1007/978-3-319-66146-9_16

The benefits of a multidisciplinary approach in patient care, for example in treating cancer but also in many chronicle diseases, has been widely recognized in the healthcare literature [22] and boosted the dissemination of Patient-Focused Care (PFC) principle [14]. In this direction, one of the most relevant initiative is the creation of hospital service lines (or product line), in which care is organized around an identifiable service (e.g. cardiac care), a segment of market (e.g., child patients), or some combination of these two [12].

Positive expectations deriving from a centralized approach to specialist services are mostly linked to the potential to reduce variations in care. Thus, innovative "patient-centric" organizational models (e.g. specialized care units and patient care pathways) were proposed and implemented in numerous healthcare organizations in the attempt to offer patients the best chance of (1) receiving high quality medical procedures, (2) being served by a team of specialists, which is able to tailor treatment, and (3) guaranteeing access to specialist counselling, supportive care and rehabilitation. Evidence of such a trend can be found in the cases of specialist Breast Centres [2] and other specialized units which are currently largely diffused in all Europe. See the German case for Prostate Cancer Units, and the British Urological Malignancies units.

This emergent trend is prompting hospitals to reorganize current activities around patients and their diagnoses rather than in discipline focused departments [13]. However, devising service lines, around a specific group of patients with similar diagnosis and needs, entails the most substantial reconfiguration of hospital operations and has a huge impact on the care process, particularly when the resources to serve these patients are physically co-located [12]. Most critical decisions belong to planning the necessary resources, organizing the workflow of activities and the coordination of internal/external processes, defining the best layout for the new unit [12]. In fact, while addressing the aim of improving the care effectiveness, other conditions call for reducing costs by improving the system efficiency. Therefore, a methodology supporting managerial decisions during the reconfiguration process is extremely valuable.

Concluding, reorganizing activities for a specialized unit or simply a pathway requires effort for process analysis and planning in order to explore the feasibility of solutions: effective/efficient methods for pathways identification and analysis, demand planning and forecasting, resource allocation and optimization systems. This work aims to support decision makers in such reconfiguration process by proposing a feasible and efficient method for patient flow identification/analysis and for resource allocation. In so doing we also set the basis for an integrate modelling and planning of activities within healthcare organizations.

2   Theoretical Background: Resource Planning and Patient-Flows

The planning of resources in healthcare organizations requires the balancing of two opposite objectives: maximizing service levels and minimizing resource/capacity costs [5]. More resources (like doctors, instruments, beds) reduce patient waiting time, staff workloads and congestion. On the other hand, more resources increase the operating costs of the unit. In balancing these two goals, the main problems are the stochastic nature of the healthcare demand and the wide diversity of patients' conditions. For this last reason, this context appears to be similar to the engineer-to-order [11] or make-to-order environment in manufacturing.

The problem of planning resources and activities could be addressed with different decision horizon length. Hans et al. [11] and Vissers et al. [26] in their healthcare planning frameworks identified the strategic level (very long term), the operational level (short term), and an intermediate level (medium/long term) recognized as the tactical level. Until now, the researches focused particularly on the operational level (OR scheduling and bed management), with very few works at strategic level and, to a lesser extent, at tactical level [9, 11].

Literature shows a lack of efficient approaches for handling resource planning at strategic-tactical level in healthcare systems. This is mostly due to the peculiar characteristics of healthcare environment like heterogeneity, patient participation in the service process, simultaneity of production and consumption [11]. Specifically, one of the main limit in the medium-long term is the poor knowledge of the patient-flows that instead is very important in order to plan resources and organize/reorganize processes [4, 6, 10, 25].

Healthcare processes are in fact difficult to study: they are deeply interconnected, dynamic, multidisciplinary in nature, numerous, and ad hoc for a single patient [1, 7, 15, 18, 27]. Thus, traditional approaches to business process analysis are low effective and scarcely efficient in this context: they are often very time-consuming and might not provide an accurate representation of the processes and patient-flows.

More recently, the pervasive adoption of information systems within healthcare organizations and the development of effective data analysis techniques (e.g. process mining) have raised data availability and contributed to overcome such problems in finding and mapping patient-flows and analyzing the related healthcare processes [17, 18, 21, 23, 24].

As a combination between data mining and traditional model-driven Business Process Management, process mining aims to discover, analyze and improve processes exploiting the event logs available in the information systems [23]. In so doing, it tries to understand how processes are actually performed and which resources are involved, rather then what is prescribed or supposed to happen [16].

Though various authors have already shown its suitability [17, 19], the use of process mining technique to support healthcare management is a relatively new and unexplored field still valuable to investigate in order to support the planning of activities and resources inside healthcare organizations.

3   Research Design

3.1   Research Objective

This research aims to develop a methodology supporting service reconfiguration in healthcare systems. It exploits the process mining approach for identifying the service workflow of activities and the related resources, which are necessary for a specific class (/group) of patients. Knowledge is extracted from event logs by using the information stored in the hospital information systems.

Specifically, supporting the process discovery, analysis and monitoring phases, the following expected contributions can be identified:

1. Improvement of healthcare services planning: e.g. reduction of services fragmentation. Space, resources and timing of services could be easily analyzed and reconsidered coherently to a patient-centric perspective.
2. Enhanced resource allocation. Resources (space and beds, equipment and staff) could be organized more effectively and efficiently to provide sufficient resources for the involved classes of patients. In a shorter view, the efficiency of the method could allow managers to dynamically update and review plans.
3. Enabling an integrated resource planning at Hospital level. An extension of this modelling approach to all classes of patients served by the hospital may support managers to develop effective and integrate planning of resources in the medium-long term.

In this paper, we report some preliminary evidence deriving from the application of the methodology for the creation of a Patient-Focused Care Units, a Lung Cancer Unit in Italy. The method was applied concurrently to the traditional approach usually adopted by the management in order to test its suitability and compare benefits/costs.

3.2   Methodology

The methodology draws on the process mining approach [3, 8, 23] and goes through the following five main phases: Problem Statement; Log Preparation; Log Inspection; Process Discovery; Activities and Resources Evaluation.

1. *Problem Statement* aims to define the unit of analysis i.e. the class(es) of patients under investigation. For example, some selected groups of patients taken over by a hospital focused unit, a specific Diagnostic Related Group (DRG) or any other patient group identifiable inside the hospital.
2. *Data Retrieval and Log Preparation* create an adequate event log by preprocessing the event data gathered from one or more information systems.

3. The *Log Inspection* offers first insights about the followed patient-flows, permits to filter incomplete and/or outlier cases, and provides some useful descriptive statistics as the number of cases, the total number of events, the number of different sequences, etc.

4. *Process Discovery* phase aims to find the patient-flows under investigation, starting from the event log. It is possible to apply many techniques to act process discovery like α-Algorithm, heuristic miner, fuzzy miner, genetic miner and region-based miner. For complex environment, the most promising approaches seems to be heuristic mining, genetic mining and fuzzy mining [23]. However, in order to point out clear patient-flows it is sometimes useful to combine more process maps obtained using various process mining algorithms. In this phase, the involvement of experts (medical staff, physicians, nurses) can support to review, refine and validate the achieved patient-flows.

5. *Activities and Resources Evaluation* aims to estimate the activities consumptions and the related resources required for a specific class(es) of patients, starting from the patient-flows and using the process mining tools. Process mining tools like ProM and Disco®, can help decision makers to identify an "average" patient-flow with the correspondent expected demand of activities. In order to connect the activities to the related resources, a Bill Of Resources (BOR) could be used. The BOR is a BOM-like structure that includes all the resources needed to perform a specific class of activity [20], including people (doctors, physicians, nurses), instrumental resources, and materials.

4 Case Study

The methodology was applied to support a real case of service reconfiguration at the University Hospital of Pisa. The management was considering to reorganize the service for the lung cancer patients, by evaluating the creation of a Lung Cancer Unit in order to provide a better care to this serious class of patients. Management agreed that before to launch the new Patient-Focused Care Unit, it is important to known the main patient-flow, the most relevant activities, their interactions and the related resources. This is also interesting in order to evaluate the best layout for providing the service.

As a first step, we started formalizing the patients with lung cancer as the class of patients under investigation in order to set up a Lung Cancer Unit. Then, we created an adequate event log with the all the relevant instances, by aggregating and refining data coming from different information systems. We collected data of lung cancer patients served in the year 2014. The generated event log comprised more than 10000 events.

After a first inspection of log, we removed the outlier cases (after this operation, the dataset contained 470 cases) and obtained some relevant statistics about the process, such as the number of different activities, the number of different sequences/paths etc. For instance, descriptive stats revealed that the patients utilized 61 different activities, but that the 10 most frequent covered more than 95% of total events.

During the process discovery phase, we applied different process mining algorithms using two process mining tools (ProM and Disco®): the heuristic miner, the fuzzy miner, and the inductive miner. Along with the medical experts, we reviewed and refined the various process models obtained in order to achieve the most correct patient-flow. Figure 1 shows the resultant patient-flow expressed in BPMN language. We excluded from the map the less frequent and not pertinent activities. Thanks to this evidence, it was possible to support the new service planning with the purpose of reducing the services fragmentation in term of time, resources and space. In fact, doctors and managers can get a tangible proof of which activities are critically interconnected, those more relevant, which could be performed on the same day, like general physic examination and CT scan, and thus they might

Fig. 1 Patient-flow resultant for lung cancer patients

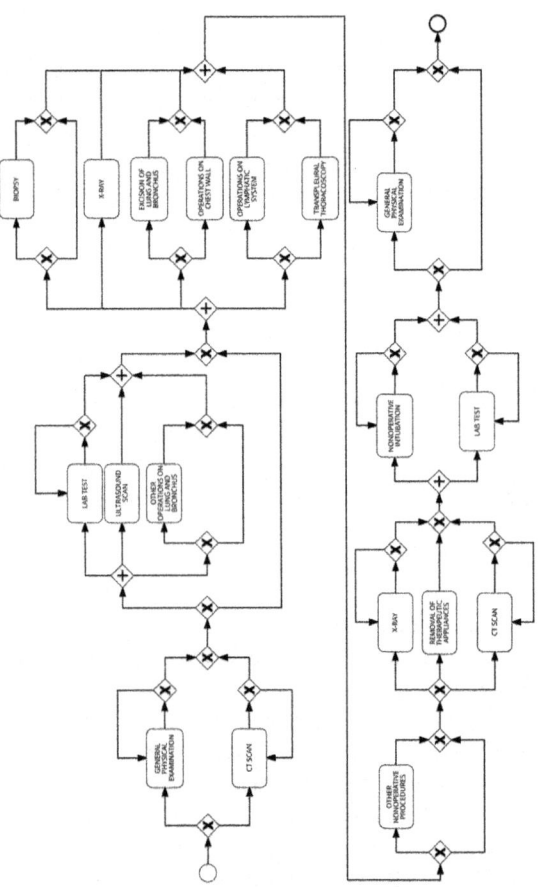

Activities	Average demand per patient
Lab test	8.14
Physical examination	4.51
X-ray	2.60
Non-operative intubation	1.29
CT scan	0.89
Excision of lung and bronchus	0.86
Other operations on lung and bronchus	0.56
Removal of therapeutic appliances	0.55
Ultrasound scan	0.49
Operations on lymphatic system	0.34
Transpleural Thoracoscopy	0.24
Operations on chest wall	0.20
Other procedures non-operative	0.09
Biopsy	0.08

Fig. 2 Activities for an average lung cancer patient

organize appointments and physical layout accordingly. Furthermore, the process map makes also possible to get relevant information on how to streamline the current patient-flow and devise an ideal one for future improvements.

The last step, activities and resources evaluation, provided us the average volume of activities required by a patient affected by lung cancer. Exploiting ProM and Disco®, it was possible to estimate the overall demand of activities requested by the lung cancer patients. Starting from the total volumes of activities and observing the patient-flow, we could calculate the activities necessary for an "average" patient. Figure 2 reports the average demand of activities by a lung cancer patient.

Showing the average demand of activities, possibly its variability in time, and related resource consumption, outcomes from this last step could really support managers in taking decisions about the creation/implementation of the new Lung Cancer Unit. For example, they can easily estimate the needing resources for the service, starting from the average demand of activities and using the number of patients as a parameter (in term of space, beds, equipment, and staff). Also, they can evaluate the feasibility of the new focused unit in respect to a number of variables as such as available space and resources, law and regulations concerning to the specialized staff etc.

On a next step of the research, we aim to finalize the BORs in order to complete the overall analysis of resources related to the patient-path. Moreover, this information could also drive decisions about new investments or resource centralization/sharing and enable what-if analysis, simulations, and other more in-depth analysis.

5 Discussion and Conclusions

This work respond to the need of a methodology supporting decision makers when a healthcare organization is considering to reorganize the delivery of cares or to introduce a new service in its portfolio. This is an increasingly frequent situation in this service context [12]. The relevance of the proposal is confirmed by a lack of efficient approaches and tools for supporting activity and resource planning in the medium-long term, able to handle the peculiar characteristics of healthcare systems [9, 11].

Findings suggest that the proposed approach can profitably support managers during the re-organization of services (e.g. a new Patient-Focused Care Unit) by estimating the patient-flows and the activities/resources which are necessary for different classes of patients. This is valuable in order to improve the overall service planning, reduce the services fragmentation and optimize resource allocation. A deep comprehension of the patient-flows, in fact, enables to streamline business processes and can support performance analysis.

Moreover, evidences from the case application also show that the proposed method is quite efficient and easy to apply in a real environment. Thus, managers could also periodically review the path in order to check if the service flow and the related demand of activities and resources have changed. This would allow to dynamically modify the service planning and resource allocation.

As an addition, the extension of the proposed approach to the entire organization could enable an integrated planning of the activities and resources within the healthcare organization. Specifically, the method would help managers firstly to investigate each class of patients and the related patient-flows, and then to aggregate their demands for activities and resources. In so doing, it could also provide interesting insights to set-up (new investments) and manage efficiently several hospital subunits (like X-ray) or other shared resources.

Finally, discovering the activities and resources linked to the different classes of patients, this method could support cost accounting in an Activity Based Costing perspective.

Clearly, some critical issues still remains, as for example the possible coexistence of two or more diseases in the same patient. The service and resource planning for these cases of comorbidity could be more difficult. Moreover, the case application we assessed concerns just a single class of patients, therefore this circumstance could affect meaningfulness and generalizability of our investigation.

Nevertheless, the positive results of this preliminary study encourage us to extend the research to a wider set of services and patient classes.

Acknowledgements We thank the Direction of the University Hospital of Pisa, and in particular Dr. Marco Nerattini, for their support and precious contribution to the research.

References

1. Anyanwu, K., Sheth, A.P., Cardoso, J., Miller, J.A., Kochut, K.J.: Healthcare Enterprise Process Development and Integration (2003)
2. Blamey, R., Blichert-Toft, M., Cataliotti, L., Costa, A., Greco, M., Holland, R., Sainsbury, R.: The requirements of a specialist breast unit. Eur. J. Cancer **36**(18), 2288–2293 (2000)
3 Bozkaya, M., Gabriels, J., Werf, J.M.E.M.: Process diagnostics: a method based on process mining. In: International Conference on Information, Process, and Knowledge Management, 2009, eKNOW'09, pp. 22–27. IEEE (2009)
4 Brailsford, S.C., Lattimer, V.A., Tarnaras, P., Turnbull, J.C.: Emergency and on-demand health care: modelling a large complex system. J. Oper. Res. Soc. **55**(1), 34–42 (2004)
5 Bretthauer, K.M., Côté, M.J.: A model for planning resource requirements in health care organizations. Decis. Sci. **29**(1), 243–270 (1998)
6 Cannavacciuolo, L., Illario, M., Ippolito, A., Ponsiglione, C.: An activity-based costing approach for detecting inefficiencies of healthcare processes. Bus. Process Manag. J. **21**(1), 55–79 (2015)
7 Cannavacciuolo, L., Iandoli, L., Ponsiglione, C., Maracine, V., Scarlat, E., Nica, A.S.: Mapping knowledge networks for organizational re-design in a rehabilitation clinic. Bus. Process Manag. J. **23**(2) (2017)
8 De Weerdt, J., Caron, F., Vanthienen, J., Baesens, B.: Getting a grasp on clinical pathway data: an approach based on process mining. In: Emerging Trends in Knowledge Discovery and Data Mining, pp. 22–35. Springer, Berlin, Heidelberg (2013)
9 Dobrzykowski, D., Deilami, V.S., Hong, P., Kim, S.C.: A structured analysis of operations and supply chain management research in healthcare (1982–2011). Int. J. Prod. Econ. **147**, 514–530 (2014)
10 Haraden, C., Resar, R.: Patient flow in hospitals: understanding and controlling it better. Front. Health Serv. Manag. **20**(4), 3 (2004)
11 Hans, E.W., Van Houdenhoven, M., Hulshof, P.J.: A framework for healthcare planning and control. In: Handbook of Healthcare System Scheduling, pp. 303–320. Springer US (2012)
12 Hyer, N.L., Wemmerlöv, U., Morris, J.A.: Performance analysis of a focused hospital unit: the case of an integrated trauma center. J. Oper. Manag. **27**(3), 203–219 (2009)
13 Kremitske, D.L., West Jr., D.J.: Patient-focused primary care: a model. Hosp. Top. **75**(4), 22–28 (1997)
14 Lathrop, J.P., Seufert, G.E., MacDonald, R.J., Brickhouse Martin, S.: The patient-focused hospital: a patient care concept. J. Soc. Health Syst. **3**(2), 33–50 (1991)
15 Lenz, R., Reichert, M.: IT support for healthcare processes—premises, challenges, perspectives. Data Knowl. Eng. **61**(1), 39–58 (2007)
16 Mans, R.S., Schonenberg, M.H., Song, M., van der Aalst, W.M., Bakker, P.J.: Application of process mining in healthcare—a case study in a dutch hospital, pp. 425–438. Springer, Berlin, Heidelberg (2009)
17 Mans, R.S., van der Aalst, W., Vanwersch, R.J.: Process Mining in Healthcare: Evaluating and Exploiting Operational Healthcare Processes. Springer (2015)
18 Rebuge, Á., Ferreira, D.R.: Business process analysis in healthcare environments: a methodology based on process mining. Inf. Syst. **37**(2), 99–116 (2012)
19 Rojas, E., Munoz-Gama, J., Sepúlveda, M., Capurro, D.: Process mining in healthcare: a literature review. J. Biomed. Inform. **61**, 224–236 (2016)
20 Roth, A.V., Dierdonck, R.: Hospital resource planning: concepts, feasibility, and framework. Prod. Oper. Manag. **4**(1), 2–29 (1995)
21 Stefanini, A., Aloini, D., Dulmin, R., Mininno, V.: Linking Diagnostic-Related Groups (DRGs) to their processes by process mining. In: HEALTHINF 2016—9th International

Conference on Health Informatics, Proceedings; Part of 9th International Joint Conference on Biomedical Engineering Systems and Technologies, BIOSTEC 2016, pp. 438–443 (2016)

22 Stitzenberg, K.B., Meropol, N.J.: Trends in centralization of cancer surgery. Ann. Surg. Oncol. **17**(11), 2824–2831 (2010)

23 Van Der Aalst, W.: Process Mining: Discovery, Conformance and Enhancement of Business Processes. Springer Science & Business Media (2011)

24 Van der Aalst, W. M., Reijers, H. A., Weijters, A. J., van Dongen, B. F., De Medeiros, A. A., Song, M., & Verbeek, H. M. W: Business process mining: An industrial application. Information Systems, 32(5), 713-732 (2007)

25 Vissers, J.M.: Patient flow-based allocation of inpatient resources: a case study. Eur. J. Oper. Res. **105**(2), 356–370 (1998)

26 Vissers, J.M., Bertrand, J.W.M., De Vries, G.: A framework for production control in health care organizations. Prod. Plan. Control **12**(6), 591–604 (2001)

27 Weske, M.: Business process management architectures. In: Business Process Management, pp. 333–371. Springer, Berlin, Heidelberg (2012)

Improving Emergency Medical Services with Time-Region-Specific Cruising Ambulances

Jiun-Yu Yu and Kwei-Long Huang

Abstract This study aims to propose a dynamic resource allocation policy for Emergency Medical Service (EMS) of New Taipei City, Taiwan. Response time is one of the major key performance indicators in EMS system since rapid ambulance response provides patients better chances of recovering or surviving. Ambulances in New Taipei City used to park in separate fire stations and wait for the calls from emergency medical dispatch center (EMD). As an alternative, a dynamic resource allocation policy is designed and proposed by dispatching ambulance to patrol on the streets or stay at specific locations in the areas with high emergency events demand time slots. These areas and time slots are determined by statistical analysis of the historical emergency medical service data. The main idea of this dynamically allocated ambulance policy is to increase the probability that an ambulance will be on stand-by at the nearby area when an emergency event occurs. In addition, this policy is investigated under the condition that no extra EMS resource is available. A simulation model is developed to evaluate this dynamic resource allocation policy, and the results of this study show that time-region-specific ambulance cruising policy can significantly reduce the EMS response times.

Keywords Response time · Resource planning · Resource allocation Healthcare

J. -Y. Yu (✉)
Department of Business Administration, National Taiwan University, Taipei, Taiwan
e-mail: jyyu@ntu.edu.tw

K. -L. Huang
Institute of Industrial Engineering, National Taiwan University, Taipei, Taiwan
e-mail: craighuang@ntu.edu.tw

© Springer International Publishing AG 2017
P. Cappanera et al. (eds.), *Health Care Systems Engineering*, Springer Proceedings in Mathematics & Statistics 210, https://doi.org/10.1007/978-3-319-66146-9_17

1 Research Objective

This study aims to achieve two goals. First, reduce the response time without substantially increasing the amount of or significantly changing the allocation of the current emergency medical service (EMS) resources. Second, propose a time-region-specific cruising ambulances system which dispatches ambulance(s) to patrol on the streets or stay at some specific locations in the high EMS demand areas and time slots. This study is conducted based on the historical EMS data provided by the fire department of New Taipei City, Taiwan.

2 Literature Review

Brotcorne et al. [1] review many OR/MS models for ambulance location and allocation. These models usually focus on minimizing the number of ambulances with the location set covering model (LSCM) or on maximizing the demand covered with the tools such as the maximal covering location problem (MCLP) model and the double standard model (DSM) subject to two covering constraints. Many models that contribute to location research afterward are introduced in this paper.

Green and Kolesar [2] conclude that emergency medical service system (EMSS) has been significantly influenced by operations research and management science (OR/MS) after reviewing the related literatures published from 1960s. Many papers on EMSS management appear in 1970s, leading to implementation of new policies in firefighting and policing. Lots of the practices and experiences from New York City are then adopted by many other cities. Green and Kolesar also discuss how original models affected EMSS, investigating the pace of new models in OR/MS area and suggesting the future potential of OR/MS application.

Ingoldsson et al. [3] develop a discrete event simulation model to evaluate the impacts of the different deployment scheme. In the simulation models, ambulances having completed their missions are to be directed to certain location according to the proposed "park search" algorithm. "Park search", consisting of a prioritized list of locations where ambulances should park, is conducted after ambulances transfer patients to hospitals. The system allows ambulances to be available near the area with high EMS risk.

3 Problem Description

New Taipei City is the most populated region in Taiwan, roughly 4 million people living in a donut-shaped area of 2,000 km^2. The north part of New Taipei City is coastal area while the south part of it is covered by mountains more than 1,000 m high. Due to the unique geographic characteristic, the inhabitants are densely living

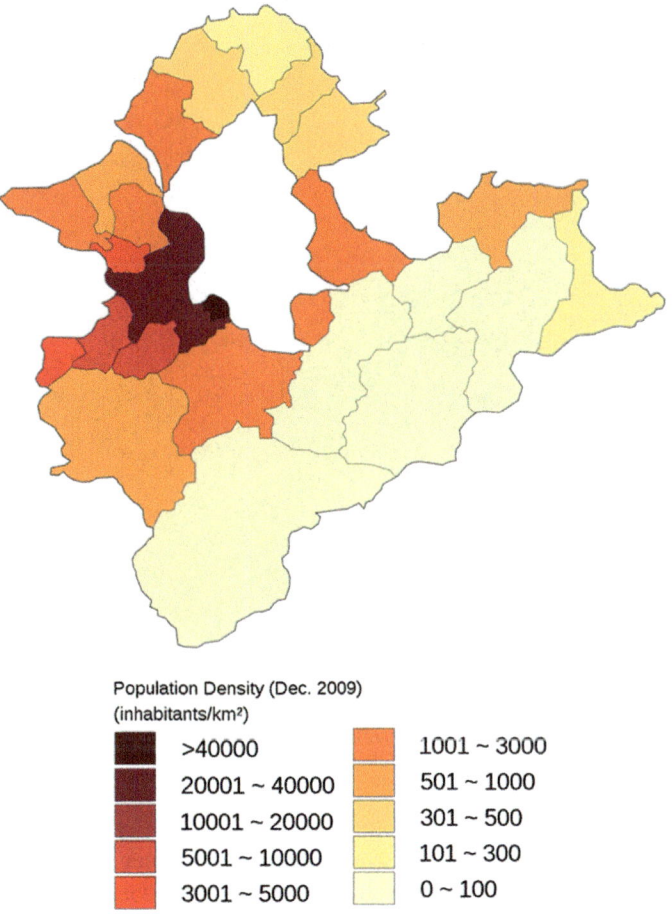

Fig. 1 Population density of New Taipei City

in the small west region, and the medical resources such as hospitals and fire stations are unevenly distributed (Figs. 1 and 2). In sum, there are currently 66 EMS units within the 29 districts in New Taipei City.

The original ambulance operation is that each ambulance is standing-by at its designated fire station. Once an emergency call is dispatched to a specific fire station, its standing-by ambulance will be sent to the scene to execute the task. After completing the task, the ambulance returns back to its designated fire station for posted cleanup and then become ready for the next task. The historical data of emergency medical service studied in this research is the record of location and time of every emergency event occurred in New Taipei City. There are seven steps of handling an incoming emergency call:

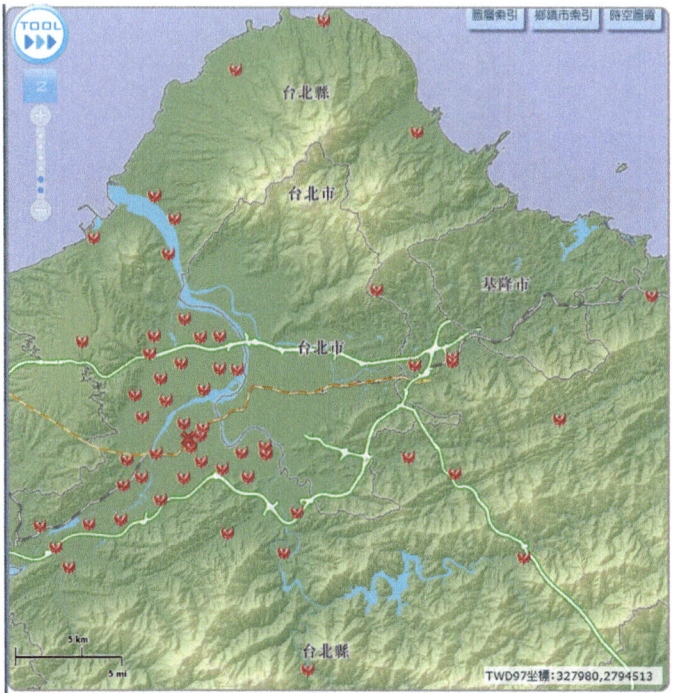

Fig. 2 Fire station locations in New Taipei City

1. Dispatch Center receives a call from an emergency scene.
2. Dispatch Center dispatches an available ambulance from the fire station which is the nearest to the scene.
3. The ambulance arrives at the emergency scene.
4. After the EMTs finish the first aid emergency treatments, the ambulance leaves for the nearest hospital or the hospital requested by the patient.
5. The ambulance arrives at the hospital.
6. The ambulance leaves the hospital and returns back to the fire station.
7. The ambulance arrived at the fire station and the task closes.

Due to the physical locations of fire stations, response times for some events may be longer than the required standard and are not improvable inherently under the current configurations and resources. Subjected to the resource constraint (e.g., without adding new ambulances and emergency medical technicians (EMTs)), new ways of dynamically deploying the ambulances must be investigated and proposed.

Under the time-region-specific cruising policy, one or more ambulances will patrol on some certain routes or stay at a specific location during a certain time slot instead of residing at its designated fire station. It is expected that the dynamically deployed ambulance(s) can more rapidly respond to the emergency events and provide instant care to patients simply through the re-allocation of the current resource.

4 Research Methods

In this study, a set of heat maps based on the Geographical Information System (GIS) is developed to show the frequency of emergency events spread over in each district in New Taipei City. Figure 3 is generated based on the data of emergency events collected from 2010 to 2011. In this figure, New Taipei City is divided into small grids, each 450 m^2. The grid colored in red denotes the higher frequency of events while the blues ones denotes the lower frequency.

After analyzing the historical data from the 29 districts, three busiest regions, Banqiao, Sanchong, and Zhonghe, are selected for further investigation with the time-region-specific cruising policy using simulation modeling. Since the EMS demand is not time invariant on hourly base, each region is only simulated for two time slots with high EMS event frequencies, namely, 17:00–19:00 and 21:00–23:00. This simulation setting in fact increases the practicality of the time-region-specific cruising policy under investigation as the ambulance(s) will only have to be re-deployed in these time slots, if this policy is proved to be significantly better and is then implemented in real practice.

The simulation model is developed using AnyLogic software, and the input parameters used for simulation settings include (1) arrival rate of EMS events, (2) location of each EMS event, (3) delay time between the call from EMD and the ambulance leave the fire station, (4) ambulance average driving speed, (5) duration

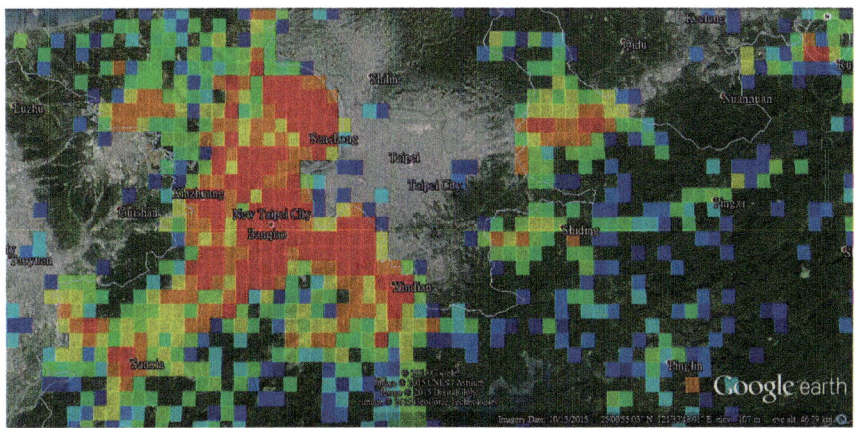

Fig. 3 Example of heat map for New Taipei City

Region & time slots	Banqiao 17:00- 19:00	Banqiao 21:00- 23:00	Sanchong 17:00- 19:00	Sanchong 21:00- 23:00	Zhonghe 17:00- 19:00	Zhonghe 21:00- 23:00
K-S test p-value	0.0068	0.0154	0.1155	0.0509	0.0507	0.4585
Paired t-test p-value	0.5190	0.8738	0.8091	0.0670	0.2354	0.8488

Fig. 4 Simulation model validation resutls

time EMT stay at the scene, (6) duration time of paper work and patient transfer at the hospital, and (7) number of ambulance in each fire station.

The simulation model is validated by comparing the simulated results with the historical EMS data. Simulation results are obtained by random generation of EMS events, the locations of which are based on historical data. Ambulances in the simulation model are operated according to the original operations: standing-by in the fire station and waiting for the calls from dispatch center. Kolmogorov-Smirnov test and paired t-test are conducted to examine the difference between the simulation results and the historical response time. The test results show that there are no significant differences between the result generated by simulation model and the historical response time (Fig. 4).

5   Simulation Analysis

In this study, two types of ambulance dynamic allocation policies are proposed and investigated. In the Guarding mode, the ambulance is on standby at the gravity location determined from the historical EMS events. In the Cruising mode, the ambulance patrols on either a short or a long routes that are specifically designed around the gravity location. Since vehicles are driving on the right-hand side in Taiwan, making right-turn is more convenient and efficient than turning left. Thus, in the Cruising mode the ambulance patrols on the designed routes clockwise. Moreover, as some of the fire stations are equipped with two ambulances, it is also good to know if the response times would be reduced if both ambulances are operated under the proposed dynamic allocation policy. Figure 5 summarizes the ambulance operation policies studied in the simulation.

Figure 6 demonstrated the simulation example of Banqiao region, in which the dotted blue line illustrates the shorter route while the solid black line the longer route. There are actually 6 fire stations in Banqiao region, locations of which are also indicated in Fig. 6.

For each of the "One" ambulance operation policies listed in Fig. 4, Banqiao region may have 6 possible arrangements as there are 6 stations; only one ambulance is needed to run on this policy. Therefore, 18 different settings are to be

Plans		Description
Original Operations		Ambulance standing-by in the fire station, waiting for the calls from dispatch center.
Dynamic Allocation One Ambulance	**Guarding Mode**	Standing-by at the gravity location
	Cruising Mode	Cruising on (shorter) Route 1 clockwise
		Cruising on (longer) Route 2 clockwise
Dynamic Allocation Two Ambulances	**Guarding Mode**	Standing-by at the gravity location 1
		Standing-by at the gravity location 2
	Cruising Mode	Cruising on (shorter) Route 1 clockwise
		Cruising on (longer) Route 2 clockwise

Fig. 5 Ambulance operations policies studied via simulation

Fig. 6 Demonstration of simulation model for Banqiao region

simulated and examined in the model. Similarly, for each of the "Two" ambulance operation policies, Banqiao region will have to choose from the 6 stations and the 4 policies, resulting in total 180 possible arrangements to be simulated.

In the time slot 17:00–19:00, the simulation results with "One" ambulance operation policy for Banqiao is shown in Fig. 7. It is obvious that the response

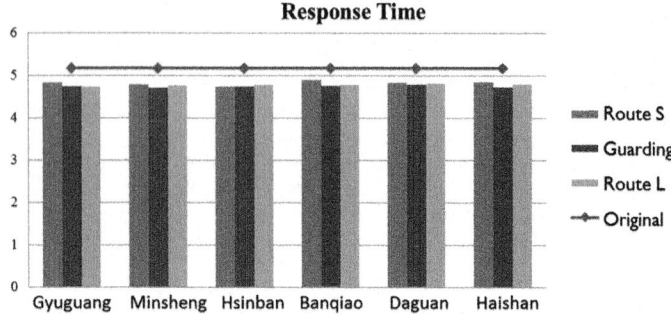

Fig. 7 Simulation results with One ambulance policy for Banqiao region, 17:00–19:00

Fig. 8 Policy combinations
for Two ambulance policy for
Banqiao region

A	Gravity 1 & Route 1
B	Gravity 1 & Route 2
C	Gravity 2 & Route 1
D	Gravity 2 & Route 2
E	Gravity 1 & Gravity 2
F	Route 1 & Route 2

times for the 3 proposed "One" ambulance policy are all shorter than those of the original practice. The average reduction in response times is 7.56%.

In the same time slot, the "Two" ambulance operation policy are examined under 6 different combinations, shown in Fig. 8. The simulation results of these 6 combinations are compared using Tukey's Honest Significant Differences (HSD) Test, which shows that the combinations C and D are significantly better than the other four combinations (Fig. 9). This result suggest that in Banqiao for the time slot 17:00–19:00, sending one ambulance to stay at gravity center 2 and the other ambulance to patrol on either shorter or longer route will significantly reduce the response time, from 5.16 to 4.36 min, in average. Further simulation investigations reveal that any two of the six stations can be chosen to implement these C or D combination as the simulated performances are similar.

The same simulation investigation is conducted for Banqiao for the time slot 21:00–23:00 as well. The results show that again combinations C and D are significantly better than the other four combinations. However, the performances are not the same across the six stations. Tukey's HSD Test shows that only the following practice produces significant reduction in response time: one ambulance

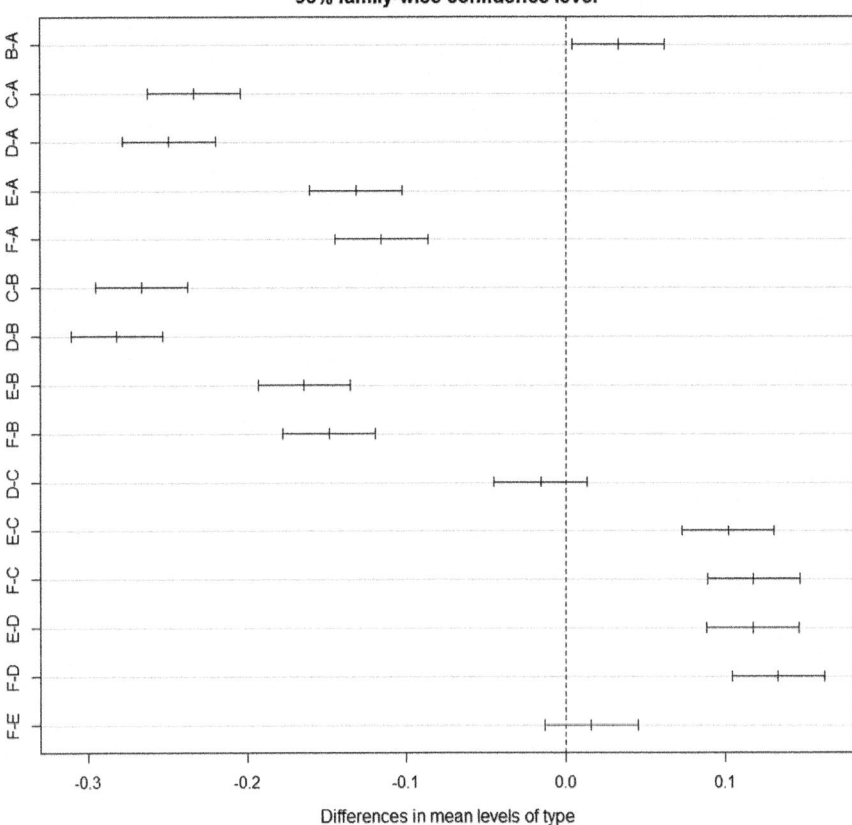

Fig. 9 Tukey's honest significant differences (HSD) test results for Two ambulance policy for Banqiao region, 17:00–19:00

from Minsheng station stay at gravity center 2 and the other ambulance from Daguan patrol on route 2 (Fig. 10). This set of collaboration reduced the average response time from 4.74 to 3.82 min, 19.35% of reduction.

In sum, the simulation study shows that for Banqiao region, both One and Two ambulance dynamic allocation policies generate significant reduction in response time. But if the Two ambulance policy is to be implemented, different time slots required different set of stations to collaborate to achieve the best result.

Similar simulation modeling and experimentations for the other two chosen regions, Sanchong and Chonghe, are conducted as well for the two chosen time slots, 17:00–19:00 and 21:00–23:00. The optimal ambulance dynamic allocation policies for these time-region-specific combinations are all different, and significant reductions in response time are all achieved.

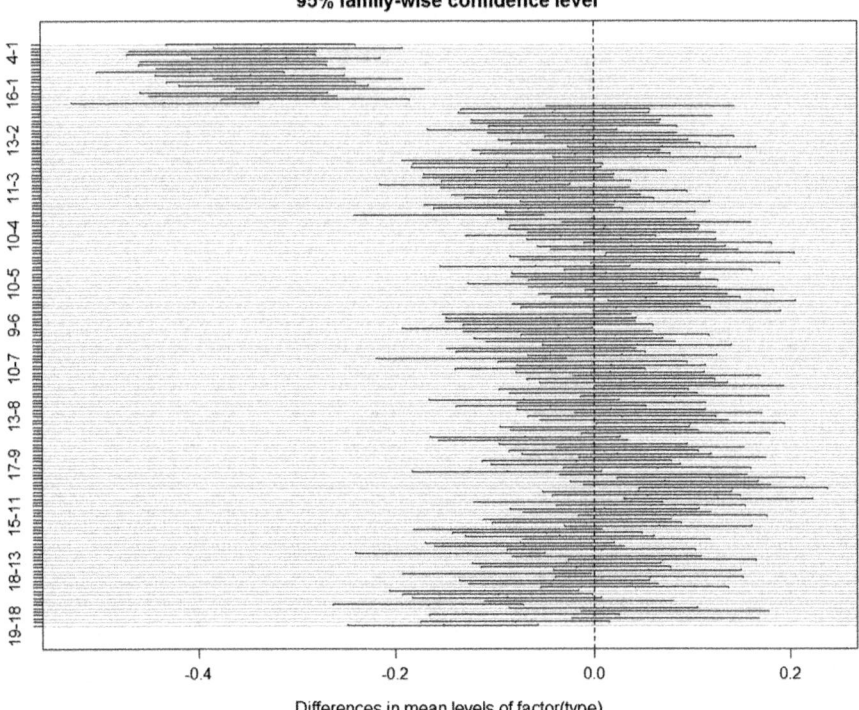

Fig. 10 Tukey's honest significant differences (HSD) test results for Two ambulance policy for Banqiao region, 21:00–23:00, six stations collaboration

6   Conclusion

This simulation study successfully achieved the two aforementioned aims. Given the same limited amount of EMS resources, the time-region-specific cruising ambulance scheme proposed and tested in this study helps to significantly reduce response times for the chosen fire stations in New Taipei City. These dynamically allocated ambulances help to reduce the EMS covering areas of every station. The simulation model developed in this study is the prototype and can be suitably extended to different time-region combinations.

References

1. Brotcorne, L., Laporte, G., Semet, F.: Ambulance location and relocation models. Eur. J. Oper. Res. **147**, 451–463 (2003)

2. Green, L.V., Kolesar, P.J.: Improving emergency responsiveness with management science. Manag. Sci. **50**(8), 1001–1014 (2004)
3. Ingolfsson, A., Erkut, E., Budge, S.: Simulation of single start station for Edmonton EMS. J. Oper. Res. Soc. **54**(7), 736–746 (2003)

A Simulation Model for Optimizing Staffing in the Emergency Department

Melik Koyuncu, Ozgur M. Araz, Wes Zeger and Paul Damien

Abstract In this study we used a nonparametric statistical approach estimate time varying arrival rates to an Emergency Department (ED). These rates differ hourly within each month in a year. The time varying arrival rates serve as a key input into a new discrete-event simulation model for patient flow at an ED of the case study hospital. Our simulation model estimated Length of Stay (LOS) and Door to Doctor Time (DTDT) for patients in three different acuity groups. The simulation model also used for optimizing the staffing allocations in three different shifts. The methodological contributions are exemplified using real ED data from a major teaching hospital.

Keywords Quality of service · Waiting time · Discrete event simulation Optimization · Time-varying arrival rates

1 Introduction

Hospital emergency departments (EDs) are one of the most critical elements of the national health system in the United States. According to the US Government Accountability Office (GAO) report, hospital ED crowding continues to increase

M. Koyuncu (✉)
Industrial Engineering Department, Cukurova University, Adana, Turkey
e-mail: mkoyuncu@cu.edu.tr

O. M. Araz
College of Business Administration, Supply Chain Management and Analytics,
University of Nebraska Lincoln, Lincoln, NE 68588, USA

W. Zeger
Emergency Medicine Department, University of Nebraska Medical Center,
Omaha, NE, USA

P. Damien
McCombs School of Business, Information Risk and Operations Management,
University of Texas at Austin, Austin, TX, USA

© Springer International Publishing AG 2017   201
P. Cappanera et al. (eds.), *Health Care Systems Engineering*, Springer Proceedings
in Mathematics & Statistics 210, https://doi.org/10.1007/978-3-319-66146-9_18

wherein some patients wait longer than the recommended times [1] thereby receiving poor quality of service [2].

A commonly used performance measure for evaluating ED operations is Length of Stay (LOS) defined as the time between the arrival to discharge of a patient from an ED. Another measure is Door to Doctor Time (DTDT), i.e., the time difference between arrival at an ED and the time a physician first sees the patient. LOS generally consists of the sum of two components (a) processing times which are related to patient acuity involving triage, physician examination, laboratory tests, medical treatment; and (b) patient waiting times for reaching these processes, often called process queue times. In some cases, LOS could be high for admitted patients due to boarding, or some patients could be held in ED until their conditions are stabilized. DTDT could be a more critical performance indicator for immediate and critical patients due to the severity of their medical conditions. In the estimation of DTDT and identification of operational strategies that could improve this measure, acuity levels of patients should be taken into account, rather than taking the average DTDT to increase service quality. Ang et al. [3] used a Q-lasso method that combines statistical learning and fluid model estimators to predict ED wait times for different patient acuity groups. Kuang and Chan [4] evaluated proactive patient diversion policies by using predicted waiting time information to improve service quality at an ED. Similar methodological approaches of integrating predictive modeling with simulation were used by Saghafian et al. [5] in order to develop patient streaming policies and sequencing to improve ED service responsiveness.

In this research we first present the analyses of patient arrivals data to an ED to better parametrize a discrete event simulation model then using simulation optimization in Arena software [6] we allocate the staff optimally in three shifts for staff balancing. Althoguh there are extensive studies in the literature using queuing theory principles and simulation-optimization approaches for EDs, most of them do not consider seasonal or monthly variations in patient arrival rates. Saunders et al. [7], Duguay and Chetouane [8], Kuo et al. [9] presented discrete event simulation models for process flow of EDs and used simulation experiments for assessing ED performance are some examples.

2   Patient Flow

Patient flow in the ED of the case study hospital involves triage, physician examination, extra testing, nurse evaluation, minor treatment and major treatment operations, and patient stabilization times. Patients arriving at the ED are evaluated at the triage then categorized and prioritized according to their medical condition wherein the acuity levels range between 1 (most critical) and 5 (least critical). If a patient arrives to the ED as a walk-in, acuity level ranges between 2 and 5 most of the time. However, if patients arrive in an ambulance, they could skip triage and are considered acuity 1. All patients are examined by a physician based on the priority rule that accounts for acuity level in the examination room. Physicians direct some patients to a test room to have their tests completed, following which they revisit the physician for assessment. Patients could follow different routes after the physician's assessment; these routes along with routing probabilities are presented in Fig. 1 based on 2014 ED visits data that was exported from the hospital's database. Total ED visits during that calendar year was 49,180—1% acuity 1 patients, 11% acuity 2, 52% acuity 3, 30% acuity 4, and 6% acuity 5.

3 Simulation Model

We developed a discrete event simulation model to analyze the patient flow in the ED based on Fig. 1. The following discussion details each of the key inputs to be used in the overall discrete-event simulation model.

Service (Processing) Times: The required service times for processes in the ED include, examination time, extra test time, nurse evaluation time, major and minor treatment times. We estimate these times from the hospital data and used feedback from experts for verification purposes. As discussed in Kuo et al. [9], service times do not have a consistent structure due to the urgency in any ED; hence it is difficult to use statistical methods to estimate each of the service steps in the ED flow. The staffing included physicians who are scheduled daily, lab technicians, (scheduled only for week days), and nurses who are scheduled for a week. We estimated the triage processing time from ED data as the average time patients spent from initial arrival to the ED until completion of triaging at the nurses' station. The other processing times are estimated based on interviews with ED staff, and are presented in Table 1. Given the inherent uncertainty in these data, we model them using probability distributions that allow for considerable and realistic variability in these input parameters.

Routing Probability: The routing probabilities are directly related to patient acuity level in the ED. Using the hospital's data these probabilities were calculated as detailed in Fig. 1.

3.1 Time-Varying Poisson Arrival Rates

A Poisson distribution was fit to hourly arrivals data from which time varying hourly arrival rates in a day were obtained: we have 24 parameters, $\lambda(t)$, $(t = 1,\ldots, 24)$, and these are shown in Table 2. The distinction between these Poisson time-varying rates and the ones detailed under time-varying rates is that no distribution assumption is made in the latter. This, we believe, may be useful in this context since, there's considerable variability in the observed data that is quite difficult to fit using a parametric model.

Table 1 Processing time estimates used in the discrete-event simulation

Process	Distribution parameters (min)	Resource(s)	Reference
Triage	TRI A(4.68:8.54; 15)	Nurse	Hospital data
Examination	UNIF(5,8) & UNIF (4,7)*	Physician, bed	Expert view
Extra test	TRIA (10,60,110)	Technician	Expert view
Stabilization time for acuity 4–5 patients	UNIF (5,80)	Bed	Expert view
Nurse last evaluation for acuity 4–5 patients	TRIA (10,15,25)	Nuxse, bed	Expert view
Major treatment	UNIF (45,90)	Physician, bed	Expert view
Minor treatment	UNIF (5,10)	Phvsician. nurse, bed	Expert view
Stabilization time for acuity 2–3 patients	UNIF (20,440)	Bed	Expert view
Nurse last evaluation for acuity 2–3 patients	UNIF (20,40)	Nurse, bed	Expert view

*For returning patients, *UNIF* Uniform Distribution, *TRIA* Triangular Distribution

3.2 Model Validation

To ensure that the model is a valid representation of real ED, we compare the real patient arrivals with the simulated patient arrivals. While real data contain 48,140 patient records, simulated data generated 45,688 patients.

In addition we analyzed the observed LOS and simulated as presented in Fig. 2 to compare the distributions of both real and the simulated LOS. The simulated LOS's mean is 245.46 min and its standard deviation (SD) is 144.88 min and observed mean of LOS is 243.81 min (SD = 148.49 min). We performed the following hypothesis for testing whether they come from the same statistical distribution or not.

Hypothesis

H_0: Observed length of stay (LOS) and the simulated LOS follow same statistical distribution.

H_1: Observed LOS and the simulated LOS do not follow same statistical distribution.

Since both real LOS and simulated LOS distributions do not fit a theoretical statistical distribution, we applied Mann Whitney U test, a non-parametric statistical test, for the stated hypothesis. As a result of the Mann Whitney U test, null hypothesis cannot be rejected with the p value of 0.19.

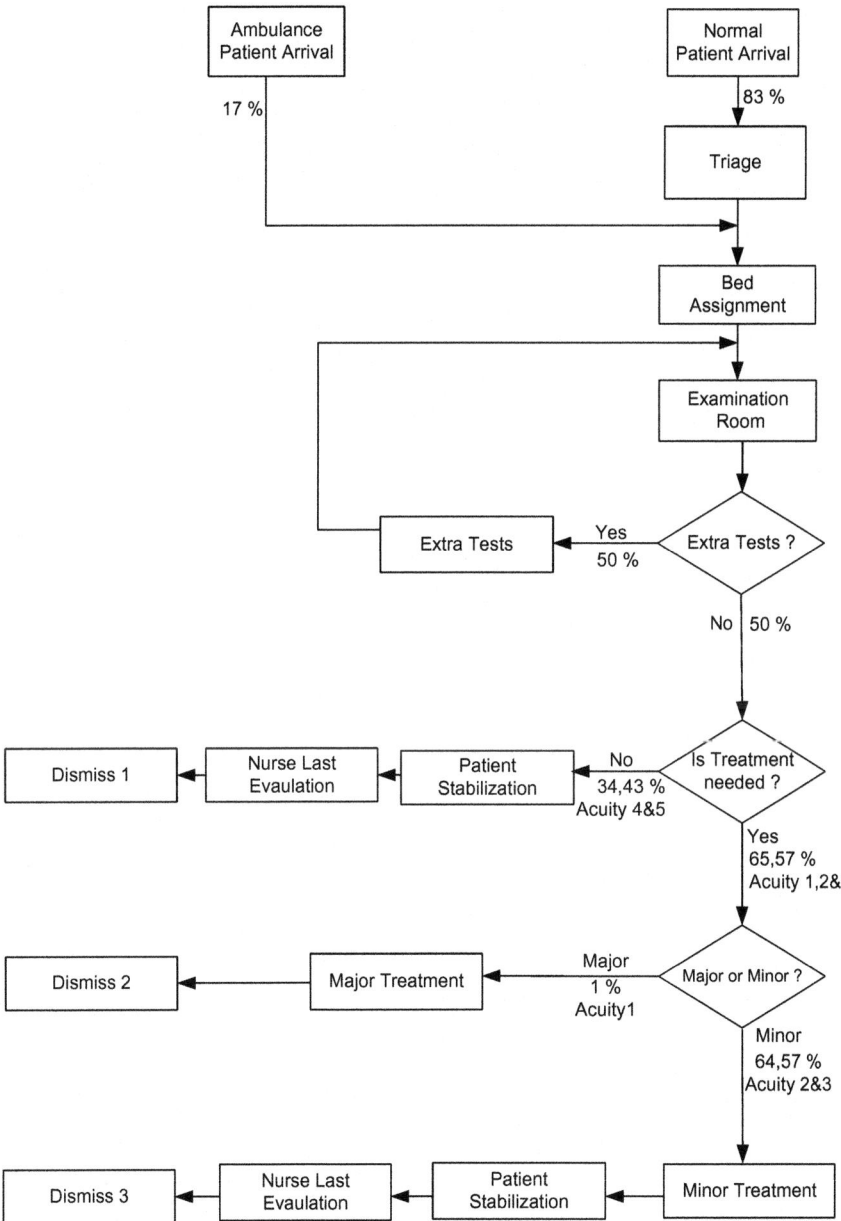

Fig. 1 Conceptual patient flow in the ED with routing probabilities

Table 2 Time varying Poisson rates

Time (h)	Poisson rate	Stand. deviation	Time (h)	Poisson rate	Stand. deviation
1	3.08	0.501	13	7.83	0.807
2	2.50	0.456	14	7.83	0.808
3	2.00	0.408	15	7.50	0.790
4	2.08	0.416	16	7.41	0.786
5	1.75	0.381	17	7.75	0.803
6	1.75	0.381	18	7.50	0.790
7	2.08	0.416	19	7.50	0.791
8	2.91	0.493	20	6.83	0.754
9	4.25	0.595	21	6.75	0.750
10	6.50	0.735	22	5.83	0.697
11	7.16	0.772	23	5.25	0.661
12	7.50	0.790	24	4.16	0.589

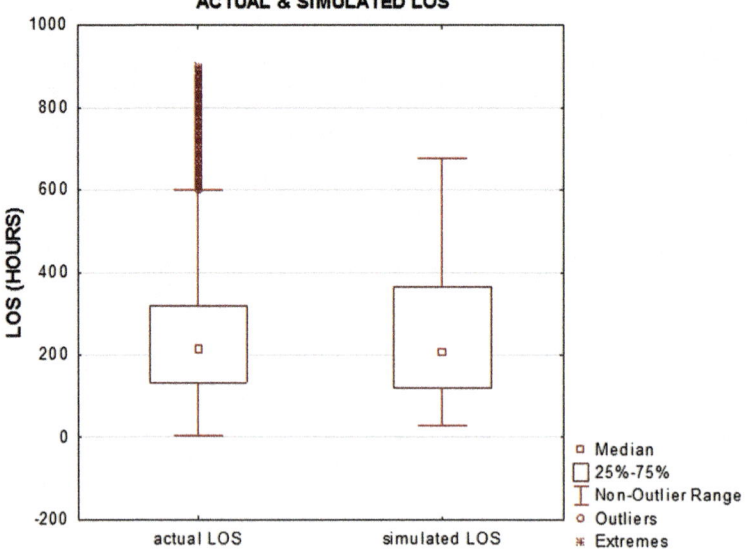

Fig. 2 Actual versus simulated LOS box plots

4   Results and Discussion

This study presented a discrete-event simulation that modeled patient arrival patterns to an ED during different times of the year.

Each day is divided into three time intervals: $00:^{00}$–$08:^{00}$, 08^{00}–$16:^{00}$ and $16:^{00}$–$23:^{59}$; labeled Shift I, Shift II and Shift III, respectively. Given predetermined (acceptable) patient waiting times in the range of 5–30 min in 5-min increments, and patient acuity levels, we aim to determine the minimum number of service providers in each of the three shifts. As the number of providers increases, the expected waiting times will obviously decrease; however the total staffing costs would likely increase, and in many cases staffing may not be feasible due to available providers.

Simulation with Time Varying (Non-Homogenous) Poisson process: Consider the hourly arrivals data with $\lambda\ (t)$, $t \geq 0$ shown in Table 2. We simulated the ED for one year using these arrival rates and we calculated the LOS and DTDTs for different waiting time in the queue levels.

Table 3 summarizes the obtained DTDTs and LOS for each patient group in all three shifts using these arrival rates in the simulation. Other than acuity 1 patients in Shift I, reductions on both DTDT and LOS can be achieved while optimizing staffing decisions. Using the validated simulation model varying wait-time goals in different stations, we used Arena OptQuest to optimize staffing in the ED. Table 4

Table 3 DTDTs and LOS in each shift for different maximum allowable waiting times in minutes in the queue (W_q^*) using *Time Varying Poisson arrival process* ($\lambda\ (t)$, $t \geq 0$)

W_q^* (min)	5	10	15	20	25	30	Current
Shift 1-Door to Doctor Time (DTDT)							
Acuity 1	0.05	0.14	0.14	0.14	0.14	0.14	0.90
Acuity 2–3	6.96	7.07	7.07	7.07	7.07	7.07	8.09
Acuity 4–5	9.46	9.43	9.43	9.43	9.43	9.43	10.62
Shift 1-Length of Stay (LOS)							
Acuity 1	106.16	113.88	113.88	113.88	111.88	113.88	101.75
Acuity 2–3	298.25	298.1	298.1	298.1	298.1	298.1	303.53
Acuity 4–5	95.52	96.08	56.08	96.08	96.08	96.08	97.71
Shift 2-Door to Doctor Time (DTDT)							
Acuity 1	0.03	0.04	0.04	0.04	0.21	0.19	1.10
Acuity 2–3	7.05	7.03	7.08	7.08	7.13	7.44	8.34
Acuity 4–5	9.41	9.54	9.54	9.54	9.59	10.01	9.85
Shift 2-Length of Stay (LOS)							
Acuity 1	107.76	107.75	107.26	107.26	108.1	107.04	116.77
Acuity 2–3	298.91	298.61	298.57	298.57	299.32	299.99	303.4
Acuity 4–5	95.4	92.26	95.57	95.57	95.78	96.65	100.84
Shift 3-Door to Doctor Time (DTDT)							
Acuity 1	0.01	0.01	0.01	0.1	0.1	0.1	0.56
Acuity 2–3	7.02	6.97	6.97	7.23	7.23	7.23	7.97
Acuity 4–5	9.38	9.51	9.51	9.85	9.85	9.85	10.72
Shift 3-Length of Stay (LOS)							
Acuity 1	107.3	107.43	107.43	103.51	103.51	103.51	117.22
Acuity 2–3	298.76	299.45	299.45	299.82	299.82	299.82	307.45
Acuity 4–5	95.04	95.77	95.77	96.50	96.50	96.5	102.60

Table 4 Staffing distribution: non-parametric time-varying arrival rates

Wq (mins.)	Shift I			Shift II			Shift III		
	N	P	T	N	P	T	N	P	T
5	3	2	3	6	3	6	7	3	6
10	3	2	3	6	3	6	6	3	6
15	3	2	3	6	3	5	6	3	5
20	3	2	3	7	2	5	7	2	5
25	3	2	3	6	2	5	7	2	5
30	3	1	3	6	2	5	6	2	5

N Nurses, *P* Physicians, *T* Technicians

summarizes the staff distribution of ED to achive these predetermined goals for waiting time in the queues.

This study presented a discrete-event simulation that modeled patient arrival patterns to an ED during different times of the year. A simulation-optimization framework using the DES optimized staffing in three different shifts after accounting for patient acuity levels in each shift which is the main contribution of this study to the current body of literature.

Acknowledgement Melik Koyuncu was supported by The Scientific and Technological Council of Turkey during this research between 06/30/2015 and 03/31/2016 as a postdoctoral fellow by grant number 1059B191401508.

References

1. United States Government Accountability Office: Hospital Emergency Departments. Report to the Chairman, Commitee on Finance U.S. Senate (2009)
2. McCarthy, M.L., Zeger, S.L., Levin, S.R., Desmond, J.S., Lee, J., Aronsky, D.: Crowding delays treatment and lengthens emergency department length of stay, even among high-acuity patients. Ann. Emer. Med. (2009)
3. Ang, E., Kwasnick, S., Bayati, M., Plambeck, E.L., Aratow, M.: Accurate emergency wait time prediction. Manuf. Serv. Oper. Manag. **18**(1), 141–156 (2016)
4. Kuang, X., Chan, C.W.: Using future information to reduce waiting times in the emergency department via diversion. Manuf. Serv. Oper. Manag. (2016) (Articles Advance)
5. Saghafian, S., Hoop, W., Van Oyen, M.P., Desmond, J.S., Kronick, S.L.: Patient streaming as a mechanism for improving responsiveness in emergency departments. Oper. Res. **60**(5), 1080–1097 (2012)
6. http://www.rockwellautomation.com/rockwellsoftware/simulation (2016)
7. Saunders, C., Makens, P., Leblanc, L.: Modeling emergency department operations using advanced computer simulation. Ann. Emerg. Med. **18**(2), 130–140 (1989)
8. Duguay, C., Chetouane, F.: Modeling and improving emergency department systems using discrete event simulation. Simul. Trans. Soc. Model. Simul. Int. **83**(4), 311–320 (2007)
9. Kuo, Y., Rado, O., Lupia, B., Leung, J.M.Y., Graham, C.A.: Improving the efficiency of a hospital emergency department: a simulation study with indirectly imputed service-time distributions. Flex. Serv. Manuf. J. **28**, 120–147 (2016)

Patient Pathways Discovery and Analysis Using Process Mining Techniques: An Emergency Department Case Study

Waleed Abo-Hamad

Abstract Acute hospitals are currently facing major pressures due to rising demand, caused by population growth, ageing and high expectations of service quality. Simulation studies have been used to overcome these challenges and to drive improvements. However, the majority of these studies derive their process models manually which is not only unrealistic but also time-consuming. Based on process mining, this paper presents a methodology to analyze the complexity of patients flow within an emergency department. Based on patients event log, process mining techniques are used in this paper to discover the actual patient pathways; understand the high variance in patient pathways taken by diverse groups of patients; and gain insights into bottlenecks and resource utilization. The work is a step forward towards minimizing the latency in the decision making process in such complex healthcare systems.

Keywords Emergency department · Patient pathways · Process mining

1 Introduction

Health-care managers are currently under constant pressure to control rapidly escalating expenses, while still responding to growing demands for both high-class patient service levels and medical treatment. Resolving such challenges requires a consistent understanding of health-care systems, which can be an overwhelming task, given the high levels of uncertainty and interdependence. Overcrowding in emergency departments (EDs) has become a significant international crisis that negatively affects patient safety, quality of care, and patient satisfaction [5, 7, 8, 16]. The research on healthcare system efficiency improvement has attracted much attention, in which Discrete Event Simulation (DES) based approach appears to be a

W. Abo-Hamad (✉)
College of Business, Dublin Institute of Technology (DIT), Aungier Street,
Dublin, Ireland
e-mail: wabohamad@dit.ie

© Springer International Publishing AG 2017   209
P. Cappanera et al. (eds.), *Health Care Systems Engineering*, Springer Proceedings
in Mathematics & Statistics 210, https://doi.org/10.1007/978-3-319-66146-9_19

dominant tool. Efforts to develop simulation models have advanced since the late 1980s when simulation was used to investigate the impact of key resources on waiting times and patient throughputs [15], and it has since been used to study the effect of a wide range of health interventions on health-care processes' performance [3, 4, 11, 12].

In spite of these advances in analyzing and improving healthcare processes, new approaches are still needed for two reasons. First, the majority of reported studies are not flexible and cannot be adapted or reused [9]. Second, due to the complexity of healthcare process, it is challenging to get a precise view of the patient flow in the health system with enough details. Therefore, the majority of DES studies develop the process model from observations which is not only unrealistic but also time-consuming. Process models are the core component of any DES projects; therefore, it is essential for any successful DES project to develop process models that are reusable and close reflection of reality. Using process mining mitigates these issues by automatically discover process models from event logs [17]. Through the application of process mining techniques, healthcare organizations can discover the actual patient pathways that are conducted in reality [14], understand the high variance in clinical pathways taken by diverse groups of patients, and gain insights into bottlenecks and resource utilization [13].

As process models form the foundation for a simulation model and based on the abovementioned merits, this paper demonstrates how process mining can be applied to support key modeling efforts by the discovery of process models from historical event. Process mining techniques are used to analyze the event log from an Emergency Department (ED), identify the critical resources and performance bottleneck.

2 Methodology

2.1 Project Background

The hospital studied in this paper is an acute, public, voluntary, and adult teaching hospital that holds a unique place in the delivery of healthcare not only to the community of North Dublin but also to the rest of Ireland. This 570-bed hospital provides primary, specialized, and tertiary healthcare services, with a 24 h "on-call" ED which services over 55,000 patients annually. According to the task force report in 2007, the overall ED physical space and infrastructure is inadequate. Additionally, the hospital is operating at approximately 99% occupancy with resultant difficulty in accommodating surges in numbers of ED admissions. This is often aggravated by delays in patients transfer to critical care (ICU/HDC) beds. Consequently, the hospital is not compliant with volume and wait time targets (6-h patient experience time target). A detailed simulation model for the ED was developed in [1, 2] and a number of improvement strategies were proposed to achieve this target. Although these strategies were effective, the model was not flexible to accommodate the constant changes in patient care pathways and to sustain improvement efforts. In order to overcome these issues, the process model of the ED (that was developed manually) has to be updated to capture the changes in patient flow. Following the manual process of developing and updating the ED process model would take 6–8 weeks [1, 2]. Given the fast changes in healthcare process, by the time the model is completed the process model will not be reflective. Therefore, in order to minimize the latency between the occurrence of events and decision making, process mining techniques were applied to automatically discover patients' pathways from historical data. A collaborative project with the hospital was established in order to achieve the following objectives: (1) discover the actual patient pathways that are conducted in reality; (2) understand the high variance in patient pathways taken by diverse groups of patients; (3) gain insights into bottlenecks and resource utilization; and (4) minimize the latency in the decision making process.

2.2 Dataset and Methods

A real-time patient tracking information system was used to track the patient's journey within the ED. A 1-year historical data with anonymous patients' records have been provided by the hospital managers. The dataset was provided in an event log structure with a total of 229,971 event logs representing 40,777 patients. Each log in the table represents an event (i.e. one process stage of the patient journey in the ED) with the following attributes (patient ID, Triage Category, Presenting Complaint, Date of Birth, Gender, Event ID, Tracking Step Name, Tracking Step Date Time, Location, By (Staff)). Events with the same name, patient ID and timestamp were removed. This resulted in 210,180 event logs.

Due to the unstructured nature of ED processes, the fuzzy miner [10] is used in this paper to discover the ED patient flow and to assess the variability of the flows within the process. The fuzzy miner allows to observe complex processes at different levels of granularity. This is achieved by applying two fundamental metrics: significance and correlation [10]. The significance metric assesses the relative importance of a precedence relation between two event classes, i.e. the more often two event classes are observed after one another, the more significant their precedence relation. On the other hand, the correlation metrics indicates how closely two events (i.e. activities) are following each other. Therefore, fuzzy mining could reduce and focus the displayed event classes by applying the two metrics on the discovered process map to achieve different levels of aggregation and abstraction.

3 Patient Pathway Discovery and Analysis

The main flow unit for the ED is the patient. A flow unit can be defined as an entity that enter the system, where various activities are performed before exiting the system. The event log for patients was analyzed to extract patients' characteristics and types. Upon their arrival, patients are assigned a clinical priority (triage category) according to the Manchester Triage System (MTS) that is widely used in UK, Europe, and Australia [6]. The MTS uses a five level scale for classifying patients according to their care requirements; immediate, very urgent, urgent, standard, and non-urgent. Patients were grouped based in their triage category. Immediate and very urgent patients represent 15%, urgent patients (triage category 3) represent the largest group of attendees to the ED annually (59% average), while standard and non-urgent patient 26% of all patients. As advised by ED consultants, the analysis of these patients' groups are critical as each group of patients can have a different journey within the ED and hence a different pathway.

3.1 Patient Pathway Discovery

The main building block of patient pathways are the activities that patients go through their journey in the ED. Twenty-two different activities within the ED were identified from the event log data. The fuzzy miner was then applied on the whole event log to construct the first top-level process map of the overall ED (Fig. 1a). The resulted map is too complex and is not interpretable due to the high variances in patients' pathways.

This confirms the perception of the complex nature of patient journeys within the ED, there will always be patients presenting to the ED with a unique characteristic

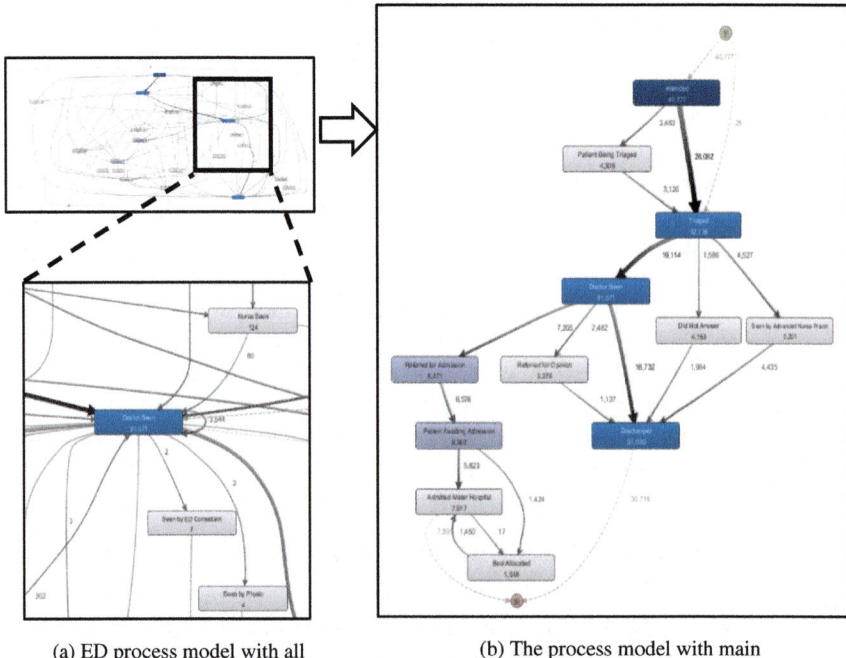

(a) ED process model with all activities and paths

(b) The process model with main activities and paths

Fig. 1 The discovered patient flow model of the emergency department

that would require the patient to follow a unique care pathway. Furthermore, this complexity is what doctors and nurses deal with on a daily basis and it is one of the main reasons they do not believe that system engineering techniques can contribute to understand the complexity of patient flow. However, the fuzzy miner allows to observe complex processes at different levels of granularity. This is achieved by applying two fundamental metrics: significance and correlation. The significance metric assesses the relative importance of a precedence relation between two event classes, i.e. the more often two event classes are observed after one another, the more significant their precedence relation. On the other hand, the correlation metrics indicates how closely two events (i.e. activities) are following each other. Therefore, fuzzy mining could reduce and focus the displayed event classes by applying the two metrics on the discovered process map to achieve different levels of aggregation and abstraction. The fuzzy process miner was therefore applied on the data to show the main highway paths for patients and to hide less frequent paths (Fig. 1b). The number inside the rectangle shows how many times an activity has been executed. For instance, activity 'Doctor Seen' occurred 31,571 times. The number on the arc represents the co-occurrence frequency between any two activities. For example, the co-occurrence frequency between 'Doctor Seen' and 'Referred for Admission' is 7,205. Due to excluding low frequent paths there are differences between the numbers of activities shown on incoming arrows and activity boxes on the process maps. Further analysis of this model revealed that there were 1,984 different patient pathway patterns (Table 1).

Over 60% of these patterns are one-off paths and only 31 of these patterns account for 80% of ED patients. Therefore, the remaining 1,951 patterns, which accounts to 20% of patients, were filtered out in order to reflect the common behavior of the ED. The fuzzy miner was applied again on the resulted 31,447 patients to drive final top-level process map of the ED (Fig. 2). The most followed paths are shown with thick arcs between activities. However, the analysis of exceptional pathways (paths with very low frequency) can give deep insights for medical professionals regarding the main factors behind these patterns.

3.2 Performance and Bottleneck Analysis

By considering the timestamp of events in the dataset, the ED performance and bottlenecks can be identified and analyzed (Fig. 3). The numbers on the arcs represents the waiting time between any two activities. The average length of stay (LOS) for all patients from arrival to departure (whether discharged or admitted to the hospital) is 9.1 h which is 3 h above the national target of 6-h average length of stay (LOS) in Ireland. However, the waiting time for admitted patients is 14 h on average. Patients have to wait 3.3 h on average to be seen by a doctor and 5.1 h afterwards to be discharged from the department. The main bottlenecks in the ED are the "Seen by Doctor" and "Patients waiting admission" activities.

Table 1 Discovered patient pathway patterns

Pathway	Cases	Relative frequency	Events	Step 1	Step 2	Step 3	Step 4	Step 5	Step 6	...	Step 13
1	9374	23%	4	Attended	Triaged	Doctor seen	Discharged				
2	3406	8%	4	Attended	Triaged	Seen by advanced nurse practitioner	Discharged				
3	3264	8%	3	Attended	Doctor seen	Discharged					
4	3005	7%	3	Attended	Triaged	Discharged					
5	1949	5%	6	Attended	Triaged	Doctor seen	Referred for admission	Patient awaiting admission	Admitted to hospital		
.	.	.	.	.	.	.	.	.	.	.	.
.	.	.	.	.	.	.	.	.	.	.	.
.	.	.	.	.	.	.	.	.	.	.	.
1984	1	0.002%	6	Attended	Did not answer	Triaged	Referred for opinion	Doctor seen	Admitted to hospital		

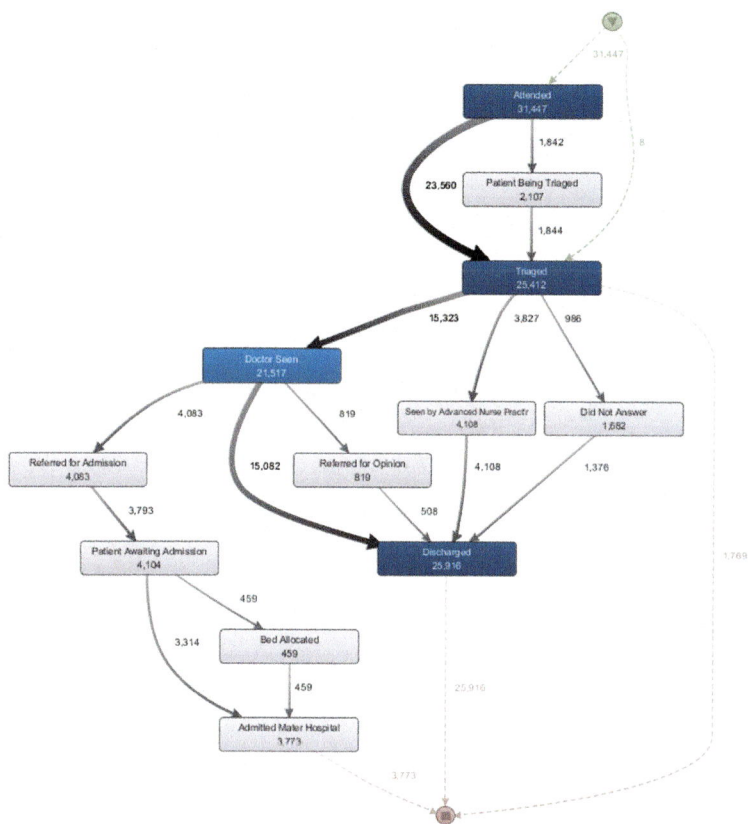

Fig. 2 The top-level process map of patient pathways

To gain a deeper understanding of the process flow of patients and the causes of these bottlenecks, the process model was analyzed at a more fine-grained level. The "Triage Category" attribute was used to divide patients into three groups; Immediate and very urgent, non-urgent and standard, and urgent patients. The process map of each patient group was constructed using the fuzzy miner and pathway patterns that reflect the common behavior for each group was analyzed (Fig. 4). There are obvious variances in the associated pathways for patients with different urgency categorization (i.e. triage). The first patient groups (Immediate and Very Urgent) represents 15% of all patients with the majority of them have been admitted to the hospital with an average waiting time of 13.7 h, for the admission process to be completed (Fig. 4a). While 26% of all patients are Standard and Non Urgent patients whom have a shorter pathway with a discharge outcome and 5.1 h average length of stay. The "Did not Answer" activity for this group represents patients who left the ED after being triaged without waiting to be investigated by a physician (whether a doctor or advanced nurse practitioner) (Fig. 4b).

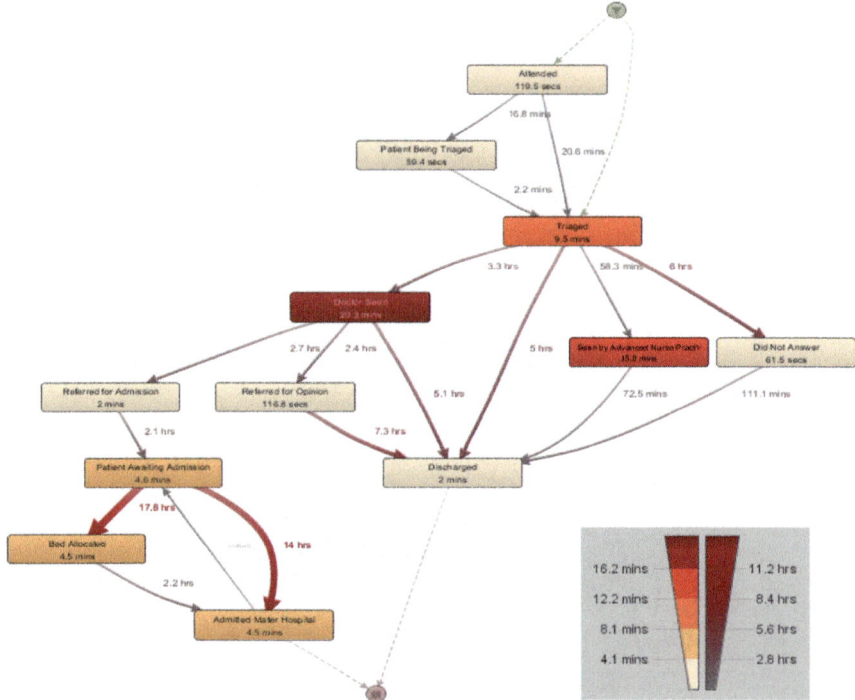

Fig. 3 Performance analysis of the top-level process map

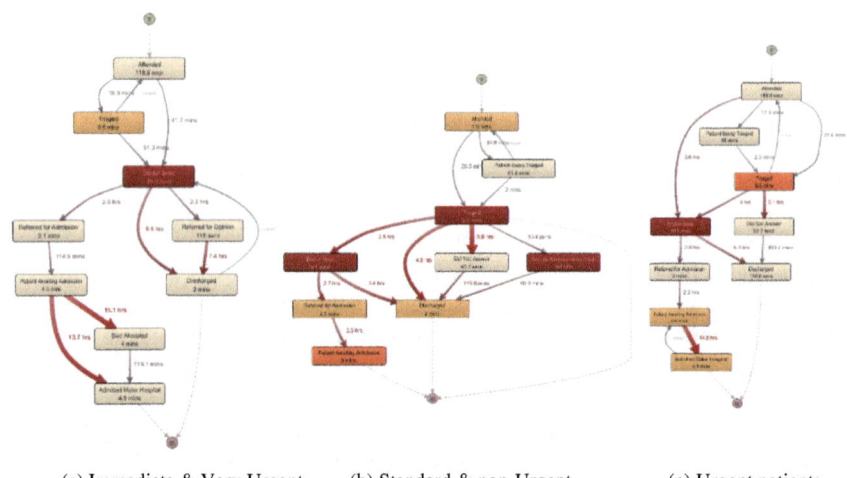

(a) Immediate & Very Urgent   (b) Standard & non-Urgent   (c) Urgent patients

Fig. 4 Performance analysis for patients with different triage categories

Urgent patients represent almost 60% of all patients with 10.1 h average LOS. This group are presented to the ED with a wide range of complaints with 27% are referred for admission and the remaining are discharged with an average waiting time of 5.2 h (Fig. 4c). The insights from this analysis enabled the ED decision makers to identify the bottlenecks for each group of patients and the challenges that they need to address.

3.3 Staff and Resources Analysis

Two types of resources were provided for each event; location and staff type. Therefor resource requirement was analyzed for different activities in patients' pathways (Table 2).

The resource analysis gives deep insights regarding the gap between the guidelines that should be followed and what is actually happening. For example, the triage activity should take place in the triage room by a registered nurse (RGN). However, the analysis reveals that 68% this activity takes place in the triage room and 77% of the times is performed by the RGN. This is a clear evidence of the overcrowding of the ED and also quantify how fare this activity from the guidelines. Similarly, the "Doctor seen" activity is performed by senior hospital officer (SHO) (58%) and registrars (20%) in the Majors area in the ED (55%) and in the Resuscitation room (18%). This highlights the actual time spent by doctors in this activity and the actual locations where this is happening. These insights helped the ED managers to understand the actual allocation of staff and resources within the ED and to identify the gaps between best practices and the actual performance.

4 Discussion and Conclusion

The different views of the analysis provided in previous sections (patient pathways, performance and bottlenecks, and resources) enabled the ED managers in the hospital partner to discover the actual patient pathways that are conducted in reality, understand the high variance in patient pathways taken by diverse groups of patients, and gain insights into bottlenecks and resource allocation and utilization. In general, healthcare processes are highly complex, dynamic, and ad hoc in nature. Therefore, there is a need for techniques that can cope with the intricate complexity of healthcare systems. Healthcare information systems generates event log data that track patient-care processes as they take place are a valuable source of data for analyzing and studying these processes. Based on these event logs, process mining techniques were used in this paper to discover patients' pathway patterns that are consistent with the observed dynamic behavior. Therefore, a more accurate process model for patient journey were developed and hence more adequate decision support. A current project with the partner hospital is in progress in order to

Table 2 Resource analysis for the main activities in patient pathways

		Triaged	Doctor seen	Seen by ANP	Referred for admission	Referred for opinion	Discharged	Admitted to hospital
Medical staff	ED consultant				**46%**		**21%**	**95%**
	ANP			24%	11%	4%	3%	3%
	SPR		15%			**10%**	10%	1%
	RGN	77%	0%					1%
	ADN			76%		13%	11%	
	SHO		58%		**36%**	51%	**30%**	
	Registrar		20%		7%	10%	12%	
	CNM	13%				5%	9%	
	Intern		7%			7%	4%	
Physical location	Majors area (15)	6%	55%		65%	57%	71%	72%
	Resuscitation room (3)	6%	18%		31%	29%	4%	24%
	Ambulatory care unit (6)	3%	11%	57%	3%	14%	23%	2%
	Rapid assessment triage (1)	17%	9%					
	Triage room (2)	68%						

ANP Advanced Nurse Practitioner; *RGN* Registered General Nurse; *CNM* Clinical Nurse Manager; *SPR* Specialist Registrar; *SHO* Senior Hospital Officer; *ADN* Associate Degree in Nursing

integrate the process mining engine with the Hospital Information System to provide a real-time tracking of the ED processes and to minimize the latency in the decision-making process.

References

1. Abo-Hamad, W., Arisha, A.: Simulation-based framework to improve patient experience in an emergency department. Eur. J. Oper. Res. **224**(1), 154–166 (2013)
2. Abo-Hamad, W., Arisha, A.: Multi-criteria approach using simulation-based balanced scorecard for supporting decisions in health-care facilities: an emergency department case study. Health Syst. **3**, 43–59 (2014)
3. Ahalt, V., Argon, N., Ziya, S., Strickler, J., Mehrotra, A.: Comparison of emergency department crowding scores: a discrete-event simulation approach. Health Care Manag. Sci. 1–12 (2016)
4. Ahmed, M., Alkhamis, T.: Simulation optimization for an emergency department healthcare unit in Kuwait. Eur. J. Oper. Res. **198**, 936–942 (2009)
5. Crawford, K., Morphet, J., Jones, T., Innes, K., Griffiths, D., Williams, A.: Initiatives to reduce overcrowding and access block in Australian emergency departments: a literature review. Collegian **21**(4), 359–366 (2014)
6. Cronin, J.: The introduction of the Manchester triage scale to an emergency department in the Republic of Ireland. Accid. Emerg. Nurs. **11**, 121–125 (2003)
7. Forero, R., Hillman, K.M., McCarthy, S., Fatovich, D.M., Joseph, A.P., Richardson, D.B.: Access block and ED overcrowding. Emerg. Med. Australas. **22**(2),119–135 (2010)
8. Graff, L.: Overcrowding in the ED: an international symptom of health care system failure. Am. J. Emerg. Med. **17**, 208–209 (1999)
9. Günal, M., Pidd, M.: Discrete event simulation for performance modelling in health care: a review of the literature. J. Simul. **4**, 42–51 (2010)
10. Günther, C., van der Aalst, W.: Fuzzy mining–adaptive process simplification based on multi-perspective metrics. In: Business Process Management, pp. 328–343. Springer, Berlin, Heidelberg (2007)
11. Ingolfsson, A., Budge, S., Erkut, E.: Optimal ambulance location with random delays and travel times. Health Care Manag. Sci. **11**, 262–274 (2008)
12. Kadri, F., Chaabane, S., Tahon, C.: A simulation-based decision support system to prevent and predict strain situations in emergency department systems. Simul. Model. Pract. Theory **42**, 32–52 (2014)
13. McGregor, C., Catley, C., James, A.: A process mining driven framework for clinical guideline improvement in critical care. In: Proceedings of the Learning from Medical Data Streams Workshop, Bled, Slovenia (2011)
14. Rovani, M., Maggi, F., de Leoni, M., van der Aalst, W.: Declarative process mining in healthcare. Expert Syst. Appl. **42**(23), 9236–9251 (2015)
15. Saunders, C., Makens, P., Leblanc, L.: Modeling emergency department operations using advanced computer simulation systems. Ann. Emerg. Med. **18**, 134–140 (1989)
16. Schafermeyer, R.W., Asplin, B.R.: Hospital and emergency department crowding in the United States. Emerg. Med. (Fremantle) **15**(1), 22–27 (2003)
17. van er Aalst, W.: Process Mining: Discovery, Conformance and Enhancement of Business Processes. Springer Science & Business Media (2011)

A Decomposition Approach for the Home Health Care Problem with Time Windows

Semih Yalçındağ and Andrea Matta

Abstract Optimization tools are necessary to efficiently plan service delivery for patients at home in the context of Home Health Care services. In the scientific literature, Periodic Vehicle Routing Problem with Time Windows (PVRPTW) is proposed to address the assignment, scheduling and routing processes with time windows. However, PVRPTW is computationally difficult and not viable for large-size problems. Thus, a practical approach is proposed to decompose the problem. Time windows are considered at the assignment level using a probabilistic model without the need of solving the routing problem. Mixed integer mathematical programming models are proposed and solved by CPLEX solver. Numerical experiments are executed to validate the performance of the proposed models with respect to the PVRPTW.

Keywords Home health care · Decomposition · Periodic vehicle routing problem · Time windows

1 Introduction

Home Health Care (HHC) services aim at satisfying people health and social needs at their home by providing appropriate and high-quality health and social services. In this work we mainly focus on the assignment, scheduling and routing problems of the HHC services. The assignment problem is used to determine which operators will deliver the care service to which patients. The scheduling problem determines the days in which operators will visit the assigned patients with the use of a pattern structure (care plan). The routing problem specifies the sequence in which the patients are

S. Yalçındağ (✉)
Industrial and Systems Engineering Department,
Yeditepe University, Istanbul, Turkey
e-mail: semih.yalcindag@yeditepe.edu.tr

A. Matta (✉)
Dipartimento di Meccanica, Politecnico di Milano, Milan, Italy
e-mail: andrea.matta@polimi.it

© Springer International Publishing AG 2017 221
P. Cappanera et al. (eds.), *Health Care Systems Engineering*, Springer Proceedings in Mathematics & Statistics 210, https://doi.org/10.1007/978-3-319-66146-9_20

visited by the assigned operator on the assigned day. In the HHC literature these problems are modeled and solved via simultaneous or two-stage approaches [7]. In the first approach, all the related decisions are held at the same time in a single model, whereas in the latter one, the decisions are considered sequentially with two models.

The HHC literature on these problems is analyzed and discussed in a recent review paper of Fikar and Hirsch [4]. Here we present a short list of some of the related existing works. Single period simultaneous approach are presented in [1, 5]. Only few works on the simultaneous approach are developed within the multiple planning period framework [2, 6]. As for the two-stage approach, only few works are available for both planning period alternatives [7, 8]. However, time window constraints are not taken into account in none of these decomposition based models.

Although the simultaneous approach is theoretically the best alternative, PVRPTW is a computationally difficult problem [3]. Since the efficiency of two-stage approaches has already been shown in the previous works, in this paper a new one based on the work of Yalçındağ et al. [7] is developed to decompose the PVRPTW. In this approach, since the assignment process is held independently from the routing process, exact visiting sequences of patients are not available. Thus, the assignment decisions might result in infeasible operator routes due to the possibility of overlapping of visits as the result of the time window constraints. To address this problem, in the assignment stage, a goal programming model is proposed to preferentially balance the utilization among operators and then avoid overlapping of visits. Time windows are considered at this level by using a novel probabilistic model without having the exact visiting sequences. The consideration of time windows at the assignment level via probabilistic method represents the main contribution of this work as well as the formulation of the goal programming model.

The rest of the paper is organized as follows. Problem definition is presented in Sect. 2. In Sect. 3, the two-stage approach and probabilistic estimation method are presented. Computational analysis for the preliminary results are reported in Sect. 4. Finally, concluding remarks and future research directions are presented in Sect. 5.

2   Problem Definition, Assumptions and Notation

The assignment problem is used to assign a set of operators to a set of patients while taking into account the care plans of each patient in the planning horizon $\mathcal{W}$ which is assumed to be composed of multiple planning periods (i.e., a week). The scheduling problem determines the days in which operators are planned to visit the assigned patients via pattern structure. Lastly, the routing problem specifies the sequence in which the patients are visited by each operator.

In this paper, we consider a complete directed network $G = (\mathcal{N}, \mathcal{A})$ that have n nodes in the set $\mathcal{N} = \{1, \ldots, n\}$, where each node j corresponds to a patient. We also have an additional node (node 0), which is used to denote the health care center of operators. For each patient, a care plan q_j is assumed to be known denoting the the number of visits required by patient j in the considered planning horizon $\mathcal{W}$.

In particular, each patient is assumed to be visited according to the associated time windows where earliest start time and latest finish time for each day d ($d \in \mathscr{W} = \{1, \ldots, W\}$) are represented by EST_j^d and LST_j^d respectively.

We consider that all the processes are held within a single category of operators (nurse or doctor) with same professional capabilities (skills). Models are developed under continuity of care where only one operator, i, from the set of all operators, $\mathcal{O}$, can be assigned to each patient, j. Each operator is characterized by a deterministic daily capacity, a_i, which is the maximum amount of regular working time according to his (her) working contract.

As described before, scheduling decision are taken with the use of pattern concept. Patterns are a priori given possible care plans that are used to satisfy the care requests of patients by selecting the most appropriate one from set $\mathscr{P}$. For example, if a patient requires two visits in a week, visits can be provided according to the pattern Monday–Wednesday or Tuesday–Thursday. For each pattern $p \in \mathscr{P}$, we define $p(d) = 0$ if no service is delivered at day d whereas it is $p(d) = 1$ if a visit is provided according to pattern p on day d.

In the following sections further notations and assumptions are described with the proposed two-stage approach and the probabilistic method.

3 Two-Stage Approach

When assigning patients to operators, fair allocation of tasks among operators is the primary objective, which is used to ensure equity among operators and improve their satisfaction. The utilization variation between operators is taken as a measure for the fairness of the task allocation.

Since routes of operators and the corresponding visiting sequences are not known in the assignment level, constructing routes in the second stage with the use of assignment decisions might not be feasible due to the time window constraints (i.e., two or more visits might overlap). To this end, a good strategy to mitigate this risk would be to minimize the number of visits that overlap in the same period. Hence, in the assignment stage, a goal programming approach (with 2 steps) is adopted to preferentially balance the utilization among operators and then avoid overlapping of visits to reach the hierarchical optimization goals. For the latter case, a new probabilistic approach is developed and integrated within the assignment stage.

3.1 Assignment Model Step (I): Workload Balancing

The considered assignment problem with the scheduling decisions consist in matching operators with patients, and patients with patterns, in a way that the utilization rates of operators are balanced. Minimization of the maximum daily operator utilization strategy is used to pursue the workload balancing.

The following mathematical formulation is a slightly modified version of the one presented in [7].

$$\min \quad m \tag{1}$$

$$\sum_{i \in \mathcal{O}} u_{ij} = 1 \qquad \forall j \in \mathcal{N} \tag{2}$$

$$\sum_{p \in \mathcal{P}} z_{jp} = 1 \qquad \forall j \in \mathcal{N} \tag{3}$$

$$\sum_{i \in \mathcal{O}} u_{ij}^d = \sum_{p\,:\,p(d)=1} z_{jp} \qquad \forall j \in \mathcal{N}, \forall d \in \mathcal{W} \tag{4}$$

$$u_{ij}^d \leq u_{ij} \qquad \forall j \in \mathcal{N}, \forall d \in \mathcal{W}, \forall i \in \mathcal{O} \tag{5}$$

$$\sum_{i \in \mathcal{O}} \sum_{d \in \mathcal{W}} u_{ij}^d = q_j \qquad \forall j \in \mathcal{N} \tag{6}$$

$$D_{id} = \sum_{j \in \mathcal{N}} (\delta_j + \tau_j) \cdot u_{ij}^d \leq a_i \qquad \forall d \in \mathcal{W}, \forall i \in \mathcal{O} \tag{7}$$

$$\frac{D_{id}}{a_i} \leq m \qquad \forall d \in \mathcal{W}, \forall i \in \mathcal{O} \tag{8}$$

$$u_{ij} \in \{0, 1\} \qquad j \in \mathcal{N}, i \in \mathcal{O} \tag{9}$$

$$u_{ij}^d \in \{0, 1\} \qquad j \in \mathcal{N}, \forall d \in \mathcal{W}, i \in \mathcal{O} \tag{10}$$

$$z_{jp} \in \{0, 1\} \qquad j \in \mathcal{N}, \forall p \in \mathcal{P} \tag{11}$$

where variable u_{ij} is a binary variable to indicate whether the operator i is assigned to the patient j or not. Similarly, the decision variable u_{ij}^d is also a binary variable to indicate if the operator i visits patient j on day d or not. The decision variable z_{jp} takes the value 1 if the pattern p is assigned to patient j and 0 otherwise. The decision variable D_{id} is a continuous variable and is used to calculate the total workload of operator i, which is based on the estimated average travel time to reach patient j (τ_j) and service time required to care patient j (δ_j). The auxiliary variable, m, is used to estimate the daily maximum utilization rate of the operators.

Constraints (2) are the assignment constraints. Constraints (3) are the scheduling constraints. Constraints (4) and (5) link together assignment and scheduling decisions; specifically, constraints (4) guarantee that, at each day, exactly one operator must visit a patient for whom a visit is scheduled on that day, while constraints (5) guarantee that an operator can visit a patient only if she/he has been assigned to that patient. Constraints (6) are used to guarantee that each patient is visited according to the associated care plan. Constraints (7) control the daily workload of the operators and ensures that the operator capacities are not exceeded. Constraints (8) express the maximum utilization rate m over the days which is minimized in the objective function (1). Finally, the remaining constraints define the domain of the variables.

3.2 Assignment Model Step (II): Avoiding Overlapping of Visits

In this step, we want to avoid the overlapping of visits when assigning patients to operators. To reach this goal, it is proposed to minimize the maximum expected number of visits among operators to prevent scheduling more than one patient in the same time unit.

As we are using goal programming approach, maximum daily utilization level from the first step is incorporated as a constraint in the second step. Since the mathematical model in this step is similar to the one of the previous step, here we only present and explain the additional ones.

$$\min \quad h \tag{12}$$

$$Constraints \quad (2) - (7) \tag{13}$$

$$\frac{D_{id}}{a_i} \leq m^* \qquad\qquad \forall d \in \mathscr{W}, \forall i \in \mathscr{O} \tag{14}$$

$$\sum_{j \in \mathscr{N}} P_j(t) \cdot u_{ij}^d \leq h \qquad\qquad \forall d \in \mathscr{W}, \forall t \in \mathscr{T}, \forall i \in \mathscr{O} \tag{15}$$

$$Constraints \quad (9) - (11) \tag{16}$$

In constraints (14), m^* is a parameter expressing the minimum value of the maximum utilization among operators that is obtained from the first assignment step. In constraints (15), $P_j(t)$ denotes the probability of patient j to be visited by an operator in period t, with $\mathscr{T} = \{1, \ldots, T\}$, denoting a time unit inside the planning horizon. Hence, constraints (15) calculate the maximum expected number of visits of operators which is minimized in the objective function (12) to avoid overlapping visits. In the following section, an analytical method for estimating the expected number of visits is provided. In other words, details for the derivation process of the parameter $P_j(t)$ are presented.

3.3 An Analytical Method to Estimate the Expected Number of Visits

As previously discussed, time window for a patient is defined as the time between the earliest start time and the latest end time of the service. As patients should be visited within their time windows, assigning two visits at the same period might create an infeasible solution. This situation might be highly observed in the two-stage approach as the result of unavailable visiting sequences of patients at the assignment level. A good strategy to mitigate this risk would be to minimize the possibility of overlapping two or more visits at the same period. For this reason, the maximum expected number of visits per period per operator is minimized at the 2nd step of the assignment stage. However, estimating the number of visits assigned to an operator in a time period is not an easy process.

In order to estimate the expected number of visits for operators, we firstly have to calculate the probability that a patient will be served at period t. It is also assumed that each choice to start the service in the time window is equally likely to occur. In the generalized case, time horizon is divided into T time slots. Time slot associated with the earliest start time (EST) of the service in the time window is expressed as l and time slot corresponding to the latest start time (LST) of the service is expressed as $r - \delta + 1$. In particular, here we also denote the earliest (EET) and latest (LET) end times of the service as $l + \delta - 1$ and r respectively. As defined previously $P(t)$ is the probability of the patient to be visited at period t and Ω is the set containing all time slots at which the service can start from in the time window (for simplicity patient index j is not included). Accordingly, here we also define the set Ω_t containing all time slots at which the service occurred in period t can start from, $\Omega_t \subseteq \Omega$. Since no other information is available, the distribution of the starting time can be assumed to be uniform with:

$$P(t) = \frac{|\Omega_t|}{|\Omega|}, \quad t \in [l, r] \tag{17}$$

For time window $[l, r]$, $\Omega = [l, r - \delta + 1]$ and there are $|\Omega| = r - l - \delta + 2$ possible service starting times. For all the services that can be delivered in the time window, earliest start time of the service cannot be earlier than EST. Hence, the earliest start time for the service being delivered at period t is $max\{t - \delta + 1, l\}$. Also, the latest start time (LST) for the service being delivered at period t is $min\{t, r\}$. In conclusion, when the service is being delivered at period t, $\Omega_t = \{max\{t - \delta + 1, l\}, ..., min\{t, r - \delta + 1\}\}$. Correspondingly $|\Omega_t|$ equals to $min\{t, r - \delta + 1\} - max\{t - \delta + 1, l\} + 1$ with the probability for the service occurred at period t as

$$P(t) = \frac{min\{t, r - \delta + 1\} - max\{t - \delta + 1, l\} + 1}{r - l - \delta + 2}, \quad t \in [l, r] \tag{18}$$

Depending on the service time interval, we can discuss two different cases: $\delta \leq \frac{r - l + 1}{2}$ and $\delta > \frac{r - l + 1}{2}$. In the first case, the inequality $l + \delta - 1 < r - \delta + 1$ is satisfied. Thus, the EET is earlier than LST. In this first case (i.e., $\delta \leq \frac{r - l + 1}{2}$), the time window $[l, r]$ is divided into three intervals with EET and LST as the break points. These three intervals are $[l, l + \delta - 1]$, $(l + \delta - 1, r - \delta + 1]$ and $(r - \delta + 1, r]$. When the service is operated before EET, by considering $t \leq EET = l + \delta - 1$, we can obtain $max\{t - \delta + 1, l\} = l$. On the other hand, if the service is operated by considering $t \leq LST = r - \delta + 1$, we can obtain $min\{t, r - \delta + 1\} = t$. Hence, we can conclude that the possible service starting choices can only be in space $[l, t]$. Therefore, the probability for the patient to be visited at period t before EET is:

$$P(t) = \frac{t - l + 1}{r - l - \delta + 2}, \quad t \in [l, l + \delta - 1]. \tag{19}$$

The same procedure can be easily adapted to the other intervals and also to the case of $\delta > \frac{r-l+1}{2}$. Due to limited space, we just provide the derived formula. When $\delta \leq \frac{r-l+1}{2}$, the probability for the patient to be visited at period t is:

$$P(t) = \begin{cases} \frac{t-l+1}{r-l-\delta+2}, & l \leq t \leq l+\delta-1 \\ \frac{\delta}{r-l-\delta+2}, & l+\delta-1 < t \leq r-\delta+1 \\ \frac{r-t+1}{r-l-\delta+2}, & r-\delta+1 < t \leq r \\ 0, & t < l, t > r \end{cases}$$

When $\delta > \frac{r-l+1}{2}$ is considered, the probability for the patient to be visited at period t is:

$$P(t) = \begin{cases} \frac{t-l+1}{r-l-\delta+2}, & l \leq t \leq r-\delta+1 \\ 1, & r-\delta+1 < t \leq l+\delta-1 \\ \frac{r-t+1}{r-l-\delta+2}, & l+\delta-1 < t \leq r \\ 0, & t < l, t > r \end{cases}$$

3.4 Routing Model

At the routing phase (in the 2nd stage), several independent periodic TSP (PTSP) models (as the number of operators) with time windows are solved on a subgraph ($G = (\mathcal{N}', \mathcal{A}')$) induced by the patients assigned to the considered operator (according to the decision provided in the second step of the first stage).

The objective function is set as to minimize the maximum utilization among days as in the first step of the assignment problem. Considered model is obtained by modifying the PTSP model that is used in the work of Yalçındağ et al. [7] and adding the following time window constraints.

$$s_{j'}^d + \delta_{j'} + tr_{jj'} - M(1 - x_{j'j}^d) \leq s_j^d \qquad \forall(j',j) \in \mathcal{A}', \forall d \in \mathcal{W} \qquad (20)$$

$$s_j^d \geq EST_j^d \qquad \forall j \in \mathcal{N}', \forall d \in \mathcal{W} \qquad (21)$$

$$s_j^d + \delta_j \leq LET_j^d \qquad \forall j \in \mathcal{N}', \forall(j',j) \in \mathcal{A}', \forall d \in \mathcal{W} \qquad (22)$$

Constraints (20) establish the relationship between the operator departure time from a patient and its immediate successor where, s_j^d is the decision variable denoting the service starting time of serving patient j on day d, $x_{j'j}^d$ is the binary variable that takes value 1 if the operator travels along the arc (j',j) (in the arch set $\mathcal{A}$) on day d, $tr_{jj'}$ is the travel time required to reach patient j from patient j' and M is a very large number. In particular, constraints (21) and (22) guarantee that time windows (the earliest start time and the latest end time of the service) restrictions are respected.

4 Computational Study: Preliminary Analysis

In this section we analyze and compare the proposed two-stage approach with the simultaneous approach (PVRPTW). The simultaneous approach is also adopted from the work of Yalçındağ et al. [7] by including time window constraints.

The models are tested under four instance groups, G-1, G-2, G-3 and G-4 which are mainly differentiated based on the average operator utilization (around %40 and %60) and time window to service time ratio ($\frac{TW}{\delta}$). Each group is composed of 5 randomly generated instances having 15 patients and 2 operators in which locations (randomly sampled from a uniform distribution between 0 and 20 in a square grid), time windows and visiting requirement (within the planning horizon) of patients are randomly decided. Table 1 presents the main instance features.

The models have been implemented in Python and solved with CPLEX for optimality on a computer with double CPU Intel Xeon Processor E5-2699 v4 2.20 GHz and 256 GB of RAM. All of the results obtained with four instance groups are presented in Table 2.

In the table, objective function values of the corresponding models are shown with Z_{OFV}. Obtaining Z_{OFV} for the simultaneous approach is trivial, whereas for the two-stage approach is calculated after the routing stage by taking the maximum daily utilization rate of all the operators. Similarly, Uti_{max} is used to show the maximum overall utilization rate (through the whole planning horizon) among all the operators. B_{rng} corresponds to the workload balancing value which is calculated by taking the difference of the maximum and minimum overall utilization values (workload ranges) of the corresponding operators. CPU times are reported as well where this value in the two-stage approach corresponds to aggregated CPU times of all the stages (including independent routing calculations). Lastly, $\%\triangle_{PVRPTW}$ corresponds to the percentage differences based on the objective function values (Z_{OFV}) between the simultaneous (PVRPTW) and the two-stage approaches.

The results in Table 2 show that, if time window is not too tight and operators are not fully utilized (i.e., instances of G-1), two-stage approach is able to provide feasible solutions for all the instances that are close to the ones of the simultaneous case (on the average 2.8% difference according to the Z_{OFV} values). It can also be observed that the computation effort required by the proposed two-stage approach is significantly low with respect to the simultaneous approach. Even in these small scaled instances, it takes up 8880 s for the simultaneous approach to provide solutions whereas these values do not exceed 22.11 s in the two-stage case. In particular, workload balancing levels, B_{rng}, of both approaches are very close to each other. Similar observations can be also seen in the instances where $\frac{TW}{\delta}$ are smaller (i.e., instances of G-2). In this case percentage differences between two approaches based on the objective function values is slightly higher with the average value of %5.1.

For the G-3 instances where operators are more utilized with respect to the previous ones and $\frac{TW}{\delta}$ value is over 3, almost all the observations are similar to the previously explained ones. In some instances, simultaneous approach requires higher computation times to reach optimality (up to 143694 s) which again indicates the

Table 1 Instance features

Instance group	Num. of patients	Num. of operators	Av. Visit duration (min)	Av. Operator capacity (min/day)	Num. of days	Num. of patterns	Av. Num. of visits (operator per day)	Av. utilization (%)	Av. $\frac{TW}{\delta}$ value
G-1	15	2	45	420	6	9	2.35	40	3.31
G-2	15	2	45	420	6	9	2.31	40	2.20
G-3	15	2	45	420	6	9	3.10	60	3.38
G-4	15	2	45	420	6	9	3.15	60	2.18

Table 2 Results with instances

Group	Instance	Z_{OFV}	Uti_{max}	B_{rng}	CPU (sec)	$\%\triangle_{PVRPTW}$
G-1 PVRPTW	1	0.614	0.491	0.014	130.47	
	2	0.481	0.403	0.022	8133.56	
	3	0.468	0.423	0.085	8880.08	
	4	0.391	0.354	0.012	139.86	
	5	0.420	0.406	0.098	82.56	
G-1 Two-Stage	1	0.646	0.519	0.033	1.44	5.12
	2	0.497	0.427	0.065	1.30	3.33
	3	0.468	0.408	0.045	1.63	0.00
	4	0.403	0.360	0.007	11.89	3.08
	5	0.432	0.406	0.081	22.11	2.75
G-2 PVRPTW	1	0.486	0.405	0.017	1972.28	
	2	0.381	0.340	0.009	43.53	
	3	0.398	0.369	0.020	14.77	
	4	0.499	0.426	0.001	449.13	
	5	0.523	0.448	0.046	241.84	
G-2 Two-Stage	1	0.503	0.399	0.002	6.77	3.51
	2	0.388	0.365	0.043	1.31	1.89
	3	0.418	0.378	0.025	1.33	5.16
	4	0.536	0.425	0.005	1.42	7.28
	5	0.563	0.439	0.001	2.09	7.67
G-3 PVRPTW	1	0.697	0.571	0.051	143694.17	
	2	0.532	0.477	0.012	983.61	
	3	0.514	0.476	0.001	1626.97	
	4	0.584	0.496	0.030	21001.91	
	5	0.593	0.491	0.015	123517.44	
G-3 Two-Stage	1	0.722	0.556	0.020	14.49	3.58
	2	0.549	0.511	0.035	13.53	3.22
	3	0.522	0.478	0.006	8.41	1.62
	4	0.594	0.493	0.026	2.47	1.71
	5	0.603	0.484	0.011	9.97	1.75
G-4 PVRPTW	1	0.611	0.545	0.041	1043.06	
	2	0.640	0.590	0.126	2418.23	
	3	0.638	0.538	0.022	13643.09	
	4	0.504	0.463	0.026	1331.25	
	5	0.631	0.578	0.106	3074.69	
G-4 Two-Stage	1	Inf.	Inf.	Inf.	N.a	N.a
	2	0.677	0.574	0.036	594.52	5.73
	3	Inf.	Inf.	Inf.	N.a	N.a
	4	0.529	0.480	0.023	473.03	5.04
	5	Inf.	Inf.	Inf.	N.a	N.a

importance of using decomposition approach (i.e., two-stage approach). In particular, the maximum value for the expected number of visits for this case is on the average (of 5 instances) 1.25. For some instances the two-stage approach results in infeasibilities (denoted with Inf.) in the routes of operators. The reason behind this observation is due to having higher operator utilization rates and relatively smaller time window lengths that result in a large maximum value for the expected number of visits (on the average 1.87).

We can conclude that proposed two-stage approach is generally very successful to decompose the PVRPTW model and obtains solutions with good qualities and low computation efforts. For the cases where infeasibilities are observed, relaxing the m^* value (in the 2nd step of the assignment stage) might be helpful.

5 Conclusions

In this work, we decompose the PVRTW problem into assignment, scheduling and routing problems via two-stages where scheduling decision is incorporated both into the assignment and routing stages. In the assignment stage, a goal programming approach is constructed to preferentially balance the utilization among operators and then avoid overlapping of visits. Time windows are considered at assignment level using a probabilistic model without the need of solving the routing problem.

Results indicate that the performance of the proposed methodology where time windows are included in the assignment stage via probabilistic model is inspiring since in most of the cases the two-stage approach provides similar solutions as the PVRPTW approach provides. However, the results, observations and conclusions reported in this paper suffer from limited experimentation, thus they have to be verified on extended experiments with larger sized instances. To this end, a future activity is to develop heuristic algorithms to be able to solve large sized instances for both approaches. In particular, the proposed methodology can also be extended by considering lunch breaks, synchronization and precedence operations.

References

1. Bredstörm, D., Rönnqvist, M.: Combined vehicle routing and scheduling with temporal precedence and synchronization constraints. Eur. J. Operation. Res. **191**(1), 19–31 (2008)
2. Cappanera, P., Scutella, M.G.: Joint assignment, scheduling, and routing models to home care optimization:a pattern-based approach. Trans. Sci. **49**(4), 830–52 (2015)
3. Christofides, N., Beasley, J.: The period routing problem. Networks **14**(2), 237–256 (1984)
4. Fikar, C., Hirsch, P.: Home health care routing and scheduling: a review. Comput. Operation. Res. **77**, 86–95 (2017)
5. Rasmussen, M.S., Justesen, T., Dohn, A., Larsen, J.: Home care crew scheduling problem: preference-based visit clustering and temporal dependencies. Eur. J. Operation. Res. **219**(3), 598–610 (2012)

6. Trautsamwieser, A., Hirsch, P.: A branch-price-and-cut approach for solving the medium-term home health care planning problem. Networks **64**(3), 143–59 (2014)
7. Yalcindag, S., Cappanera, P., Matta, A., Scutellá, M.G., Sahin, E.: : Pattern-based decompositions for human resource planning in home health care services. Comput. OR, **73**, 12–26 (2016)
8. Yalcindag, S., Matta, A., Sahin, E., Shanthikumar, J.: The patient assignment problem in home health care: using a data-driven method to estimate the travel times of care givers. Flex. Serv. Manuf. **28**, 1 (2016)

Discrete-Event Simulation of an Intrahospital Transportation Service

Maxime Painchaud, Valérie Bélanger and Angel Ruiz

Abstract Several types of transportation are required daily to support medical activities and services to the patient at a hospital. Intrahospital transportation activities range from transporting inpatients with limited mobility from one location to another to moving samples from one care unit to the hospital laboratory. In practice, patient and material transportations are often managed independently. However, dealing with both types of requests in a centralized framework can help improve the system performance and ensure a better use of resources. It also results in a more complex decision-making process. Given the high level of uncertainty related to intrahospital transportation, especially regarding patients, adequate tools are required to help hospital managers plan, organize, and operate such a service. In this study, we propose a discrete-event simulation tool to support the management of an intrahospital transportation service. The tool is validated in the context of a Canadian hospital, and used to evaluate and compare the system performance under several decisions.

Keywords Healthcare · Simulation · Patient transportation · Decision-support

1 Introduction

On a daily basis, intrahospital transportation activities are required so that every person, material, furniture needed to perform or support clinical activities are available at the right place and at the right time. Transportation activities range from picking mail from a care unit and delivering them to the correct department to moving patients in critical situations from their care units to the operating room. Intrahospital

M. Painchaud · A. Ruiz
Department of Operations and Decision Systems, Université Laval, Québec, Canada
e-mail: maxime.painchaud.1@ulaval.ca

A. Ruiz
e-mail: angel.ruiz@osd.ulaval.ca

V. Bélanger (✉)
Department of Logistics and Operations Management, HEC Montréal, Montréal, Canada
e-mail: valerie.belanger@cirrelt.ca

© Springer International Publishing AG 2017
P. Cappanera et al. (eds.), *Health Care Systems Engineering*, Springer Proceedings in Mathematics & Statistics 210, https://doi.org/10.1007/978-3-319-66146-9_21

transportation activities can be divided in two main categories: *patient* transportation and *material* transportation. They both include a combination of planned and unplanned requests with specific requirements, characteristics and expected delays.

Patient transportation encompasses the movement of patients with limited mobility that require safe and supervised transports from one location to another for diagnosis or therapeutic reasons. When transportation is needed between units in the same building, patients are generally transported by trained personnel using stretchers or wheelchairs. When transportation is required between buildings of a same institution, patients are transported by ambulances, owned by the organization or by an ambulance transportation provider. Patient transportation requests show several characteristics, including the level of priority, the need for particular equipment (oxygen tanks, etc.), and the need for cleaning after transportation. In addition, they need to be planned carefully to ensure the comfort and well-being of the patient, an efficient use of the granted resources (stretchers, vehicles, etc.), and a reliable medical activities execution (e.g. schedule of operating theatres or magnetic resonance imaging services).

Material transportation covers the movement of mail, laboratory samples, medical items, laundry and furniture performed within a hospital to support a patient clinical pathway. When transportation has to be performed in the same building, material is generally transported by personnel on foot. When transportation is required between buildings, material are moved by cars or trucks, either owned by the organization or by an external transportation service provider. Material transportation needs to be performed efficiently and without unwanted delay to ensure a smooth clinical pathway.

Typically, patient and material transportations are managed independently, and sometimes in a decentralized manner. Indeed, not all hospitals can actually rely on a centralized transportation service, meaning that trained personnel, often nursing assistants, are still responsible for performing transportation activities originating from their care units or departments. While a centralized system offer potential for improved performance, implementing and operating a centralized transportation service that deals with both types of requests yields a complex decision-making, including the number of resources needed to provide such a service, the schedule of the personnel, and the policy that rules the assignment of requests to porters. Given the high level of uncertainty, especially with respect to patient transportation, it is important to provide decision makers with adequate tools to assess the impact of each decision on the system performance.

In this paper, we present a generic discrete-event simulation tool to support the management of a centralized intrahospital transportation service. We focus on a transportation service that operates within the same building and is responsible for the transportation of patients as well as the transportation of small medical items, mail and laboratory samples. In the first part of this paper, the intrahospital transportation process, from the initial request to the release of a porter, is described, followed by a short literature review. The proposed simulation model is then presented and computation experiments are reported along with some results. Finally, a discussion on future research concludes the paper.

2 Intrahospital Transportation Process

This paper focuses on an intrahospital transportation service providing both the transportation of patients and the transportation of material, including small medical items, mail and laboratory samples. Such a transportation service receives several requests daily, mostly unplanned requests with several characteristics and levels of priority. When a specific clinical activity, e.g. X-ray scans, needs to be performed, a nurse or a clerk from the patient's care unit generally sends the appointment request to the corresponding department. The latter plans the activity and informs the care unit of the expected schedule. At the right moment, a nurse or a nurse assistant prepares the patient, and if needed, sends the transportation request, which will be assigned a priority level. When a porter is dispatched to serve a request, the porter moves to reach the patient, and transports him to his final destination. The porter is finally released. When the patient is ready to go back, a transportation request is sent again to the transportation service and a similar process is followed.

In the case of material transportation, a similar process is also followed except that only one request, from the origin to the destination, is needed. Figs. 1 and 2 present an overview of the transportation process from the initial request to the release of a porter.

The design, planning and management of a centralized transportation service involve several decisions, which can be classified according to the classical three decision-making levels: strategic, tactical and operational. Strategic decisions include the choice of services to be provided, the hiring of staff, the definition of each actor's roles and responsibilities, and the choice of communication and information system. The tactical level addresses decisions such as the working hours of the transportation service as well as the scheduling of staff. Finally, operational decisions, often taken in real-time, are concerned with short term decisions such as the dispatching of requests. Each decision generally impacts each other, but more importantly, they affect the system performance from the efficiency, patient experience, quality of care and cost standpoints.

3 Literature Review

Logistics, including transportation, plays an important role in healthcare systems as it ensures that everything or everyone that is needed to perform or support clinical activities at the right place, at the right time. Facing a continuous pressure to deliver high quality care at the least possible cost, healthcare organizations should not neglect the improvement of transportation practices. Despite a growth in interest in the past decade, research dedicated to hospital transportation, especially with respect to patient transportation, remains limited. In this section, we review the main works related to intrahospital patient transportation. From a methodological standpoint, we focus on papers from or close to the field of operations research. Table 1 summa-

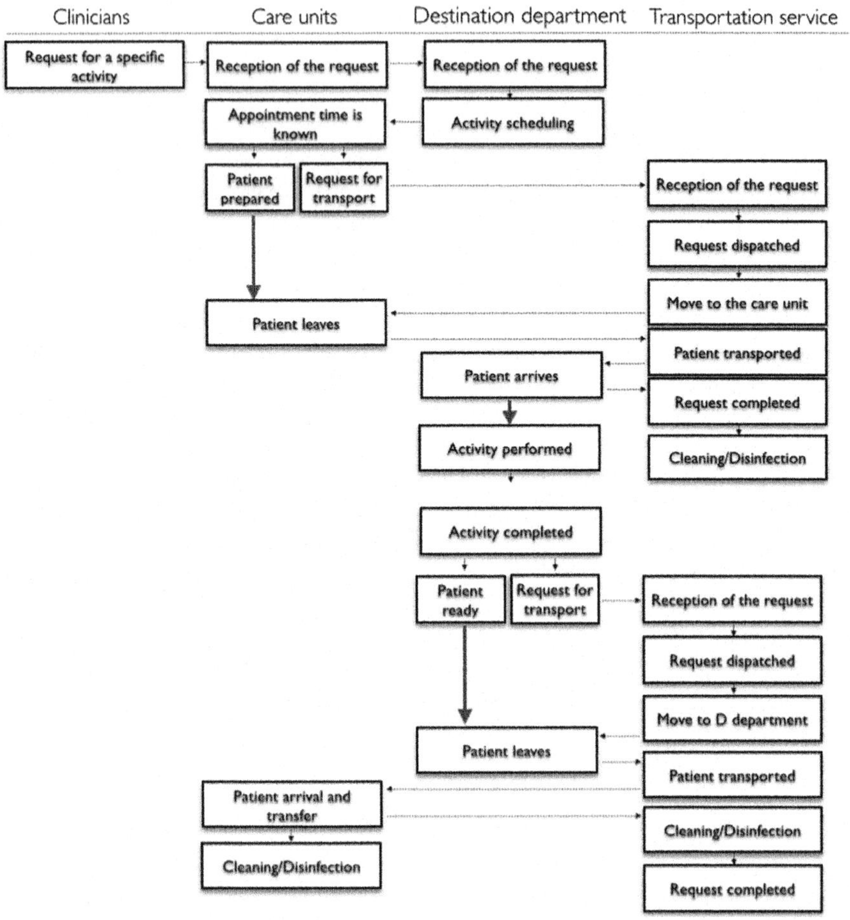

Clinicians	Care units	Destination department	Transportation service

Request for a specific activity	→	Reception of the request	←	Reception of the request		
		Appointment time is known	←	Activity scheduling		
		Patient prepared	Request for transport	······→		Reception of the request
						Request dispatched
						Move to the care unit
		Patient leaves	←			Patient transported
				Patient arrives	←	Request completed
				Activity performed		Cleaning/Disinfection
				Activity completed		
				Patient ready	Request for transport	······→ Reception of the request
						Request dispatched
						Move to D department
				Patient leaves	←	Patient transported
		Patient arrival and transfer	←			Cleaning/Disinfection
		Cleaning/Disinfection				Request completed

Fig. 1 Patient transportation process

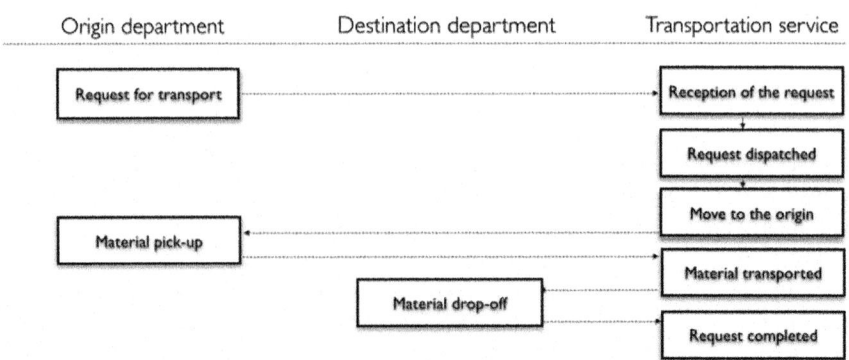

Origin department	Destination department	Transportation service
Request for transport	······→	Reception of the request
		Request dispatched
		Move to the origin
Material pick-up	←	Material transported
	Material drop-off	← Request completed

Fig. 2 Material transportation process

Table 1 Literature review

	Hanne et al. [4]	Fiegl and Pontow [3]	Beaudry et al. [1]	Turan et al. [12]	Kergosien et al. [6]	Parragh et al. [9]	Bowers et al. [2]	Schmid and Doerner [11]	Zhang et al. [13]	Henshaw [5]
A. Types of transport										
A.1 Patient	X	X	X	X	X	X	X	X	X	
A.2 Material		X								X
A.3 Intra-building	X	X	X	X				X		
A.4 Inter-building	X		X		X	X	X		X	X
B. Decision problem										
B.1 Routing	X		X		X	X	X	X	X	
B.2 Scheduling		X						X		
B.3 Workforce planning										X
C. Characteristics										
C.1 Static				X		X	X	X	X	
C.2 Dynamic	X	X	X		X					

(continued)

Table 1 (continued)

	Hanne et al. [4]	Fiegl and Pontow [3]	Beaudry et al. [1]	Turan et al. [12]	Kergosien et al. [6]	Parragh et al. [9]	Bowers et al. [2]	Schmid and Doerner [11]	Zhang et al. [13]	Henshaw [5]
C.3 Deterministic	X	X	X	X	X			X	X	
C.4 Stochastic							X			
D. Methodology										
D.1 Simulation	X						X			X
D.2 Heuristic method	X	X					X			
D.3 Graph theory		X								
D.4 Tabu Search			X		X					
D.5 VNS						X				
D.6 Column generation						X				
D.7 Memetic alg.									X	
D.8 Coop. meta.								X		
D.9 Commercial solver				X						

rizes the main studies reviewed. To categorize each article, we identify four main characteristics: (1) the type of transport, (2) which decision problem the paper focus on, (3) how the study deals with uncertainty, and (4) the methodology. Although not presented in the table, it is important to mention that several constraints and characteristics, such as the need for disinfection, the mix of vehicles or equipment, and level of priorities, are generally incorporated into classical problems to account for the specific aspects of intrahospital transportation.

As highlighted in Table 1, most studies focus on operational decision-making, assuming decisions like the number of porters or teams and their work shifts as already fixed. In particular, several studies address the problem of determining the ambulance routes in the case of inter-building transportation. For intra-building transportation, operational decisions that have been considered frequently are the scheduling and the dispatching of requests to porters. In both cases, static and dynamic versions of the underlying problems are studied, mostly in a deterministic framework, assuming that all information about requests (e.g. the service time, travel time, the duration of the corresponding appointment) is known when decisions are taken, which is generally not the case in healthcare.

From a methodological standpoint, optimization models are generally proposed and solved by means of heuristic approaches. In addition to optimization, Hanne et al. [4] and Bowers et al. [2] also relied on simulation techniques to assess the impact of using the proposed dispatching policies. Contrarily to previous studies, Henshaw [5] adresses tactical decisions. Henshaw [5] developed a simulation model to test several decisions such as adding porters at specific moments over a day.

In this paper, we aim to develop generic discrete-event simulation tool that will provide a realistic representation of intrahospital transportation activities, allowing the evaluation and comparison of several alternatives. One specific aspect of this tool is that it will be flexible and generic enough to consider several types of requests in an integrated manner and will allow the study of the interaction between several levels of decision, including strategic, tactical, and operational decisions. In particular, as compared to the work of [5], the proposed simulation model will be able to consider many dispatching strategies, ranging from simple heuristics to more sophisticated optimization tools. We strongly believe that simulation provides a powerful management tool to support decision-making at several points in a transportation service lifetime, from its design to its continuous improvement.

4   Simulation Model

Simulation has been used for several years in healthcare to analyze, evaluate and compare several types of decisions and scenarios. In this study, we propose a generic discrete-event simulation tool to support the management of an intrahospital transportation service with the aim of reducing patient wait time and improving the use of resources. The proposed simulation model is built around five main components: the input data, the demand and random variables generator, the simulation engine, the

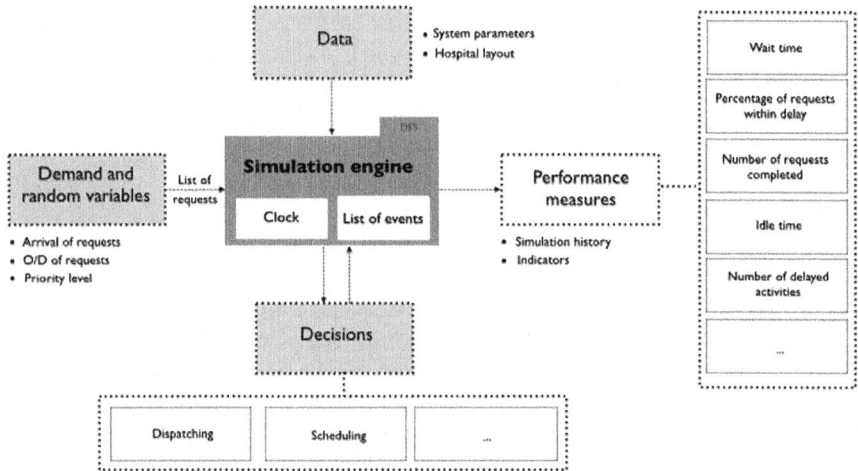

Fig. 3 Simulation architecture. [adapted from [7]]

decisions module, and the performance measures. Figure 3 illustrates the proposed simulation framework followed by a short description of each of its components.

- *Input data.* Input data contains information about the hospital configuration as well as the system itself. In particular, it defines each possible origin and destination, along with their precision location. Input data also includes a matrix, which defines times between each origin and each destination under normal operating conditions. Finally, it comprises information related to the workforce, including work shifts and breaks.
- *Demand and Random Variables.* Simulation models uncertain events using probability distributions. The demand and random variables generator samples these probability distributions to determine specific values for uncertain events to be used during the simulation execution. In the proposed simulation, the arrival of calls and their origin/destination, the time spent by the porter at the origin and at the destination, and the request priority level, are assumed to follow given probability distributions (see Sect. 5).
- *Simulation Engine.* The proposed simulation model is based on discrete-event simulation (DES). The simulation includes two main types of entities, namely the porters and the requests, both with their respective sets of possible states. The simulation engine was developed based on the three-phase approach [10]. The latter allows the simulation clock to advance asynchronously from one event to the next. The simulation engine works with a list of events sorted by their execution time. Each time an event is executed, the system state is modified accordingly and the simulation clock moves to the following event in the list. The three-phase approach distinguishes two types of events: bounded (B) and conditional (C). B-type events are those that unconditionally occur at a given time, generated according

to previous events. C-type events conditionally occur at a given time, depending on the current system.

- *Decisions module.* During the simulation execution, some events trigger decisions. Therefore, the simulator needs to reproduce as faithfully as possible the decision processes carried out by operators or dispatchers, either based on current practices or on potential ones.
- *Performance measures.* At the end of the simulation execution, performance measurement is required to compare and analyse the tested scenarios. The performance measures module records a complete history of the simulation, including all movements and requests performed by each porter, and all the times at which the entities states changed. The simulation history is then used to calculate several performance indicators allowing verification, validation and analysis [8]. Among others, the patient wait time, the porters' busy time, and the number of delayed activities are computed.

The simulation model is implemented using a generic programming language (Java) to allow more flexibility and easier integration of decision routines.

5 Computational Experiments

The proposed simulation tool is validated and used to test several improvement alternatives in the context of a Canadian hospital. More precisely, the hospital under consideration contains over 650 beds and performs more than 250,000 medical imaging examinations, 500,000 surgeries, and 6,500,000 laboratory procedures yearly, which will often result in transportation requests. Overall, the hospital centralized transportation service, which operates 24/7, received approximately 385,000 transportation requests also yearly. On average, the service performs 1,200 transports daily on weekday and about 700 on weekend day, including both patient and material transportation requests. Four levels of priority are actually considered by the hospital, ranging from P1 to P4. The first three correspond to patient transportation, P1 being the most urgent one, and the fourth one, P4, is used only in the case of material transportation. Each level of priority is characterized by an expected delay or, in other words, a due time. For instance, the hospital aims to serve all P1 requests right away, P2 requests within 7 min, P3 requests within 20 min and P4 within 60 min.

Based on historical data, probability distributions were fitted to characterize uncertain events (see Table 2). During the experiment, the simulation model can be fed either by historical data or by values drawn from the uncertain event specific probability distribution.

Once the number of porters and their schedule are fixed, transportation requests need to be assigned to porter. The hospital transportation service currently uses a specific software to support the real-time dispatching of requests. The latter considers the priority level: each request having a given level of priority runs for a fixed delay before being considered as a due request. When a porter completes a request, the due

M. Painchaud et al.

Table 2 Probability distributions

	Distribution
Interarrival time	Exponential (non-stationnary)
Origin	Discrete
Destination	Discrete
Priority level	Discrete
Time at origin	Triangular
Travel time	Based on the distance matrix
Time at destination	Triangular

Table 3 Computational results by work shift

	S1			S2		
	Night	Day	Evening	Night	Day	Evening
Nb. porters	7	19	6	8	20	7
Nb. transports	799	4268	1450	699	4319	1499
Transports/Porter	114	225	242	87	216	214
Busy time (%)	45	91	96	29	86	85

request with the higher level of priority is assigned. If there is no due request, the closest one is assigned to the porter.

The objective of the current experiment is to verify and validate the simulation model, and to show how this simulation model can be used to evaluate and compare potential improvement solutions. For validation purpose, simulation runs that replicate the current process of the hospital transportation service were performed, and the results obtained were compared with historical data. Results indicate that the proposed simulation model constitutes a faithful representation of the current situation. Several scenarios with varying number of porters were then created to demonstrate how this model can be used to evaluate and compare potential solutions. Tables 3 and 4 present some of the results obtained for two of these scenarios, which highlight important issues. The first scenario presented includes 32 porters working each week day (7 on night shifts, 19 on day shifts and 6 on evening shifts). As compared to the first scenario, the second one includes one additional porter on each work shift.

Unsurprisingly, results presented in Tables 3 and 4 show that the service performance increases with the number of porters. However, performance measures for each priority level do not improve in the same proportion. When the number of porters increases, the service provided for the most urgent patient requests (P1) improves in a greater proportion. Then, the priority level that encounters the best improvement is P4, which corresponds to material transportation requests. Therefore, increasing the number of porters seems to benefit more requests with low priority level. This should be explained, at least partially, by the dispatching policies and the concept of due request that rules it. This raises an important question, how to de-

Table 4 Computational results by priority level

	S1				S2			
	P1	P2	P3	P4	P1	P2	P3	P4
Nb. transports	19	2620	3472	406	19	2620	3472	406
Transports over delay	9	946	1286	122	2	688	764	65
Av. wait time (min)	1.4	6.2	30.2	85.9	0.2	4.4	12.4	27.3

termine the delay to attribute to each priority level? Our observations suggest that the expected delay (or due time) influences significantly the overall performance as it impacts the dispatching. Currently, the expected delay (or due time) is given as a parameter in the software, but in practice, it is much more important than this. The choice and definition of each priority level is an important issue that should be further investigated.

6 Conclusion

The planning and management of a centralized intrahospital transportation services involves many challenging decisions. Given the high level of randomness and dynamism with respect to the demand, the development of a simulation-based analysis tool can be very useful to help managers in their decision-making process. In this paper, we presented a flexible and generic discrete-event simulation model that can represent all main operations related to intrahospital transportation activities. One interesting aspect of this proposed tool is that it explicitly considers both patient and material transportation requests, and can be adapted to incorporate easily more sophisticated dispatching policies. Computation experiments highlight the importance of the definition of each priority level and the dispatching policy that rules the assignment of requests to porters, issues that should be further investigated.

References

1. Beaudry, A., Laporte, G., Melo, T., Nickel, S.: Dynamic transportation of patients in hospitals. OR Spect. **32**, 77–107 (2009)
2. Bowers, J., Lyons, B., Mould, G.: Developing a resource allocation model for the scottish patient transport service. Oper. Res. Health Care **1**, 84–94 (2012)
3. Fiegl, C., Pontow, C.: Online scheduling of pick-up and delivery tasks in hospitals. J. Biomed. Informat. **42**, 624–632 (2009)
4. Hanne, T., Melo, T., Nickel, S.: Bringing robustness to patient flow management through optimized patient transport in hospitals. Interfaces **39**, 241–255 (2009)

5. Henshaw, C.: Improving patient transportation performance by developing and implementing a generic simulation model. Master's thesis, University of Toronto (2015)
6. Kergosien, Y., Lenté, C., Piton, D., Billaut, J.C.: A tabu search heuristic fot the dynamic transportation of patiens between care units. Eur. J. Oper. Res. **214**, 442–452 (2011)
7. Kergosien, Y., Bélanger, V., Soriano, P., Gendreau, M., Ruiz, A.: A generic and flexible simulation-based analysis tool for EMS management. Int. J. Prod. Res. **53**, 7299–7316 (2015)
8. Law, A.M.: Simulation Modeling & Analysis. McGraw-Hill, Boston, MA (2006)
9. Parragh, S., Cordeau, J.F., Doerner, K.F., Hartl, R.F.: Models and algorithms for the heterogeneous dial-a-ride problem with driver-related constraints. OR Spect. **34**, 593–633 (2012)
10. Pidd, M.: Computer Simulation in Management Science. Wiley, New York, N.Y. (2006)
11. Schmid, V., Doerner, K.F.: Examination and operating room scheduling including optimization of intrahospital routing. Trans. Sci. **48**, 59–77 (2014)
12. Turan, B., Schmid, V., Doerner, K.F.: Models for intra-hospital patient routing. In: 3rd IEEE International Symposium of Logistics and Industrial Informatics, pp. 51–60 (2011)
13. Zhang, Z., Liu, M., Lim, A.: A memetic algorithm for the patient transportation problem. Omega **54**, 60–71 (2015)

Appointment Overbooking and Scheduling: Tradeoffs Between Schedule Efficiency and Timely Access to Service

Yan Chen, Hari Balasubramanian and Yong-Hong Kuo

Abstract Appointment overbooking is one of the most popular ways to mitigate the risk of no-shows. Although appointment overbooking can improve resource utilization, it may also lead to adverse consequences such as increases in resource overtime and patient waiting time. While the majority of research on appointment overbooking focuses on the system performance within the session, our work aims to study the tradeoffs between schedule efficiency and timely access to service. We integrate a stochastic mixed-integer linear program, which assigns patients to appointment time slots, into a queueing model, which evaluates the timeliness of access to service. Our computational results suggest that when the session capacity is less than the appointment request rate, overbooking can greatly reduce appointment lead time and patient abandonment rate and only slightly increases resource overtime and patient waiting time. However, the benefits of overbooking become mild when the session capacity is larger than the appointment request rate.

Keywords Appointment scheduling · Overbooking · Healthcare management Stochastic programming · Simulation · Waiting time

1 Introduction

Scheduling of patients' appointment requests has been an important subject for healthcare service delivery and is critical for the efficiency of the appointment sys-

Y. Chen
Department of Decision Sciences, Macau University of Science and Technology,
Avenida Wai Long, Taipa, Macau, China
e-mail: yachen@must.edu.mo

H. Balasubramanian
Department of Mechanical and Industrial Engineering, University of Massachusetts,
Amherst, United States
e-mail: hbalasubraman@ecs.umass.edu

Y. -H. Kuo (✉)
Stanley Ho Big Data Decision Analytics Research Centre, The Chinese University
of Hong Kong, Shatin, New Territories, Hong Kong
e-mail: yhkuo@cuhk.edu.hk

© Springer International Publishing AG 2017
P. Cappanera et al. (eds.), *Health Care Systems Engineering*, Springer Proceedings
in Mathematics & Statistics 210, https://doi.org/10.1007/978-3-319-66146-9_22

tem. Due to the uncertainty arising from various components in an appointment system (e.g., non-deterministic service time), the events and activities may not be actualized as planned and this problem creates challenges when designing an efficient appointment schedule. When scheduling the appointment requests, one may need to consider the interests of both the service providers (e.g., physicians) and the patients [10]; a healthcare organization may wish that the patients have a shorter waiting time (WT) for a better satisfaction with the service and the resources have a lower idle time (IT) and overtime (OT) for a higher utilization and a lower operating cost. However, the interests of the service providers and the patients can be conflicting; a higher utilization of resources can lead to a longer patient waiting time.

The no-show phenomenon (i.e., patients not attending the scheduled appointments) has been observed in different types of healthcare units around the globe. The unattended time slots increase resource idleness and, as a results, lead to lower utilization and productivity. Furthermore, these no-show cases prevent patients from accessing immediate medical services due to the long wait list. Different from the time for physically waiting at a healthcare facility, this waiting time between the referral date and the appointment date is kind of "indirect". In this study, we refer it to the appointment lead time (LT). In practice, this appointment LT is of particular importance and can be critical to an early diagnosis of diseases, which could result in more effective medical treatments. Therefore, such no-show phenomenon not only is an issue regarding the profit or cost for a healthcare organization and patient experience when attending their appointments, but also has a crucial impact on the effectiveness of health services provided to patients. To mitigate its adverse impacts, appointment overbooking is one of the most popular schemes suggested.

Although overbooking can reduce the resource IT and appointment LT, the overbooked time slots can increase resource OT and patient WT (within the session) if the patients show up more than expected. While the research on appointment overbooking and scheduling is inclined to the performance of the appointment schedule within the session, our paper aims to study the tradeoffs between the schedule efficiency (in terms of resource OT and IT and patient WT) and the timely access to appointment (in terms of the appointment LT). In this paper, we integrate a stochastic mixed-integer linear program (SMILP), which determines the optimal appointment schedule within the session, into a queueing model, which evaluates the timeliness of access to appointment with different levels of session capacity. (i.e., number of patients scheduled per session).

2 Brief Literature Review

There has been vast research on appointment scheduling in healthcare settings; for literature reviews or discussion, we refer the reader to Cayirli and Veral [2], Gupta and Denton [7], and Ahmadi-Javid et al. [1].

Patient no-shows have been a prevalent and important issue for healthcare units which offer appointment times slots for patients to access their services. Several studies (e.g., [3, 8, 9]) suggest that patient no-shows have a great impact on appointment system performance and is one of the most important factors that should be considered when designing the appointment scheme.

LaGanga and Lawrence [11] propose appointment overbooking to reduce the adverse consequences of no-shows. They find that overbooking is more beneficial when there are a larger number of patients, a higher no-show rate, and a less variated service time. Muthuraman and Lawley [17] develop a stochastic overbooking model to compensate for patient no-shows. They consider a setting where a schedule assigns patients to appointment time slots through a sequential call-in process and propose a sequential booking policy that is unimodal. Later, several extensions of their work on sequential clinical scheduling have been studied, e.g., [4], [19] and [12]. LaGanga and Lawrence [11] develop an appointment scheduling and overbooking model which trades off the benefit of servicing additional patients, reducing patient WT costs, and clinic OT costs. Zacharias and Pinedo [18] consider the case that patients have heterogeneous no-show probabilities and different weights, characterize features of optimal solutions, and develop a heuristic procedure for sequential scheduling. Chen et al. [5] develop a simulation model to assess the effects of different overbooking schemes and conclude that overbooking may not necessarily increase resource OT and the way of assigning overbooked patients does have a significant impact. According to Gupta and Denton [7], patients delays can be classified into two types: direct (the time difference between arrival at the healthcare facility and service start) and indirect (the difference between the time the request is made and the appointment time). Most of these studies along this line of research considers direct patient WT, while indirect WT (i.e., appointment LT) has drawn relatively less attention.

There are some studies which consider indirect WT. Liu et al. [14] consider patient no-shows and cancellation and propose dynamic policies for assigning an appointment date to each patient sequentially. Their objective is to maximize the long-run average expected net reward. Liu and Ziya [15] consider that both panel size and service capacity are decision variables and illustrate the effects of behavioral interventions to reduce no-show probabilities. Liu [16] studies the choice of appointment scheduling window, where patient no-shows are present.

While there has been much research in the area of appointment scheduling and overbooking, the majority of studies consider either the performance within the session or the delay from the time the request is made to the appointment time. The research that addresses the relationship between these two types of performance measures appears to be inadequate in the literature. In this paper, we aim to study the tradeoffs between schedule efficiency and timely access to service.

3 Mathematical Models

Our methodology integrates a SMILP for an optimal assignment of patients to slots into a queueing model for evaluation of appointment lead time. Our approach first computes optimal resource OT and IT and patient WT for different numbers of patients scheduled per session, and these values will be used to calculate the key performance metrics for the queueing model.

3.1 Optimization of Appointment Times Within the Session

We first present our mathematical model that optimizes the appointment schedule within the session. This optimization model is based on the appointment scheduling model with fixed-length time slots presented in [6].

The mathematical model requires the following *sets and parameters:*

n = the number of patients scheduled per session;
J = set of appointment time slots in the session;
N = set of patients, i.e., $\{1, 2, ..., n\}$;
S = set of scenarios;
N_s^1 = set of patients who show up at the healthcare unit under scenario s;
N_s^2 = set of patients who do not show up at the healthcare unit under scenario s;
b_j = beginning time of slot j;
d_{is} = service duration for the ith patient under scenario s;
w^{OT} = penalty for each unit of resource OT;
w^{WT} = penalty for each unit of patient WT;
w^{IT} = penalty for each unit of resource IT; and
E = end time of the session.

In this model, the total number of patients scheduled per session, i.e., n, is fixed. We do not impose any restriction on n and $|J|$. When $n > |J|$, overbooking is allowed. We assume that all the patients are punctual. In other words, they arrive at the healthcare facility according to the appointment times. Within a session, we consider two dimensions of stochasticity for the appointment system: the presence of each patient and his/her service duration. We adopt a deterministic-equivalent approach to model the uncertainty; i.e., the outcomes of the stochastic components are realized under each independent and identically distributed scenario in the finite set S. Under a scenario s, each patient i has a chance of not showing up, according to the no-show rate, and, if the patient shows up, the service duration d_{is} is realized via random sampling from its probability distribution. If a patient i is not present under scenario s, the value of d_{is} is then set to 0. All scenarios are realized prior to solving the optimization problem.

Our model determines the following *decision variables:*

$x_{ij} = \begin{cases} 1, & \text{if the } i\text{th patient in the schedule is assigned to the } j\text{th time slot}; \\ 0, & \text{otherwise} \end{cases}$

a_i = appointment time of the ith patient in the schedule;
z_{is}^{start} = start time of the service provided to the ith patient under scenario s;
z_{is}^{end} = end time of the service provided to the ith patient under scenario s;

WT_{is} = WT of the *ith* patient for receiving the medical service under scenario s;
IT_{is} = resource IT between the *ith* and the $(i + 1)th$ services under scenario s; and
OT_s = OT of the resource under scenario s.

We aim to determine the optimal appointment times $\{a_i : 1 \le i \le n\}$ in a session such that the key performance metrics—resource OT and IT, and patient WT—are optimized. Other decision variables act to ensure that the key events take place and the performance metrics are measured properly.

Mathematical Model:

$$SMILP(n): \quad \min w^{OT} \sum_{s \in S} OT_s + w^{IT} \sum_{i \in N, s \in S} IT_{is} + w^{WT} \sum_{i \in N_s^1, s \in S} \frac{WT_{is}}{N_s^1} \quad (1)$$

subject to

$$a_i = \sum_{j \in J} b_j x_{ij} \quad \forall i \in N \quad (2)$$

$$\sum_{j \in J} x_{ij} = 1 \quad \forall i \in N \quad (3)$$

$$a_{i+1} \ge a_i \quad \forall i \in N\backslash\{n\} \quad (4)$$

$$z_{1s}^{start} = a_1 = 0 \quad \forall s \in S \quad (5)$$

$$z_{is}^{start} = a_i + WT_{is} \quad \forall i \in N, s \in S \quad (6)$$

$$z_{is}^{start} + d_{is} - z_{is}^{end} \quad \forall i \in N, s \in S \quad (7)$$

$$z_{is}^{start} = z_{(i-1)s}^{end} + IT_{(i-1)s} \quad \forall i \in N\backslash\{1\}, s \in S \quad (8)$$

$$z_{ns}^{end} + IT_{ns} - OT_s = E \quad \forall s \in S \quad (9)$$

$$a_i, z_{is}^{start}, z_{is}^{end}, WT_{is}, IT_{is}, OT_s \ge 0 \quad \forall i \in N, s \in S \quad (10)$$

$$x_{ij} \in \{0, 1\} \quad \forall i \in N, j \in J \quad (11)$$

Objective (1) aims to minimize a weighted sum of the resource OT and IT and average patient WT among all the scenarios. Constraints (2) assign the appointment times to the patients. Constraints (3) ensure that exactly one time slot is assigned to each patient. Constraints (4) ensure that the appointments are in chronological order. Without loss of generality, Constraints (5) initialize the appointment time and service start time of the first patient to the start time of the session. Constraints (6) and (7) ensure that the patients are seen chronologically in the order in which they are scheduled, start the services no earlier than their arrivals, and compute their WTs. Constraints (8) measure the resource IT between the $(i - 1)th$ and the *ith* patients. Constraints (9) calculate the IT after the last appointment and OT of the resource. Constraints (10) and (11) respectively impose non-negativity and integrality condition on the corresponding decision variables.

By solving $SMILP(n)$, we obtain optimal $\{OT_{is}, IT_{is}, WT_{is} : i \in N, s \in S\}$ so that the average OT, IT and WT can be used for the queueing model in Sect. 3.2.

3.2 Queueing Model for the Appointment System

In this queueing model, we consider a finite time horizon, discretized into sessions $\{1, 2, ..., T\}$ in which the patients are scheduled. We focus on the steady-state performance and consider a large T such that the queueing model will run over many successive sessions. Each session can accommodate at most C requests, i.e., maximum allowable appointments scheduled per session. For each session t, we assume that the number of appointment requests R_t is a Poisson random variable with rate λ, i.e., $R_t \sim Poi(\lambda)$, and if $R_t > C$, those requests which cannot be accommodated for session t are scheduled for the earliest available session(s). Let Q_t be the number of requests that have not been accommodated immediately after session t (i.e., those that have made a request but still are unable to access the service in session t or earlier). Suppose that, between two consecutive sessions, each unaccommodated request in the wait list has a probability of α that the patient will cancel the appointment and leave the wait list and that each cancellation of appointment is independent of each others. Thus, the number of requests in the wait list immediately before session $(t + 1)$, denoted by $\bar{Q}_{t+1}$, is a binomial random variable, i.e., $\bar{Q}_{t+1} \sim Bin(Q_t, 1 - \alpha)$. We have the following relationship:

$$Q_t = \bar{Q}_t + R_t - \min\{\bar{Q}_t + R_t, C\} \quad \forall t = 1, 2, ..., T \tag{12}$$

To analyze the queueing model, we adopt a simulation approach which starts with an empty system, i.e., $\bar{Q}_1 = 0$. We set a warm-up period t_0 such that the sessions on or before t_0 are excluded from the reported performance measures LT and patient abandonment rate (AR). We approximate LT and AR with the following equations:

$$LT = \sum_{t=t_0+1}^{T} Q_t / \sum_{t=t_0+1}^{T} R_t \tag{13}$$

$$AR = 1 - \sum_{t=t_0+1}^{T} \min\{\bar{Q}_t + R_t, C\} / \sum_{t=t_0+1}^{T} R_t \tag{14}$$

Equation (13) is an estimate of LT based on Little's law [13]. We note that if a patient leaves the wait list before the appointment date, LT is not defined for this patient. We note that Eq. (13) is an overestimate of LT. In other words, it is a conservative proxy of the actual LT.

For each session t, with $SMILP(\min\{\bar{Q}_t + R_t, C\})$, OT_t, IT_t, and WT_t can be predetermined. Therefore, the overall OT, IT, WT across the planning horizon can be evaluated by the following equations:

$$OT = \sum_{t=t_0+1}^{T} OT_t / (T - t_0) \tag{15}$$

$$IT = \sum_{t=t_0+1}^{T} IT_t/(T - t_0) \tag{16}$$

$$WT = \sum_{t=t_0+1}^{T} (\min\{\bar{Q}_t + R_t, C\} \times WT_t)/ \sum_{t=t_0+1}^{T} \min\{\bar{Q}_t + R_t, C\} \tag{17}$$

4 Computational Study

We conduct a computational study to examine the tradeoffs between schedule efficiency and timeliness of access to service. We consider a setting of the medical imaging center presented in [6]. Currently, the medical center offers 12 appointment slots for each session. Each session is of 240 min, i.e., $E = 240$, and each time slot is of 20 min. We collected data from the medical imaging center about the appointment activities during the period of January to March, 2015. The dataset includes the records about the date and time each appointment request was made, the scheduled appointment date and time of the request, whether the patient showed up or not, and the start times and end times of set up and service provided to the patient (if showed up). In this computational study, we consider that the service duration is the total of the set up time for the service and the duration of service provided to the patient. Different from [6] which presents the problem that patients did not fast as instructed such that scheduled examinations could not be performed, we only consider the traditional no-show definition (those who do not show up on the scheduled appointment dates). The required information for our mathematical models is summarized in Table 1. In this computational study, we adopt IBM ILOG CPLEX 12.6.2 as our mixed-integer linear programming solver for solving $SMILP(n)$ for optimal OT, IT, and OT for different values of n. The weights in the objective function are chosen such that the resource time is more valuable than the patients': $w^{OT} = 6$, $w^{IT} = 6$, and $w^{WT} = 1$. We conduct our simulation experiments for the queueing model for the appointment system, introduced in Sect. 3.2, on Microsoft Excel 2013. In the simulation experiments, we run over 11,000 sessions and set the warm-up period to 1,000 sessions (i.e., $t_0 = 1,000$ and $T = 11,000$). We consider the cancellation rate $\alpha = 0.01$ and evaluate the performance metrics for $C = 10, 11, ..., 18$.

We calculated from the data that the patient appointment request rate was 13.19 request per session. To analyze the performance metrics for different levels of patient demand, we consider the empirical patient appointment request rate as the baseline, and an increase and a decrease of 10% of this request rate. That is, we examine the cases of which $\lambda = 11.871$, 13.19, and 14.509.

Figures 1, 2 and 3 respectively show the key performance metrics for $\lambda = 11.871$, 13.19, and 14.509. In the three figures, the primary vertical axis indicates the minutes spent for average OT, IT and WT within a single session and the secondary vertical axis denotes the number of sessions for appointment lead time and AR in percentage. For each of these metrics, the lower the value, the better the performance it is. From

Table 1 Summary statistics from empirical data

Random variable or parameter	Mean (μ)	Standard deviation (σ)	Distribution
Examination time per test (min.)	12.70	8.09	0.5 + 87· BETA(2.3, 12.7)
Setup time for each test (min.)	6.40	5.17	-0.5 + LOGN(7.01, 6.43)
No-show probability	0.176	-	-

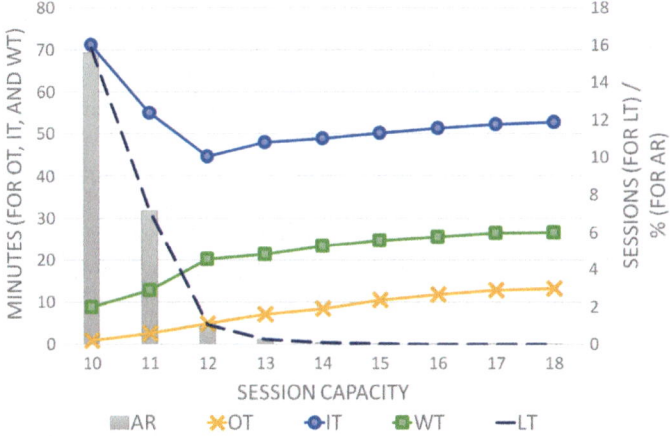

Fig. 1 Key performance metrics when $\lambda = 11.871$

the three figures, we observe the tradeoffs between schedule efficiency and timeliness of access to service; as C increases, LT and AR decrease while OT and WT increase. An interesting observation is that, the changes in the performance metrics appear to be more significant when $C < \lambda$. The reasons are that when $C > \lambda$, it is less likely to receive a sufficient number of appointment requests for filling all the C appointment slots within the session and the wait list for accessing the appointment slots would be kept at a short length (e.g., see LT at the points which $C > \lambda$). In this case, increasing C further from the request rate λ does not impact too much on the performance metrics. Thus, the benefits of increasing C to a level beyond λ become mild and this may lead to long resource OT and WT in occasional cases. We also observe that IT decreases as C increases to a certain point where C is slightly greater than λ; beyond that point, IT increases again. The reason is that, for large C, it is less likely to form a wait list and the idle time will be longer in sessions having new requests fewer than expected. This suggests that C being too large may not necessarily reduce but may increase IT.

Our solution methodology also facilitates to estimate the effects of different over-booking levels for determining the right balance between schedule efficiency and timeliness of access to service. Taking the current setting of the medical imaging

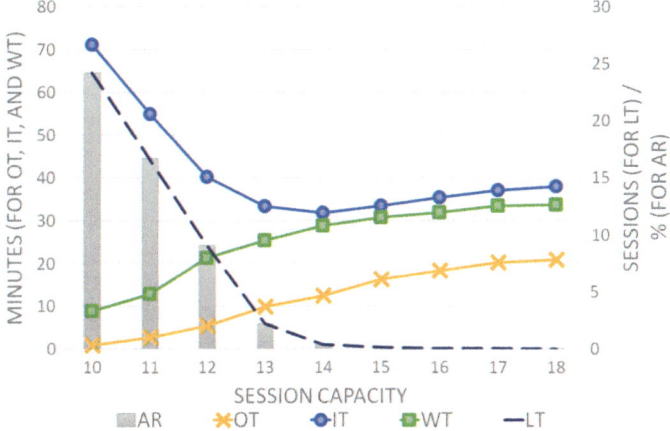

Fig. 2 Key performance metrics when $\lambda = 13.19$

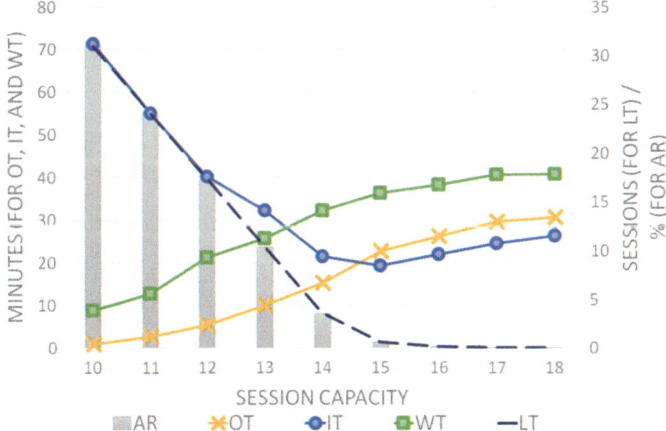

Fig. 3 Key performance metrics when $\lambda = 14.509$

center as an example (e.g., see $\lambda = 13.19$ in Fig. 2 and $C = 12$) as an example, an increment of C from 12 to 13 slots only increases the patient WT from 21.32 to 25.39 min and the resource OT from 9.91 to 12.43 min, but its benefits can be substantial; the appointment LT and AR drop from 9.08 to 0.41 sessions, and 9.10 to 2.23%. This suggests that it may be beneficial to increment C from the current level 12 to 13, as the increases in WT and OT are both mild but the patients can receive the service much earlier (almost 9 sessions earlier) and a higher percentage of patients can be seen.

5 Conclusion

The problem of patient no-shows has been reported in numerous healthcare units in different parts of the world. Appointment overbooking is one of the popular ways to mitigate the adverse effects of unattended time slots. While the majority of research on appointment scheduling and overbooking focuses on the performance measures within the session, our paper studies the tradeoffs between the schedule efficiency and timely access to service. We develop a methodology that integrates a stochastic program, which optimizes the appointment schedule of a session, into a queueing model, which evaluates the appointment lead time and patient abandonment rate. With the data collected from a medical imaging center, we conduct a computational study to evaluate the tradeoffs by adjusting the number of patients scheduled per session. The computational results suggest that overbooking can reduce the appointment lead time and abandonment rate significantly when the session capacity is less than the appointment request rate. When the session capacity is greater than then appointment request rate, the benefits become mild. Our solution methodology also enables practitioners to assess the effects of overbooking and decide the optimal overbooking level in the way which the increases in resource overtime and patient waiting time are acceptable.

Acknowledgements This research is supported by Macao Science and Technology Development Fund under Grant No. 088/2013/A3 and GRF grant 14202115 from the Hong Kong Research Grants Council. The authors would like to thank Ms. Valencia Chang, the Administrative Director of the Macau University of Science and Technology Hospital, for providing us with her concerns about appointment scheduling for medical diagnostic tests from a hospital administration point of view. The authors also wish to acknowledge Dr. Chaobai Wen for providing us with the data for this study. The authors are indebted to the anonymous referees for their valuable comments and suggestions.

References

1. Ahmadi-Javid, A., Jalali, Z., Klassen, K.J.: Outpatient appointment systems in healthcare: a review of optimization studies. Eur. J. Oper. Res. **258**(1), 3–34 (2017)
2. Cayirli, T., Veral, E.: Outpatient scheduling in health care: a review of literature. Prod. Oper. Manag. **12**(4), 519–549 (2003)
3. Cayirli, T., Veral, E., Rosen, H.: Designing appointment scheduling systems for ambulatory care services. Health Care Manag. Sci. **9**(1), 47–58 (2006)
4. Chakraborty, S., Muthuraman, K., Lawley, M.: Sequential clinical scheduling with patient no-shows and general service time distributions. IIE Trans. **42**(5), 354–366 (2010)
5. Chen, Y., Kuo, Y.H., Balasubramanian, H., Wen, C.: Using simulation to examine appointment overbooking schemes for a medical imaging center. In: Proceedings of the 2015 Winter Simulation Conference, pp. 1307-1318 (2015)
6. Chen, Y., Kuo, Y.H., Fan, P., Balasubramanian, H.: Appointment Overbooking with Different Time Slot Structures (2017) (Working paper)
7. Gupta, D., Denton, B.: Appointment scheduling in health care: challenges and opportunities. IIE Trans. **40**(9), 800–819 (2008)
8. Hassin, R., Mendel, S.: Scheduling arrivals to queues: a single-server model with no-shows. Manag. Sci. **54**(3), 565–572 (2008)

9. Ho, C.J., Lau, H.S.: Minimizing total cost in scheduling outpatient appointments. Manag. Sci. **38**(12), 1750–1764 (1992)
10. Kaandorp, G.C., Koole, G.: Optimal outpatient appointment scheduling. Health Care Manag. Sci. **10**(3), 217–229 (2007)
11. LaGanga, L.R., Lawrence, S.R.: Clinic overbooking to improve patient access and increase provider productivity. Decis. Sci. **38**(2), 251–276 (2007)
12. Lin, J., Muthuraman, K., Lawley, M.: Optimal and approximate algorithms for sequential clinical scheduling with no-shows. IIE Trans. Healthc. Syst. Eng. **1**(1), 20–36 (2011)
13. Little, J.D.: A proof for the queuing formula: $L = \lambda W$. Oper. Res. **9**(3), 383–387 (1961)
14. Liu, N., Ziya, S., Kulkarni, V.G.: Dynamic scheduling of outpatient appointments under patient no-shows and cancellations. Manuf. Serv. Oper. Manag. **12**(2), 347–364 (2010)
15. Liu, N., Ziya, S.: Panel size and overbooking decisions for appointment-based services under patient no-shows. Prod. Oper. Manag. **23**(12), 2209–2223 (2014)
16. Liu, N.: Optimal choice for appointment scheduling window under patient no-show behavior. Prod. Oper. Manag. **25**(1), 128–142 (2016)
17. Muthuraman, K., Lawley, M.: A stochastic overbooking model for outpatient clinical scheduling with no-shows. IIE Trans. **40**(9), 820–837 (2008)
18. Zacharias, C., Pinedo, M.: Appointment scheduling with no-shows and overbooking. Prod. Oper. Manag. **23**(5), 788–801 (2014)
19. Zeng, B., Turkcan, A., Lin, J., Lawley, M.: Clinic scheduling models with overbooking for patients with heterogeneous no-show probabilities. Ann. Oper. Res. **178**(1), 121–144 (2010)

Pattern Generation Policies to Cope with Robustness in Home Care

Paola Cappanera and Maria Grazia Scutellà

Abstract We consider the Robust Home Care problem, where caregiver-to-patient assignment, scheduling of patient requests and caregiver routing must be taken jointly in a given planning horizon, and patient demand is subject to uncertainty. We propose four alternative policies to fix scheduling decisions and experiment their impact when used as a building block of a decomposition approach. Preliminary experiments on large size instances show that such policies allow to efficiently compute robust solutions of good quality in terms of balancing caregivers' workload and in terms of number of satisfied uncertain requests.

Keywords Home care · Patient demand uncertainty · Robust optimization

1 Introduction

Increasing age of population, societal changes occurred in taking care of elderly people and the consequent increased hospitalization costs, compel health care providers to circulate home care services. Home care services refer to medical, paramedical and social services to be delivered at patient homes instead of in the hospital. They involve several decisions such as caregiver-to-patient assignment, scheduling of patient requests and caregiver routing, which must be taken jointly in a given planning horizon.

Although the related literature is growing at a high rate, very few works address the issue of uncertainty in problem data. Specifically, service time uncertainty is addressed in the assignment of a set of patients to a set of caregivers over a time horizon [4], stochastic service times are considered both in home care, for scheduling and routing problems [9] as well as for staff dimensioning problems [6], and in

P. Cappanera (✉)
DINFO, University of Florence, Florence, Italy
e-mail: paola.cappanera@unifi.it

M. G. Scutellà
Dipartimento di Informatica, University of Pisa, Pisa, Italy
e-mail: scut@di.unipi.it

© Springer International Publishing AG 2017   257
P. Cappanera et al. (eds.), *Health Care Systems Engineering*, Springer Proceedings
in Mathematics & Statistics 210, https://doi.org/10.1007/978-3-319-66146-9_23

the field service context [7]. Finally, uncertainty on the availability of the caregivers
is investigated in [5]. Recently, a cardinality-constrained robust framework [1], taking into account the uncertainty of the patient demand, has been addressed in [3] by
extending the deterministic problem proposed in [2] to simultaneously take assignment, scheduling and routing decisions in a multiple-day planning horizon. To cope
with the computational burden experimented by the robust approach especially on
large scale instances, in [3] a decomposition approach has been suggested along the
lines of the two-phase decomposition approaches proposed in [8] for the deterministic problem. The results of a wide set of experiments, reported in [8], reveal the
importance of anticipating in the first phase of the approach scheduling issues that
are then considered again in the second phase. Scheduling decisions, indeed, play
a preeminent role. Thus, the tool used to manage such decisions, namely *pattern
generation*, is crucial, where a pattern represents a template for scheduling patient
requests along the planning horizon.

The goal of this paper is to further investigate this algorithmic idea. Specifically,
we propose and experiment a decomposition approach where scheduling decisions
are fixed according to four alternative policies used to generate patterns. The pattern
generation policies incorporate information on uncertainty of patient demand and
adopt different strategies to pursue the common objective of spreading uncertain
visits along the planning horizon so as to increase robustness. Preliminary experiments, also on instances of large size, show that the use of these policies allows
to efficiently compute robust solutions of good quality in terms of balancing of the
caregiver workload and in terms of number of satisfied uncertain requests.

2 Nominal Home Care

The *nominal* Home Care Problem (HCP), as introduced in [2], is defined on a complete directed network $G = (N \cup \{0\}, A)$, where each node $j \in N = \{1, \ldots, n\}$ corresponds to a patient, and node 0 denotes the "depot". All caregivers start their tour
from the depot and come back to the depot at the end of their working day, for each
day of the planning horizon W. O_d denotes the subset of the caregivers available on
day d, for each $d \in W$. Each caregiver $\omega \in O_d$ is characterized by a "skill" s_ω representing the level of treatment the caregiver can provide when assisting a patient.
A hierarchical structure of the skills is assumed, so that a caregiver with skill k can
work requests with required skill up to k. For each patient $j \in N$, the *care plan* r_j
specifies the number of visits of each skill required by j in the planning horizon.

In [2], *assignment*, *scheduling* and *routing* decisions have been coordinated by
means of the concept of pattern. More precisely, it is assumed that, for each patient
j, the requests expressed by the care plan r_j can be operated according to a set P_j of
a priori given patterns. Formally, denoting by $P = \cup_{j=1}^{n} P_j$ the set of all patterns, and
assuming that each patient requires at most one visit per day, each pattern $p \in P$ is
such that $p(d) = 0$ if no service is offered on day d, while $p(d) = k$ indicates that a

visit of skill k is operated according to pattern p on day d. The pattern selected for patient j thus determines the scheduling of his requests and it has to be compatible with the care plan of j, guaranteeing the exact number of visits of appropriate skill.

Given the input data above, HCP consists of three types of decisions: (i) assigning a pattern from P_j to each patient j, so scheduling the requests of j, defined by the care plan r_j, during the planning horizon (*care plan scheduling*); (ii) assigning a caregiver to each patient j for each day in which a request of j has been scheduled (*caregiver assignment*); (iii) determining the tour of each caregiver for each scheduled day (*routing decisions*).

In addressing these decisions, the *skill constraints* (i.e. the compatibility between the skills associated with the patient requests and the skills of the caregivers), the *continuity of care* (i.e., at most T different caregivers can be assigned to each patient in W, for a given T), and caregiver *workday length constraints*, taking into account the travel time along the links of the network, and the service time at the patients, must be satisfied. Alternative objective functions, based on operating cost minimization or social equity criteria (e.g., balancing the caregiver workloads) can be considered.

3 Robust Home Care

In HCP the input data are assumed to be certain. In particular, all patient care plans are assumed to be known in advance. However this assumption may be unrealistic, since patient requests are usually subject to uncertainty in the considered time horizon. To deal with this issue, we assume that at the moment of planning some requests are *certain*, i.e. they will need to be served, while others might be cancelled. We refer to the latter as *uncertain* requests.

The mathematical formulation of the robust model is given in details in [3], while in this section the robust problem is described at a higher level with the aim of stating which kind of robustness is considered. The graphical example given in Fig. 1 supports this informal description. For each patient j, we are given both a *certain* care plan $\bar{r}_j$, corresponding to the certain skilled requests of j, and an *uncertain* care plan $\tilde{r}_j$, corresponding to the uncertain skilled requests of j. Each patient node in G is thus duplicated, by associating each patient $j \in \{1, \ldots, n\}$ with a corresponding uncertain "copy", denoted by $j + n$. Certain care plan $\bar{r}_j$ is associated with node j, while uncertain care plan $\tilde{r}_j$ is associated with node $j + n$. Certain and uncertain patterns are then used to schedule certain and uncertain requests, respectively.

In HCP all patient requests, being certain, must be scheduled over the planning horizon and served by the caregivers. Therefore, a solution to HCP is a collection of tours, for each day $d \in W$ and for each caregiver $\omega \in O_d$, such that each patient request belongs to exactly one tour. Now, consider a set of tours, over W, which includes all the certain and the uncertain requests. These tours, called a *complete set of tours*, represent a very conservative solution, since they correspond to a scenario where all the uncertain requests, in addition to the certain ones, are scheduled. But

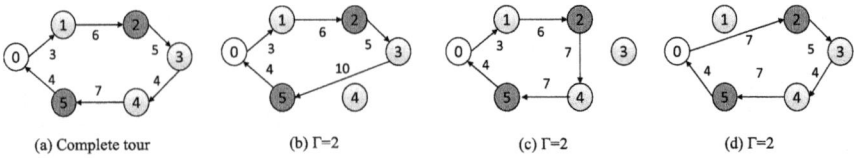

Fig. 1 A complete tour and some induced Γ-tours

this is unlikely to happen, and only a subset of the uncertain requests will realize, making this kind of solution unnecessarily expensive.

To achieve a compromise between degree of robustness and solution cost, we consider scenarios where at most Γ uncertain requests in each tour may realize, where Γ is a given parameter. Notice that any complete tour, say τ, induces a set of tours of this type, also called Γ-tours. These Γ-tours differ for the subset of (at most Γ) uncertain nodes of τ that they include. Specifically, robustness is achieved by considering *sequence-preserving Γ-tours*, i.e. Γ-tours whose nodes represent a subsequence of the corresponding complete tours. A sequence-preserving Γ-tour is thus obtained from a complete tour by deleting some uncertain nodes, without changing the order of the remaining nodes. Figure 1 shows an example of complete tour (a) and a subset of induced Γ-tours, for $\Gamma = 2$ (b), (c), (d). The dark nodes are certain, while the gray nodes are uncertain. Sequence-preserving Γ-tours are desirable from an operational perspective. In fact, usually caregivers prefer not to alter the order of the patient visits, as planned via the complete set of tours, in case some uncertain requests will be cancelled. We say that a complete set of tours C is a *robust set of tours* with respect to Γ if, for each tour $\tau \in C$, all sequence-preserving Γ-tours induced by τ satisfy the workday length of the caregiver assigned to τ. Coming back to Fig. 1, this happens when the maximum of the length of tours in (b), (c), (d) does not exceed the workday length of the caregiver, where the length of the tour is given by the sum of the travel times and of the service times along the Γ-tour. According to the definition above, C is a robust solution, with respect to Γ, if for each tour $\tau \in C$ the caregiver assigned to τ is able to perform the tour whatever the (at most Γ) uncertain requests of τ will possibly realize.

The resulting Robust Home Care problem, named the *sequence-preserving Γ-Robust Home Care Problem ($sRHC_\Gamma$)* [3], thus consists in jointly addressing: (i) assigning a pattern to each patient j that is compatible with $\bar{r}_j$, and a pattern to its uncertain "copy" $j + n$ that is compatible with $\tilde{r}_j$ (*care plan scheduling*); (ii) assigning caregivers to each patient j and to its uncertain "copy" $j + n$, for each day where a request, respectively certain and uncertain, has been scheduled (*caregiver assignment*); (iii) determining the tour of each caregiver for each scheduled day, so that the resulting complete set of tours be a *robust set of tours* with respect to Γ (*daily caregiver routing*). As for the nominal case, the skill constraints and the continuity of care have to be taken into account.

4   Pattern Generation Policies

As previously remarked, the scheduling decisions rely on a set of patterns. Both exact and decomposed approaches in [2, 8] are pattern based, i.e. they schedule the patient visits according to patterns which are generated a priori. Specifically, a pattern is generated for each patient j. However, to gain flexibility, at the end of the generation process all those patterns that are compatible with the care plan of j become eligible for j, i.e. they are used to form the set P_j, even if they have been generated for patients other than j. Such sets of patterns are then given in input to the solver that decides the optimal pattern-patient assignment while respecting the feasibility constraints.

When facing with robustness, problem solving may be both time and memory consuming, especially on large instances. To limit this complexity, in [3] we preliminary experimented a decomposition approach where each patient j is only associated with exactly the pattern determined for j in the pattern generation phase. Therefore, the scheduling decisions are fixed. Here we move some further steps along this line of research, by proposing and experimenting four alternative pattern generation approaches, one of them already introduced in [2] for the nominal problem. All the approaches involve multicommodity flows on an auxiliary layered network $G_W = (N_W, A_W)$, with $|N_W| = n_w$, having one layer for each considered day in the planning horizon W, plus a source node (denoted by 1) and a destination node (denoted by n_w). If K denotes the set of skills and $\bar{k}$ its cardinality, each layer in G_W is composed of $\bar{k} + 1$ nodes: node 0, which indicates that no visit is scheduled on the day corresponding to that layer, and a node k, for each $k \in K$, which represents the scheduling of a visit of skill k. Let L_d denote the set of nodes in the layer corresponding to day $d \in W$. Then, in G_W there exists a directed arc from the source node to the nodes in first layer, from each node in the last layer to the destination node, and from each node in L_d to each node in L_{d+1}, for each d in the planning horizon but the last, as depicted in Fig. 2.

Fig. 2   The layered graph G_W and a pattern example

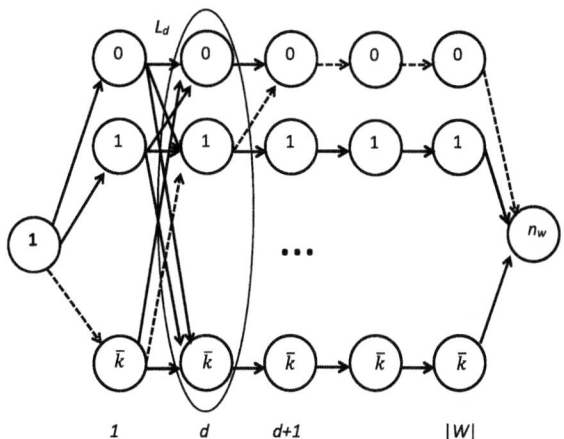

Any directed source-destination path in G_W thus corresponds to a potential pattern: if the node in L_d corresponding to skill k is visited, then a visit of skill k will occur on day d; otherwise, i.e. node 0 is visited, then no visits will occur on d. As an example, the path consisting of dashed arcs in Fig. 2, corresponds to pattern $(\bar{k}, 1, 0, 0, 0)$ according to which exactly two visits will occur: one with maximum skill on Monday and one with skill 1 on Tuesday.

Let $\overline{N}$ be the set of nodes corresponding to certain patients, and $\tilde{N}$ be the set of their uncertain copies. The four alternative pattern generation policies, named Colored flow based (*Col*), Flow Based (*FB*), Balanced (*Bal*) and Balanced w.r.t. Uncertain visits (*BalU*), are based on multicommodity flow models on G_W, where a binary commodity is introduced for each patient $j \in \overline{N} \cup \tilde{N}$ and the related flow variables $\{f_{hi}^j\}$ model a directed path in the layered graph, from its origin to its destination. Specifically, *Col* aims at minimizing the number of generated patterns for certain patients and maximizing the number of generated patterns for the uncertain ones. That is, it pursues the twofold aim of reducing the computational complexity due to a high number of patterns and spreading the uncertain visits across the planning period. *FB* is the particular case of *Col* arising when all the patients are considered as certain, and therefore it tends to minimize the overall number of generated patterns. Notice that *Col* generalizes the policy used in [2, 8] for the deterministic case, where patient demand is known in advance. The last two policies are inspired by balancing criteria. In fact, they take into account the number of visits occurring each day of the planning period, and minimize the difference between the maximum and the minimum such values over the whole planning horizon. The difference between the two policies concerns the type of visits that are considered. Specifically, in *Bal* all types of visits occurring in a day are considered, i.e. both certain and uncertain. On the contrary, in *BalU* only uncertain visits are considered.

In the following the four auxiliary multicommodity flow models, each relevant to a pattern generator policy, are presented in details.

4.1 Colored Multicommodity Flow Problem

The pattern generation policy *Col* is based on the following model:

$$(\text{Col}) \quad \min \quad \sum_{(h,i) \in A_W} (\bar{q}_{hi} - \tilde{q}_{hi})$$

$$\sum_{(h,i) \in A_W} f_{hi}^j - \sum_{(i,h) \in A_W} f_{ih}^j = \begin{cases} -1, & \text{if } i = 1, \\ 1, & \text{if } i = n_w, \\ 0, & \text{otherwise} \end{cases} \quad \forall i \in N_W, \forall j \in \overline{N} \cup \tilde{N}$$

$$\tag{1}$$

$$\sum_{d \in W} \sum_{(h,k):k \in L_d} f_{hk}^j = \bar{r}_{jk} \qquad\qquad \forall j \in \overline{N}, \forall k \in K \tag{2}$$

$$\sum_{d \in W} \sum_{(h,k):k \in L_d} f_{hk}^j = \tilde{r}_{jk} \qquad\qquad \forall j \in \tilde{N}, \forall k \in K \qquad (3)$$

$$\sum_{j \in \overline{N} \cup \tilde{N}} t_j' \sum_{k' \geq k} \sum_{(h,k'):k' \in L_d} f_{hk'}^j \leq \epsilon \cdot \sum_{\omega \in O_d: s_\omega \geq k} D_\omega \quad \forall d \in W, \forall k \in K \qquad (4)$$

$$\sum_{j \in \overline{N}} f_{hi}^j \leq |\overline{N}| \cdot \overline{q}_{hi} \qquad\qquad \forall (h,i) \in A_W \qquad (5)$$

$$\sum_{j \in \tilde{N}} f_{hi}^j \leq |\tilde{N}| \cdot \tilde{q}_{hi} \qquad\qquad \forall (h,i) \in A_W \qquad (6)$$

$$f_{hi}^j \in \{0,1\} \qquad\qquad \forall (h,i) \in A_W, \forall j \in \overline{N} \cup \tilde{N} \qquad (7)$$

$$\overline{q}_{hi} \in \{0,1\} \qquad\qquad \forall (h,i) \in A_W \qquad (8)$$

$$\tilde{q}_{hi} \in \{0,1\} \qquad\qquad \forall (h,i) \in A_W. \qquad (9)$$

Constraints (1) are the classical commodity-wise flow conservation constraints. Flow variables are used to guarantee the proper number of visits for each skill thanks to constraints (2) and (3), and therefore model a pattern which is compatible with the patient care plan. Constraints (4) take into account, skill by skill, the operators availability on each day of the planning horizon. In fact they impose, separately for each day and each skill k, that the sum of service time (t_j') of scheduled visits does not exceed the daily availability (D_ω) of the set of the operators of skill at least k, properly reduced by a parameter ϵ. This parameter reduces the overall caregiver day availability thus preventing infeasibilities that might occur when routing is considered: in fact, pattern generation neglects the traveling times. The auxiliary variables $\overline{q}_{hi}$ and $\tilde{q}_{hi}$ are introduced to detect which arcs are used to design patterns: specifically, for a given arc (h,i), $\overline{q}_{hi}$ will be set to one if the arc is used by a patient in $\overline{N}$, while $\tilde{q}_{hi}$ will be set to one when (h,i) is used by a patient in $\tilde{N}$. Finally, constraints (5) and (6) link together the flow variables f_{hi}^j respectively with the design variables $\overline{q}_{hi}$ and $\tilde{q}_{hi}$, and they guarantee that if arc (h,i) is used, i.e. $\overline{q}_{hi} = 1$ (respectively $\tilde{q}_{hi} = 1$), it can be crossed by any number of certain (uncertain) patients, whereas if that arc is not used, i.e. $\overline{q}_{hi} = 0$ (respectively $\tilde{q}_{hi} = 0$), then it cannot be traversed.

By minimizing the total number of arcs used by certain patients and maximizing the number of arcs used by uncertain patients, the model will tend to minimize, in an implicit way, the number of generated patterns for certain patients, and to maximize the number of generated patterns for uncertain ones, as previously observed.

Since *FB* is the particular case of *Col* where all the patients are considered as certain, in this case the objective function minimizes the total number of arcs used, thus minimizing indirectly the total number of generated patterns.

4.2 Balanced Multicommodity Flow Problem

The two additional models, both based on a balancing criterion, consider the number of visits occurring each day, and minimize the difference between the maximum and the minimum such values over W. This is done by means of two variables, namely, $z_{\min}$ and $z_{\max}$, which represent respectively the minimum and the maximum number of visits occurring daily in the planning horizon. The difference between the two models concerns the type of visits that are considered: specifically, in *Bal* both certain and uncertain visits occurring in a day are considered, whereas in *BalU* only uncertain visits are taken into account. The first of the two models is the following:

$$(Bal) \quad \min \quad z_{\max} - z_{\min}$$

$$\text{s.t.} (1), (2), (3), (4)$$

$$z_{\min} \le \sum_{\substack{j \in \overline{N} \cup \tilde{N}}} \sum_{k \in K} \sum_{\substack{(h,k) \in A_w \\ \text{s.t. } k \in L_d}} f_{hk}^j \le z_{\max} \quad \forall d \in W \tag{10}$$

$$f_{hi}^j \in \{0,1\} \qquad\qquad \forall (h,i) \in A_W, \forall j \in \overline{N} \cup \tilde{N}. \tag{7}$$

As in the previous models, the block of constraints (1), (2), (3) and (4) is related to the flow variables $\{f_{hi}^j\}$ for each patient j. Then in (10), for each day, the total number of visits performed on that day, independently of the skill, is bounded between the lower bound $z_{\min}$ and the upper bound $z_{\max}$. The difference between such lower and upper bound is minimized in an attempt to spread the visits equally among the days.

The model *BalU* is exactly as *Bal* but for constraints (10), which are replaced by the following counterpart, where the daily number of visits is computed by considering only uncertain patients:

$$(BalU) \quad z_{\min} \le \sum_{j \in \tilde{N}} \sum_{k \in K} \sum_{\substack{(h,k) \in A_w \\ \text{s.t. } k \in L_d}} f_{hk}^j \le z_{\max} \qquad \forall d \in W. \tag{11}$$

5 Computational Results

We generated 26 robust instances starting from the data set used in [2] and in [8], which is publicly available at http://www.di.unipi.it/optimize/. The robust data set comprises 9 instances with 60 patients on a 5-day horizon, and 17 instances with 100 patients on a 6-day horizon. All the instances are characterized by 2 types of hierarchical skills, 3 operators, travelling times obtained via Google Maps, a service time at the patients equal to 45 min, a percentage of uncertain visits equal to 20% of all the visits and $\Gamma = 1$. Instance names contain information on the name of the corresponding original nominal instance and the percentage of certain visits (0.8). The experiments, consisting in solving the robust mathematical formulation under the pattern generation policies presented in Sect. 4, have been performed on an Intel(R) Core(TM) i7-4770 CPU @ 3.40 GHz with 4 processors using Cplex 12.6, by imposing a time limit (2 h) and a memory limit (2 GB), under a balancing criterion, i.e. minimizing the maximum caregiver workload.

The number of patterns used to fix scheduling decisions greatly depends on the approach used to generate them, namely *Col*, *FB*, *Bal* or *BalU*. However, as the computational results in this section show, all the approaches are able to identify a set of patterns which successfully allows obtaining solutions of good quality within reasonable computational times.

In regards to computational efficiency, we report that, on the 9 instances with 60 patients, the average optimality gap is stable around 3.49% regardless of the pattern generation tool used, and the average computational time is 1787.5 s with a number of terminations due to memory limit equal to 34 over 36 runs. On the 17 instances with 100 patients, the average optimality gap is still good (3.24%), and the average computational time increases to 4449.33 s essentially due to a drastic reduction of the occurrences of memory limit (38 on 68 runs) w.r.t the istances with 60 patients. This phenomenon is basically due to the different structure of the two groups of instances. In fact, although characterized by a lower number of patients, the instances with 60 patients proved to be more difficult to solve than the others.

In regards to solution quality, in the rest of the section we report information on the impact of the pattern generation approaches on the distribution of the workload among the caregivers and on the number of uncertain visits selected in robust solutions. Specifically, Tables 1 and 2 report, separately for the two groups of instances, the metric Δ_{UF}, i.e. the average percentage difference between the maximum and the minimum caregiver utilization factor. For the instances with 60 patients, the role of the pattern generation tools guided by a balancing criterion is crucial to fairly distribute the workload, whereas on the instances with 100 patients all the 4 approaches allow to obtain good results.

Figures 3 and 4, respectively for instances with 60 and 100 patients, report the number of uncertain visits that are selected in the robust tours. The maximum

Table 1   60 patients: Δ_{UF}

Instance	FB	Bal	BalU	Col
0106-0-0	1.14	1.26	1.25	0.81
0106-0-1	0.37	0.33	0.03	3.30
0106-0-2	0.50	0.30	0.07	0.63
0106-1-0	0.72	0.60	1.72	2.72
0106-1-1	8.53	0.33	0.13	10.17
0106-1-2	2.77	0.70	0.83	6.80
0106-2-0	0.63	0.54	0.69	2.03
0106-2-1	8.53	0.33	0.13	10.17
0106-2-2	2.77	0.70	0.83	6.80

Table 2 100 patients: Δ_{UF}

Instance	FB	Bal	BalU	Col
INS11-0.8-0	0.03	0.03	0.06	0.25
INS11-0.8-1	0.33	0.36	0.39	1.03
INS11-0.8-2	0.00	0.75	1.14	0.11
INS12-0.8-0	1.03	0.72	0.00	0.11
INS12-0.8-1	0.22	0.72	0.78	0.42
INS12-0.8-2	0.25	0.22	0.78	2.58
INS13-0.8-0	0.08	0.31	0.00	5.67
INS13-0.8-1	0.33	0.53	0.47	0.42
INS13-0.8-2	0.00	0.39	0.56	0.17
INS14-0.8-0	0.08	0.28	0.11	2.11
INS14-0.8-1	0.11	0.39	0.06	1.14
INS14-0.8-2	0.08	0.11	0.06	0.31
INS15-0.8-0	n.a.	n.a.	10.92	n.a.
INS21-0.8-0	n.a.	0.00	n.a.	n.a.
INS22-0.8-0	0.31	1.11	0.56	0.11
INS22-0.8-1	0.47	0.00	0.19	0.61
INS22-0.8-2	0.92	0.72	0.03	0.06

theoretical number of uncertain visits that can be scheduled is given by $\Gamma \cdot |W| \cdot |O|$ where $|O|$ is the number of caregivers, i.e. it is equal to 15 and 18 respectively for instances with 60 and 100 patients. We observe that, on instances with 60 patients, the approaches which take into account the uncertain visits explicitly, namely *BalU* and *Col*, allow to obtain a number of selected visits greater than the one reported by policies not discriminating between certain and uncertain visits. On instances with 100 patients, the tool *Col* is still the one that reports almost everywhere the highest number of selected uncertain visits, whereas the tools based on balancing criteria can allow to obtain solutions where the other tools fail. Also, *FB* exhibits a good performance on this set of instances. Again, the different behaviour of the pattern generation approaches on the two groups of instances is essentially due to the different instance structure, which appears to be computationally more complex in the case of 60 patients.

Our conclusion, although preliminary, is that fixing the scheduling decisions allows to efficiently compute robust solutions of good quality in terms of balancing of the caregiver workload and in terms of number of satisfied uncertain requests. Pattern generation policies taking into account the uncertain visits explicitly are particularly effective in pursuing such objectives. We plan to extend the computational experiments to deeper understand the impact of the different pattern generation policies in guiding the robust Home Care decisions. In particular, we plan to better

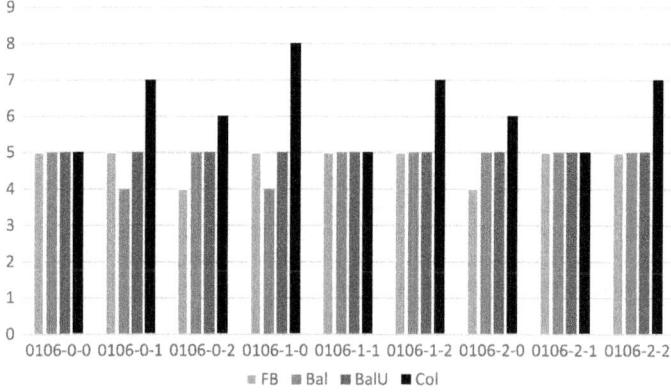

Fig. 3 60 patients: number of selected uncertain visits

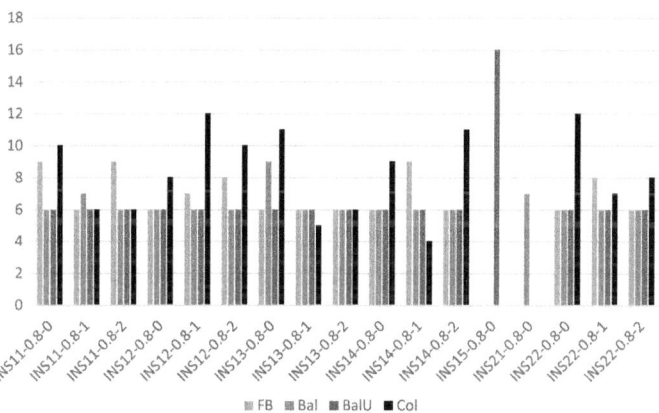

Fig. 4 100 patients: number of selected uncertain visits

investigate the role of parameter Γ which represents the maximum number of satisfiable uncertain requests per tour, also addressing values of Γ greater than 1.

References

1. Bertsimas, D., Sim, M.: The price of robustness. Oper. Res. **52**, 35–53 (2004)
2. Cappanera, P., Scutellà, M.G.: Joint assignment, scheduling, and routing models to home care optimization: a pattern-based approach (with related online supplement). Trans. Sci. **49**(4), 830–852 (2015)
3. Cappanera, P., Scutellà, M.G., Nervi, F., Galli, L.: Demand uncertainty in robust home care optimization. Omega (2017). https://doi.org/10.1016/j.omega.2017.08.012
4. Carello, G., Lanzarone, E.: A cardinality-constrained robust model for the assignment problem in home care services. Eur. J. Oper. Res. **236**, 748–762 (2014)

5. Nguyen, T.V.L., Toklu, N.E., Montemanni, R.: Matheuristic optimization for robust home health care services. Lect. Notes Manag. Sci. **7**, 1–7 (2015)
6. Rodriguez, C., Garaix, T., Xie, X., Augusto, V.: Staff dimensioning in homecare services with uncertain demands. Int. J. Prod. Res. **53**(24), 7396–7410 (2015)
7. Souyris, S., Cortés, C.E., Ordóñez, F., Weintraub, A.: A robust optimization approach to dispatching technicians under stochastic service times. Optimization Letters **7**, 1549–1568 (2013)
8. Yalçındağ, S., Cappanera, P., Scutellà, M.G., Şahin, E., Matta, A.: Pattern-based decompositions for human resource planning in home health care services. Comput. Oper. Res. **73**, 12–26 (2016)
9. Yuan, B., Liu, R., Jiang, Z.: Home health care crew scheduling and routing problem with stochastic service times. In: Proceedings of 2014 IEEE International Conference on Automation Science and Engineering, pp. 564–560 (2014)

Outpatient Day Service Operations: A Case Study Within Rheumatology Diseases Management

Giuseppe Ielpa, Rosita Guido and Domenico Conforti

Abstract This work presents an innovative quantitative approach, based on an optimization model, to manage outpatient Day Service operations. The main objective of the research work is to maximize admitted patient flow in order to use available resources efficiently. The model is designed with the help of Answer Set Programming and implemented in Datalog with Disjunction (DLV) system. As significant case study, we consider the Rheumatology domain. Preliminary results are presented.

Keywords Patient admission planning · Access rules · Priority · Appointment scheduling · Answer set programming · Day service

1 Introduction

Improving efficiency and value in healthcare systems consists in introducing new health care models to reduce gaps in quality, safety, and equity. Primary care practices often have long waits for appointments. Improving health care delivery could require process redesigns and innovative strategies to provide patients with health care quality in an effective and efficient manner.

The *Short Stay Unit* (SSU), for instance, was introduced in the 1970s to meet special needs of pediatric and surgical patients as an alternative to ordinary hospital wards. Patients require less than 5 days as hospitalization period and the same levels of medical care as an ordinary ward are provided. Shorter hospitalization periods, increased patient satisfaction, and improved resource utilization are some of the advantages of this innovative system [1]. Similar to the SSU is the *week hospital*, an innovative inpatient healthcare organization where hospital stay services

G. Ielpa (✉) · R. Guido · D. Conforti
DIMEG, University of Calabria, Rende, Italy
e-mail: giuseppe.ielpa@unical.it

R. Guido
e-mail: rosita.guido@unical.it

D. Conforti
e-mail: domenico.conforti@unical.it

© Springer International Publishing AG 2017   269
P. Cappanera et al. (eds.), *Health Care Systems Engineering*, Springer Proceedings in Mathematics & Statistics 210, https://doi.org/10.1007/978-3-319-66146-9_24

are planned in advance and delivered on week-time basis to elective patients. A week hospital model, which performs accurate diagnosis and appropriate treatments for a Rheumatology department, was defined in [2]. An integer linear programming model is formulated and solved to maximize the number of admitted patients, patients' appointments in a given planning horizon, and reduce patient waiting time as much as possible. In literature, innovative services were defined to improve service delivery, like day surgery systems, also known as ambulatory surgery or outpatient surgery system: its main characteristic is that patients are admitted for surgical procedures and overnight hospital stay is not required [3]. These characteristics belong also to the day hospital system. The *Day Service* system derives from the day hospital system as model for clinical activities. More specifically, clinical day service cannot be activated to treat emergencies, as it is suitable only for some specific diseases that could request provision of multidisciplinary clinical services. Appointment scheduling problems are broadly covered in literature, but approaches concerning Day Service are very limited.

The need to use available resources efficiently led health systems to assess the quality of the performed healthcare services in terms of hospital stay in order to detect clinical problems inconvenient for hospital stay. For this reason many services, formerly performed during hospital stay, are progressively transferred to Day Hospital, Day Service, and Day Surgery care models, which have the same effectiveness and can provide advantage in terms of efficiency for healthcare systems. A short hospitalization and a day service allow to gain many social and economic advantages, because they dramatically reduce hospitalization costs, determine hospitalization appropriately, and foster a better availability of resources for patients affected by severe diseases. Day service developments are due also to technologies innovation in the healthcare field: advantages are shorter services time and reduced care service needs thanks to new diagnostic and therapeutic techniques, which have been becoming less risky and invasive. This results in demand decreasing of long stay in hospital. Adopting of such healthcare models allows to gain efficiency in health care systems, as a suitable use of available resources is guaranteed and, at the same time, high quality of services is performed, by paying attention to individual specific care needs.

This work proposes a Day Service model management for patients affected by rheumatic diseases. The model takes into account of patient priority and patient waiting time as key performance indicators, in order to manage Day Service. The objective of this research is on a scheduling method for Day Service Unit by providing the basis for a decision support tool to make decisions, considering all appointment requests over a given planning horizon. Our model is implemented in *Datalog with Disjunction* [4] (DLV) system [5], a state-of-the-art system for Answer Set Programming (ASP). ASP is a declarative Logic Programming paradigm, developed in the area of Logic Programming and Nonmonotonic Reasoning, providing fast prototyping and very high flexibility in the implementation of new requirement specifications. ASP is suitable for academic research, as well as for industrial applications, and services operations [6]. Preliminary results on realistic instances show good performance of the proposed model.

2 Hospital Day Service

Hospital day service (HDS) is an organizational model to deliver diagnostic tests and therapeutic treatments. This model does not require hospital stay and allows patients to be followed up by a clinical doctor until diagnosis process ends and/or an appropriate therapy is prescribed. The HDS is tailored on patient needs. All multidisciplinary clinical services are distributed in the form of *complex ambulatory packages* (CAP)s. A CAP should be completed in thirty days. Since a CAP may refer to heterogeneous clinical sectors, a healthcare point-of-service provides a high clinical level and organizational management in order to allow patients to access a suitable care process. Access planning and scheduling appointments are crucial aspects of HDS. A well managed access to services can impact on quality of health care in terms of reduction of waiting time and reduction of number of accesses to hospital.

From an organizational point of view HDS can be seen as a development of the Day Hospital model. HDS relieves the pressure on inpatient wards by providing an alternative pathway for a certain case mix. Some benefits are elimination of lost, misplaced, or duplicate scheduling forms, and a unified form that significantly simplifies scheduling processes.

The main goals of HDS are:

- Simplifying patient admissions to health care services and reduce administrative burdens.
- Guaranteeing care continuity once a patient is followed up by a medical doctor.
- Enhancing level of management among different clinical departments.
- Defining a diagnosis or a therapy by performing clinical tests and treatments in only one day or in a reduced number of admissions.
- Reducing either the need of long-term hospitalization or inappropriate delivery of Day Hospital.

2.1 A Day Services Management Model

The proposed model manages a HDS system to schedule patient appointments by taking into account of patient priority, patient waiting time, patient requirements, and hospital resources availability. This model is patient-centered. A set of clinical services, prescribed to a patient, has to be performed within a given number of weeks, from the first admission. A patient admission begins every day a clinical service, prescribed to the patient, is scheduled. Patient admissions are allowed during a working day. Each working day consists in a number of time slots. Each time slot has the same time length. For each waiting patient is known:

- a priority value, defined by his/her medical specialist on basis of patient status, determining the time expected to complete the CAP,

- a set of clinical services prescribed by the medical specialist, and
- the time the patient is still waiting for his/her first appointment.

An appointment is given as tuple (*patient, clinical service, time slot*). The following constraints are defined:

- An appointment is planned for each prescribed clinical service s to patient p only once during the planning horizon.
- A patient has at most one appointment per time slot.
- First appointment date defines patient first admission date to hospital and starting date of PAC.
- Elapsed time between patient admission date and last appointment date should be minimised.

Our problem definition is inspired to the timetabling problem representation [7]. We assume that the reader is familiar with basic concepts of deductive databases and logic programming. The programming language [8] used to represent the model is an extension of Disjunctive Datalog by integrity constructs (so-called strong constraints), optimal constructs (so-called weak constraints), and aggregate functions [9]. Given any solution of the problem $\mathscr{P}$, which is implemented on basis of the designed model, has to satisfy all strong constraints. If a strong constraint is violated, the related solution will be infeasible. If any of weak constraints is violated, only those solutions, which minimise the sum of the weights of the violated constraints, according to their priority level, will be considered.

This can be expressed by an objective function H for problem $\mathscr{P}$ and a candidate solution A, for $\mathscr{P}$, and an auxiliary function $f_\mathscr{P}$ to map weak constraints weights in function of their priority level:

$$f_\mathscr{P}(1) = 1$$
$$f_\mathscr{P}(n) = f_\mathscr{P}(n-1) \times WC^\mathscr{P} \times w^\mathscr{P}_{max} + 1, n > 1$$
$$H^\mathscr{P}_A = \sum_{i=1}^{l^\mathscr{P}_{max}} (f_\mathscr{P}(n) \cdot \sum_{N \in N^{A,\mathscr{P}}_i} w_N)$$

where WC^p denotes the total number of weak constraints in $\mathscr{P}$, $w^\mathscr{P}_{max}$ and $l^\mathscr{P}_{max}$ denotes the maximum weight and maximum priority level of a weak constraint in $\mathscr{P}$, respectively; $N^{A,\mathscr{P}}_i$ denotes the set of violated weak constraints in layer i; w_N denotes the weight of the weak constraint N. All candidate models, which satisfy all strong constraints, and for which the objective function $H^\mathscr{P}_M$ is minimal among all candidate models, are considered the preferred models w.r.t. the candidate model semantics.

The input is a set of m patients $P = \{p_1, \ldots, p_m\}$, a set of n clinical exams or tests $T = \{t_1, \ldots, t_n\}$, a set of k time slots $S = \{s_1, \ldots, s_k\}$, and a set of prescriptions $R = \{(p_x; t_y; r_{xy}) \mid p \in P, t \in T\}$ modeling that a patient p, who is waiting to be admitted since a certain number of days, has to receive clinical service t with some priority. In addition, a test may be unavailable during certain periods, which is represented as a set of periods for each test ($\forall t \in T : U_t \subseteq S$).

The problem now is to assign prescribed tests to periods in such a way that: no patient is involved in more that one test during any period; that all of prescription requirements are met; all unavailabilities are honored.

More formally: Find a set $A \subseteq \{(p, t, s) \mid p \in P, t \in T, s \in S\}$, such that:

$$\forall p \in P; s \in S : (p, t_x, s) \in A \wedge (p, t_y, s) \in A \Rightarrow x = y$$
$$\forall t \in T, s \in S : (p_x, t, s) \in A \wedge (p_y, t, s) \in A \Rightarrow x = y$$
$$\forall (p_x, t_y, r_{xy}) \in R : \sum_{s \in S} \chi_A(p_x, t_y, s) = r_{xy}$$
$$\forall t \in T, p \in P : p \in U_t(p, t, s) \notin A$$

where χ_A is the characteristic function for A.

2.2 Disjunctive Logic Programming Implementation of the Model

Now that we have introduced the problem in an abstract way, we can provide a concrete formalisation in our framework, with the help of DLV. A program can be used to model a problem to be solved: the problem solutions correspond to the answer sets of the program (which are computed by DLV). Therefore, a program may have no answer set (if the problem has no solution), one answer (if the problem has a unique solution) or several ones (if the problem has more than one possible solutions).

DLV system is the product of more than twenty years of research and development, and it is nowadays one of the most relevant state-of-the art implementations of ASP [6]. In the last few years its language has been exploited for industrial use, and further improved and extended, in order to be compliant to *ASP-Core-2* standard language [10]. The following is the main ASP-Core-2 implementation of a *Guess/Check/Optimise* program [11]. The program consists in a set of rules and constraints, composed by literals with variables. The implementation of the program is fully declarative [12], i.e. focused on knowledge representation and on properties expected by the solution of the problem itself, rather than on designing and coding solving algorithms.

```
% Guess an allocation for a prescription of service T in slot S for patient P.
{ in(P,T,S) } :- prescription(P,Priority,DaysWaiting,T), slot(S),
                 not period_unavailability(S), not test_unavailability(T,S).

% Check 1. Patient P can access service T only once in the planned period.
:- in(P,T,S), in(P,T,S1), S != S1.

% Check 2. Patient P can access at most one service T during slot S.
:- in(P,T,S), in(P,T1,S), T != T1.

% Check 3. Service Capacity is the upper bound for the number
% of patients, who access service T during slot K of day D.

% Check 3.1. Monthly test_capacity(Test,Capacity,xmonth).
:- #count{P : in(P,T,S)} > C, test_capacity(T,C,xmonth).

% Check 3.2. Weekly test_capacity(Test,Capacity,xweek).
:- #count{P : in(P,T,S), slt(S,W,D,H)} > C, week(W), test_capacity(T,C,xweek).

% Check 4. Only one admission per service T is allowed, for each patient.
:- in(P,T,S), in(P1,T,S), P != P1.

% Check 5. All prescribed tests must be scheduled in the allowed period.
scheduled(P,T) :- in(P,T,S).
:~ prescription(P,Priority,DaysWaiting,T), not scheduled(P,T).
```

```
% Optimize 1. Makespan.
:~ makespan(X,Priority). [X@Priority,X,Priority]
makespan(X,Priority) :- priority(Priority,Baseline),
                        X = #max{Delta : absolute_day(Delta,W,D),
                                 slt(S,W,D,H), in(P,T,S),
                                 prescription(P,Priority,DW,T)
                                }.

% Optimize 2. Possibly schedule first patients with a higher priority.
:~ prescription(P,Priority,DaysWaiting,T), in(P,T,S),
   slt(S,W,D,H), absolute_day(AbsoluteDay,W,D), priority(Priority,Baseline),
   Delay=DaysWaiting+AbsoluteDay, Delay>Baseline,
   Weight=Delay-Baseline. [ Weight@Priority,
                            P,T,S,W,D,H,AbsoluteDay,Delay,
                            Baseline,DaysWaiting,Priority,Weight
                          ]

% Optimize 3. Possibly minimize patients admissions, in the same day or week.
:~ in(P,T,S), in(P,T1,S1), slt(S,W,D,H), slt(S1,W,D1,H1),
   T!=T1, S!=S1, D!=D1, prescription(P,Priority,DaysWaiting,T). [ 1@Priority,
   Priority,DaysWaiting,P,T,S,T1,S1,W,D,H,D1,H1
                          ]

:~ in(P,T,S), in(P,T1,S1), slt(S,W,D,H), slt(S1,W1,D1,H1),
   T!=T1, S!=S1, W!=W1, prescription(P,Priority,DaysWaiting,T). [ 1@Priority,
   Priority,DaysWaiting,P,T,S,T1,S1,W,W1,D,H,D1,H1
                          ]
```

The program below represents the *Intensional* part of the problem, which needs to be *instantiated* with input data. Input data are also referred as the *Extensional* part of the program, so-called *facts*, representing tuples of deductive databases. State of the art ASP systems take into account of the instantiation process and of the reasoning process, which lead to none, one or more solutions. For details on the instantiation of the program the reader may refer to [13], while the reasoning process refers to [14].

3 A Day Service System for Ankylosing Spondylitis: A Case Study

The efficiency of the model is assessed on a set of real data provided by the Rheumatology Department of Careggi Hospital in Florence, Italy. Data consist in a combination of clinical services, prescribed to 15 patients under the form of CAPs, for a total amount of 70 prescribed services. *Ankylosing Spondylitis* (AS) is a chronic rheumatic disease affecting bones, muscles, and backbone ligaments.

A waiting list of patients, affected by AS, has to be admitted to HDS. Appointments for laboratory tests and diagnostic tests are managed accordingly to the abovementioned requirements. Tables 1 and 2 reports the set of diagnostic tests and laboratory tests in CAPs, with the related service capacity. Along with service capacity was considered also a real calendar of services availability. In order to test the proposed model in a certain case mix of patients, we use several instances of waiting patients with different PACs, time slot length is set to 60 min, which leads to 8 time slots per day, from Monday to Thursday, only 4 time slots for Friday. The planning horizon is set to 4 weeks. Current patient appointment scheduling at Careggi

Table 1 Diagnostic and laboratory services capacity

Diagnostic tests			Laboratory tests		
Code	Description	Capacity	Code	Description	Capacity
T1	Sacroiliac pelvic radiograph	10/month	T13	Complete blood counts, erythrocyte sedimentation rate, chain reaction	10/week
T2	Standard backbone radiograph	10/month	T14	SPE-serum protein electrophoresis	10/week
T3	Peripheral parts of the body radiograph	10/month	T15	HLA-B27 tissue typing	10/week
T4	Chest radiograph	2/month	T16	Total and fractioned immunoglobulin	10/week
T5	Electrocardiogram test	10/week	T17	Phospho-calcium metabolism	10/week
T6	Limbs ultrasonography	10/week			
T7	Computed axial tomography	2/month			
T8	Sacroiliac nuclear magnetic resonance	2/month			
T9	Bone densitometry	2/month			
T10	Pulmonary function test	2/month			
T11	Echocardiogram	2/month			
T12	Colonoscopy	1/month			

Hospital has been done manually by one planner, who receives admission requests from physicians of the Rheumatology department.

The scheduling is done in such a way that patients can perform clinical services within a certain time. This time is called the *required access time* and is determined by the physician. The objective function can be considered as a weighted sum of patient and hospital performance measures: waiting time, clinical services capacity, slots selected for patient appointments. Priority levels define the maximum number of days a patient can wait to complete his/her PAC.

The data shown in the tables, along with time slots covering 4 weeks, 5 working days, for a total amount of 36 working time slots per week, represent the *Extensional* input needed to instantiate the declarative program, implemented in ASP-Core-2 language. For program execution we refer to DLV 2.0. This version of the DLV system makes use of the I-DLV system [15] to perform the instantiation process; the instantiated program is then solved with the help of the WASP system [16], an ASP solver handling disjunctive logic programs under the stable model semantics.

Table 2 Waiting patients list

Patient	Priority	Expected time (days)	Waited time (days)	Prescriptions
P1	3	10	20	T1, T7, T13, T14, T16
P2	3	10	10	T1, T3, T5, T10, T13, T15
P3	2	30	24	T2, T3, T5, T7, T13
P4	1	60	12	T3, T8, T10, T15, T16
P5	1	60	18	T2, T3, T13, T14
P6	2	30	24	T1, T6, T13
P7	2	30	24	T1, T3, T6, T13, T15
P8	1	60	45	T4, T8, T12, T14
P9	3	10	11	T6, T7, T11
P10	2	30	44	T2, T6, T11, T12, T13, T17
P11	3	10	44	T5, T7, T9, T10, T13
P12	1	60	26	T4, T5, T14, T15
P13	1	60	17	T3, T4, T10, T15
P14	3	10	29	T1, T5, T9, T10, T11, T13, T14, T16
P15	1	60	17	T8, T13, T15

We ran three different experiments in order to test the model behaviour with respect to three different inputs to the program:

1. Input with unlimited service capacity, and with service availability calendar, which defines whether a service is available in given slots.
2. Input with unlimited service capacity, and without service availability calendar (program is free to schedule the test to the best slot available).
3. Input with limited service capacity, and without service availability calendar (the real case scenario).

For the third experiment the 5th strong constraint of the *Check* part of the program was relaxed to a weak constraint, hence moved to the *Optimization* part of the program, in order to allow postponements for appointments that cannot be scheduled, when maximum service capacity is reached. Here the solutions of the problem for the three different cases are shown in the charts reported in Figs. 1, 2, and 3.

Bubbles represent patient appointments scheduled over a period of 4 weeks. Bubble size is determined in function of the value of the cost to take into account in the

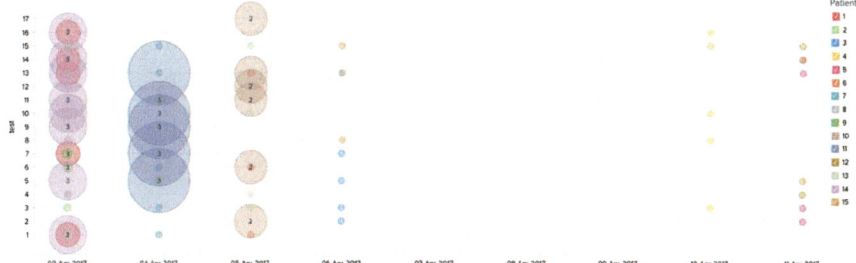

Fig. 1 Solution with unlimited service capacity, and with service availability calendar

Fig. 2 Solution with limited service capacity, and without service availability calendar

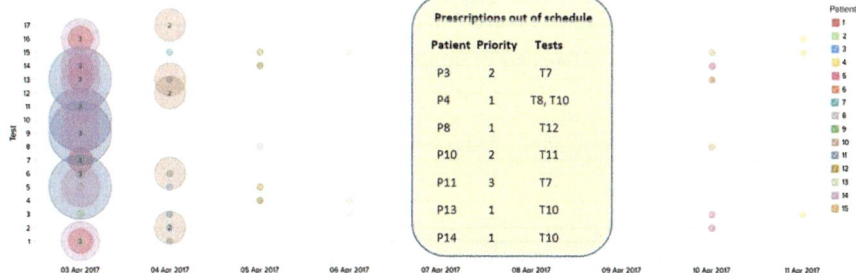

Fig. 3 Solution with limited service capacity, and without service availability calendar (real case scenario)

event an appointment is scheduled in a slot, which is over the time limit a patient is expected to complete his/her CAP. The bigger is the bubble, the higher is the influence of the allocation on the overall cost of the solution. Labels in bubbles report higher priority levels (i.e. 1, and 2), representing patients expected to complete the CAP in 10, and 20 days, respectively. Figure 2 shows that in absence of availability limitation, and a free allocation calendar, the program overlaps patients with higher priority to the very beginning of the scheduling period, in order to minimize the influence of their allocation on the cost of the solution. The majority of patients completes the CAP within one admission only. Figure 3 shows that the solution has been affected by the real service capacity, since only 62 appointments were scheduled versus the 70 candidate prescriptions. The prescriptions left-out of capacity either reflect the low priority level of the patient, or the lack of services available, and are reported in the note, which is visible in the figure.

4 Conclusions

In conclusion, Day Service represents an innovative and efficient health care model and it may be adopted in many health care systems. Day Service application allows to reduce costs dramatically.

This work mainly focused on an innovative approach to the design of patient appointment scheduling using declarative languages of Answer Sets Programming, such as the language supported by DLV. It is worth reminding that high expressiveness of ASP comes at the price of a high computational cost in the worst case, which makes the implementation of efficient systems a difficult task. ASP Programming models were developed in order to maximize patient flow of admissions to Day Service, such that each patient could do all the prescribed examinations. Results obtained by the solution of the problem developed on basis of the designed model were satisfactory. It is worth noting that, dealing with real case inputs, some exams were not scheduled, since the admission is allowed only to patients with a high level of priority, which is related to the severity of the disease, and with a longer waiting time. The selection of patients that are going to be admitted is done according to the availability of the resources needed to do the laboratory tests and the examinations.

Acknowledgements This work benefited of the discussion with Francesco Calimeri, Simona Perri, and Giovambattista Ianni, Department of Mathematics and Computer Science of University of Calabria. Thanks to Davide Fusca and Jessica Zangari for their kind support on I-DLV and WASP. Real case scenario provided by MD Maria Letizia Conforti, and the Rheumatology department of the Careggi University Hospital, Florence, Italy.

References

1. Damiani, G., Pinnarelli, L., Sommella, L., Vena, V., Magrini, P., Ricciardi, W.: The short stay unit as a new option for hospitals: a review of the scientific literature. Med. Sci. Monit. Int. Med. J. Exp. Clin. Res. **17**(6), SR15–SR19 (2011)
2. Conforti, D., Guerriero, F., Guido, R., Matucci Cerinic, M., Conforti, M.L.: An optimal decision making model for supporting Week-Hospital management. Health Care Manag. Sci. **14**, 74–88 (2011)
3. Castoro, C., Bertinato, L., Baccaglini, U., Drace, C., McKee, M., Day surgery: making it happen. Policy Brief: European Observatory on Health Systems (2007). http://www.euro.who.int/document/e90295.pdf
4. Leone, N., Faber, W., Bria, A., Calimeri, F., Catalano, G., Cozza, S., Dell'Armi, T., Greco, G., Ianni, G., Ielpa, G., Maratea, M., Panetta, C., Perri, S., Ricca, F., Scarcello, F., Terracina, G.,

Pfeifer, G., Eiter, T., Gottlob, G.: DLV: an advanced system for knowledge representation and reasoning. In: Pontelli, E. (ed.) ALP Newsletter, vol. 20(3/4) (2007)
5. DLV system 2.0 website. https://www.mat.unical.it/DLV2
6. Calimeri, F., Ricca, F.: On the application of the answer set programming system DLV in industry: a report from the field. ALP Newsl. **3**, 1–16 (2012)
7. Faber, W., Leone, N., Pfeifer, G.: Representing school timetabling in a disjunctive logic programming language. In: WLP 1998, pp. 43–52 (1998)
8. Grasso, G., Leone, N., Manna, M., Ricca, F.: ASP at work: spin-off and applications of the DLV system. In: Logic Programming, Knowledge Representation, and Nonmonotonic Reasoning. Lecture Notes in Computer Science, vol. 6565, pp. 432–451. Springer (2011)
9. Faber, W., Pfeifer, G., Leone, N., DellArmi, T., Ielpa, G.: Design and implementation of aggregate functions in the DLV system. Theory Pract. Logic Program. **8**(5–6), 545–580 (2008). https://doi.org/10.1017/S1471068408003323
10. Calimeri, F., Faber, W., Gebser, M., Ianni, G., Kaminski, R., Krennwallner, T., Leone, N., Ricca, F., Schaub, T.: ASP-Core-2: input language format. In: ASP Standardization Working Group, Technical report (2012)
11. Leone, N., Pfeifer, G., Faber, W., Eiter, T., Gottlob, G., Perri, S., Scarcello., F., The DLV system for knowledge representation and reasoning. ACM Trans. Comput. Logic **7**(3), 499–562 (2006)
12. Fuse, D., Germano, S., Zangari, J., Calimeri, F., Perri, S.: Answer set programming and declarative problem solving in game AIs. In: XIII Conference of the Italian Association for Artificial Intelligence Popularize Artificial Intelligence 2013 (PAI 2013), Turin (Italy). Torino (2013)
13. Calimeri, F., Fusca, D., Perri, S., Zangari, J.: I-DLV: the new Intelligent Grounder of DLV. Artif. Intell. (2017). To appear (In Press)
14. Alviano, M., Dodaro, C., Leone, N., Ricca, F.: Advances in WASP. In: LPNMR. LNCS, vol. 9345, pp. 40–54. Springer (2015)
15. I-DLV System website. https://github.com/DeMaCS-UNICAL/I-DLV/wiki
16. WASP System website. https://github.com/alviano/wasp

Can Performance Monitoring Identify Any Effect of Hospital for Improvement/Worsening: Case of Heart Failure Patients

Roshanghalb Afsaneh, Mazzali Cristina, Lettieri Emanuele and Paganoni Anna Maria

Abstract This study offers original insights to further the ongoing debate about the stability of 'hospital effect' on performance with administrative data as a source of evidence. The overall assumption behind is investigating the debate on organizing the hospitals by learning from their previous performance to transform change. In this way, one available evidence is the administrative data that are collected in healthcare systems to document their activity and have had an administrative rational behind. Using statistical tests (multi-level hierarchical logistic models) on the variation of performance of the hospital, we used 78,907 records for Adult Heart Failure (HF) patients from 117 hospitals in Lombardy Region considering only Incident cases from 2010 to 2012 to answer to our research questions. what we observed was that mortality and readmission were explained by a few different hospital characteristics like the percentage of surgical DRG and type of hospital in mortality model and mean length of stay in readmission one. One explanation could be the case of connection between hospital managers' capabilities to implement short-term improvement.

Keywords Administrative data · Hospital performance
Hierarchical linear model

R. Afsaneh (✉) · M. Cristina · L. Emanuele
Department of Management, Economics & Industrial Engineering,
Politecnico di Milano, Via Lambruschini 4/b, Milan, Italy
e-mail: afsaneh.roshanghalb@polimi.it

P. A. Maria
Department of Mathematics, Politecnico Di Milano,
La Nave, via Edoardo Bonardi, 9, Milan, Italy

1 Introduction

Nowadays, ranking organizations based on their performance measurement is attracting attention from both business and academic sides. Although there may come a variety of purposes from decision-makers in measuring and monitoring performance in different fields (e.g., case mix and patients' severity in healthcare), there still a debate on existing of any impact from the organizations as a unique factor affecting their performance to identify best practices and processes. Such impact could be related to management practices and processes that may came out successful/fail in improving performance over time. By having a longitudinal view, many organizations like hospitals can organize themselves to pick to a better performance in terms of effectiveness, safeness, and efficiency in healthcare. In healthcare as many studies argue—e.g., [1]—there might be a variety of key indicators need to be considered for measuring performance including a decrease in the hospital length of stay (LoS), inside or outside hospital death, and readmission cases. Considering these indicators, identifying and learning from higher performers through benchmarking hospitals, will enable health managers, regulators and consumers with a scientific base evidence to inform their decisions over time [2, 3].

Moreover, lessons can transform from other professional industries like schools or universities. By choosing and comparing a university, students or their families are expecting to receive a superior education comparing to the others. This is what the school efficiency literate call it as 'school effect' and it is based on the combination of the schools and students characteristics [4]. While, respect to the literature in health care sector, such a terminology as 'hospital effect' is still not diffused among scholars of health policy and health care management. Concerning the hospital effect, having best available evidence—with the best level of confidence—plays a key role. Both policy-makers and practitioners need reliable information arisen from the best and worst performers. Following the literature behind the use of administrative data for decision making processes, these data have been employed for a variety of purposes; examples are epidemiological studies [5], outcomes evaluation [6], and identification of risk factors [7, 8]. Within this background, this study aims at furthering the ongoing debate on the use of administrative data to identify if there is any 'hospital effect' on performance to inform improvement strategies. In particular, this study offers original insights gathered from benchmarking hospitals that deliver care to Heart Failure (HF) patients in the Lombardy Region (Northern Italy).

Despite the richness of past studies, four main identified limitations still puzzled our mind for further analyses. First, many studies were anecdotal with limited generalizability. Second, many studies collected perceptions and not actual performances. Third, many studies used patient-level variables to explain different performance, neglecting the hospital-level ones. Thus, this could be critical while the aim of analyses is to gathering best information for shaping improvement strategies on performances. Hospital-level variables like type of structure, number of hospitalizations, and percentage of DRG are potential key inputs for changing the future decisions on hospital standards in a better way. In this study, by adopting hierarchical statistical models, we tried to disentangle the contribution to performance on behalf of both patients' and hospitals' characteristics. Multilevel or hierarchical modeling is an analytic technique designed for complex nested data structures [9], frequently used in different fields like educational, social and health services research. Fourth, a limited number of studies explored the potential informative power of administrative data. These is while, some un-official reports are misleading and confusing healthcare payers, providers and citizens [10–12], toward resource allocation and patient choice. Against this background, this study, by leveraging on the opportunity to access to administrative health data, aims at answering to four main research questions: (I) is there evidence of a 'hospital effect' on performances. (II) Which are the determinants of the 'hospital effect'? (III) Is the 'hospital effect' stable over time? (IV) Which variables do drive performance improvement or worsening over time?

2   Literature Review

2.1   Use of Administrative Health Data

Historically performance measurements aim at variety of purposes like marketing, quality-improvement, accreditation, and comparisons between providers and plans [13]. This is while, measuring and reporting current performances, by using administrative data, aims at more specific decisions regarding to the both patient and hospital. Administrative data by definition are large repositories of data collected by healthcare providers for reimbursement or surveillance activities. These datasets may contain variety of information like hospital admissions, outpatient care services, emergency room services, and drug prescriptions [14]. Based on the literature, using administrative datasets includes both advantages and disadvantages. The most important critical point facing these data is about the 'quality' of these data, which is likely due to differences in data collection methodologies. Comparing to the other datasets, like clinical registries, many studies argue that these data need to be linked with the other official datasets (like chart reviews). This also will allow researchers and practitioners to consider other patients' level data in their analyses. In our case, thanks to the vital statistics datasets, we include the date of the death for mortality studies. The availability without any additional cost of such large repositories of data, is another benefit in line with using the best available evidence to emphasis on evidence-based decision-making, and brought an intense debate about whether and how to use these data for informing political and managerial decisions [15, 16]. In this study, we considered the well-established measures of quality of treatment on short-term hard outcomes for HF patients, [17]. Considering these two main metrics for hospital performances, some advantages of using such outcomes are: their availability without additional costs for their collection and their use, their large volume, the presence of long follow-up periods with stable coding procedures and population coverage [14]. As told, these data are collected during an under way process, which allows researchers to work with very up-to-date data. Moreover, administrative data offer the opportunity to perform studies based on the so-called 'real world' data, with altering the order between identification of patients and data collection they solved the unfavorable issues of small-size clinical trials or observational studies.

One significant feature of administrative data is that they are collected continuously without any pre-adjusted political or academicals purposes. This can enable researchers to formalize their interested questions and hypotheses then seeking the investigation through the already available datasets. Respectively, the first step is to extract the right dataset that might carry the most relevant evidence needed to answer to specific policy/research questions. The second step deals with the identification of the cases of interest. Typically, the process of choosing a specific patients includes using codes of diagnoses or procedures in hospital discharge forms or out-patient care datasets [14]. Each hospital is coding diagnoses and procedures through the well-established International Classification of Diseases, 9th Revision, Clinical Modification (ICD-9-CM). The ICD-9-CM is a classification system that assigns codes to diagnoses and procedures commonly adopted in different countries. A good example of these two steps is the McCoy et al. [21] study on 'diabetes care quality' by using and analyzing datasets from 'patients in a multispecialty integrated health system' in chosen year while including patients with the help of the specific code (ICD-9-CM code 250.xx). Based on clinical similarities and resource absorption, hospital cases like discharge forms, will be classified into one of the 'originally 467 groups' as so called Diagnosis Related Group (DRG) system. The third step deals with checking the quality. As indicated previously, as part of the administrative purposes the quality of these data may affected. Typically, part of these data are subjected to quality controls by the providers themselves, agencies or companies that are in charge for reimbursements or providers' evaluation like healthcare quality controls in Italy which are part of the regional government responsibilities. In this view, basing a research on these data needs an extensively knowledgeable researcher about the datasets in terms of considering variety of details, such as 'the metrics employed for each variable' and the 'consistency of data collection process'.

Another aim of using administrative data in the literature is for proposing prediction or performance evaluation models regarding the specific diseases. Many studies like Bottle et al. [17], when studying hospital performances, are using patient level variables for adjusting statistical models. One issue, while performing statistical risk adjustment models, is the lack of clinical information; for example, Tabak et al. [22] were using administrative data combined with medical records for predicting mortality. Typically, past contributions used advanced algorithms for comorbidity detection (e.g., [18]). As result of these algorithms plus the evaluation of scores, researchers and practitioners can work on patients' diseases and conditions that might affect the survival rates thus identify the need for supplementary care. Considering performance measurements through administrative data, many recent studies generally focused on so called 'hard clinical outcomes' like patient survival, non-programmed hospital readmissions, and mean hospital length of stay. Others which have more connection with the training, support or guidance interventions such as quality adjusted life years or patient satisfaction less studied using only administrative data, and thus specific information from interviews or surveys have to be added.

3 Methodology

In this section, we will explain the definition of performance for each hospital and their related variables then the kind of relevant data for our work will be described. Moreover, in measuring performance the suitable statistical or mathematical models should be considered. In our case, the choice of multilevel logistic model was based on combining both patients' and hospitals' characteristics and making it coherent to the hierarchical nature of our data. In fact, one of our expectation is that similar outcomes will be arise from treating cases in the same hospital. Outlier detection is an effort to go beyond ranking hospitals and put some further steps on identifying the characteristics and skills related to each hospital that could transform the worst performer to the first. To this end we used funnel plots as a correlative method for displaying outliers are suggested for having 'disaggregated outcomes at provider level' [19], and less biased in labelling outliers.

3.1 Hospital Performance

We started our work by choosing the 30-day mortality and 30-day readmission as the most relevant outcomes suggested in the literature. By definition, 30-day mortality refers to the total number of deaths for any cause within 30 days after the incident HF admission and 30-day unplanned readmission refers to the total number of non-programmed hospitalizations for any cause within 30 days after the incident HF admission. In fact, HF patients have a high risk of mortality and a high probability of incurring multiple urgent admissions. Additionally, using HF data, it needs considering the incident cases. By incident case, we mean the first ever admission for any patient in each hospital for HF-related complications, which will, resulted in a far better fitted comparison by excluding patients with terminal comorbidities. Respectively, 30-day mortality was measured considering intra-hospital and out–of–hospital mortality for all causes. Also using the Lombardy Region's registries about deaths; 30-day unplanned readmissions were measured excluding the cases of patient being transferred from one hospital to another, planned readmissions, and readmissions occurred more than 30 days after discharge. Additionally, to evaluate non-programmed readmissions, patients died during the incident admission or within 7 days from discharge were excluded. The latter choice was made to exclude patients who have decided, for personal reasons, to die at home rather than in hospital. Finally, hospitals located outside the Lombardy Region or with less than 100 HF hospitalizations were excluded from our analysis.

3.2 Collection of Administrative Data

After targeting our outcomes, we used administrative data from hospital discharge forms provided by Lombardy Region over 2010–2012 period. Generally, in Lombardy, administrative data contain information on patient characteristics (e.g., sex and age) and hospital admission (e.g., date of admission and discharge, principal diagnosis and comorbidities). Later on, the data combined with death statistics and regional reports on hospitals' activity. It helped us for gathering further information like percentage of surgical DRGs from regional reports and information about the date of death for patients who died. This enabled us to evaluate mortality for both outside and inside of the hospital. By selecting HF data, we decided to work on a leading cause of death in the most developed Countries, which is a key priority for policy-makers and hospital managers on how to plan, administer and pay for healthcare services. As result of the inclusion/exclusion criteria, 78,907 adult residents in the Lombardy Region and aged at least 18 were hospitalized for the first time for HF in 117 hospitals were chosen for our analyses. As such, the overall patient related variables are age, sex, length of stay, comorbidities score, and number of **hospitalizations** within the previous six months, and type of admission ward. Total hospital related variables are: number of ordinary hospitalizations, mean length of hospital stay, percentage of surgical DRG, type of hospital (ownership and teaching status), number of admissions from other local healthcare agencies, number of admissions from other Italian Regions or abroad.

4 Statistical Analysis

Using SAS statistical software (SAS 9.4 TS Level 1M3), our **statistical** analyses started with combining two-level hierarchical logistic regressions and funnel plots to identify outlier hospitals and track the trend of their improvement/worsening. In this context, we consider outliers as those hospitals that differ from other hospitals respecting the ratios of observed mortality/readmission cases out of expected ones by the same pathology. The result could provide triggers to convert leads to main decision makers, who are in charge of health care planning to set up their improvement strategies in line with understanding the causes of these occurrences [7]. First, we developed a multilevel logistic regression model that took into account the different characteristics of the data for both patient and hospital. Then, we introduced each patient-level and hospital-level covariates consecutively as first and second level of our hierarchical model. The choice of the explanatory variables have been made on the basis of our available data and what literature were contribute so far. Based on [20], we test our "null" model while evaluating the Inter-class Correlation Coefficient (ICC) by introducing the first (patient) level variables and the second (hospital) level variables respectively. Finally, at the end of the first step in our statistical stage, the overall variables were included in our final statistical model through a backward selection method. Second, the final model have been used to shape the creation of funnel plots. The identified best and worst performers are based in creating a ratio of expected cases (of mortality and readmissions) vs. observed cases, as stated in the following formula:

$$Y = \frac{\sum_{i=1}^{n_j} y_{ij}^{obs}}{\sum_{i=1}^{n_j} \widehat{p}_{ij}} = \frac{O_j}{E_j}$$

where y_{ij}^{obs} is the observed outcome for patient 'i' treated in the hospital 'j', n_j is the number of patients treated in hospital 'j' and $\widehat{p}_{ij}$ is the corresponding expected value for patient 'i' treated in hospital 'j'. By including the hospital level variables, we tried to understand if different hospital characteristics could shape the effectiveness of each hospital in case of having less 30-day mortality and readmission.

5 Findings

As result of our statistical stage, we create the funnel plots for 2010, 2011, and 2013 respectively, showing the outliers of both readmission and mortality cases. With respect to our first and second research questions, data confirm that some hospital-level variables are significant in explaining the variance of hospital performances, suggesting that there is a 'hospital effect' on performance. In this regard, our data show that mortality and readmission are explained by different variables. On the one hand, considering '30-day mortality', all patient-level variables emerged as significant together with two hospital-level variables, namely percentage of surgical DRG and type of structure. On the other hand, considering '30-day readmission', our results identified as significant only the mean length of stay from hospital-level and it also excluded the sex variable from the patient-level ones. One possible explanation is that these variables are widely intertwined with managerial choices. Our results show also that none of the outliers had different structural characteristics, confirming that managerial practices more than structural characteristics are significant in shaping hospital performance. This means that regardless of the special characteristics of admission cases, hospital managers have the opportunity to improve hospital performance. Although the 'readmission' model has identified more outliers rather than the 'mortality' one, in both cases the trend of outlier movements during time is not fully captured by our set of explanatory variables. Which was expected from the statistical point of view while doing analyses with hospital level would probably reduce the probability of identifying outliers in the final model. Still, despite within our timespan (2010 until 2012), in both cases some hospitals improved their performance while some decreased their position and failed in improving them. The overall mortality increased 32.5% while in readmission, it ranged at 48.3% and the worsening over time was 34.2% for mortality and 19.8% for readmissions. Finally, in the short-term none of the structural variables explains such differences, meaning that the improvement is led by hospital managers' capability to implement change apart from the kind ownership.

6 Discussion

This study offers original insights to further the debate about the use of 'real-world' data to measure hospital performance and drive improvements. Administrative data offers the opportunity to crystallize the 'hospital effect' and point-out the hospital-level variables that affect performance. In this regard, scholars of operation management in healthcare should take advantage of administrative data to further explore those variables that differ between best and worst performers. Best practices and managerial choices in place in best performers must be translated to the worst performers to inform change while improving the situation for the worst performers to the first. Administrative data offer large amount of 'real-world' data; however, they proved to be not enough to fully explain performance evolution in the short-term. In this regard, further research should take into account wider periods (at least ten years) or integrate administrative data with clinical registries or surveys to hospital managers.

References

1. Raghupathi, V., Raghupathi, W.: Benchmarking hospital performance using health analytics. J Heal Med. Inf. 6(2) (2015)
2. Elg, M., Broryd, K.P., Kollberg, B.: Performance measurement to drive improvements in healthcare practice. Int. J. Oper. Prod. Manag. 33(11/12), 1623–1651 (2013)
3. Kohn, M.K.: Evidence Based Strategic Decision Making in Ontario Public Hospitals (2013)

4. Konstantopoulos, S., Miller, S.R., van der Ploeg, A., Li, W.: Effects of interim assessments on student achievement: evidence from a large-scale experiment. J. Res. Educ. Eff. **9**(sup1), 188–208 (2016). https://doi.org/10.1080/19345747.2015.1116031
5. Giorgio, Lovaglio P.: Hospital effectiveness from administrative data: the Lombardy case. TQM J. **22**(5), 474–486 (2010). https://doi.org/10.1108/17542731011072829
6. Sun, R., Van Ryzin, G.G.: Are performance management practices associated with better outcomes? empirical evidence from New York public schools. Am. Rev. Public Adm. **44**(3), 324–338 (2012). https://doi.org/10.1177/0275074012468058
7. Ieva, F., Paganoni, A.M.: Detecting and visualizing outliers in provider profiling via funnel plots and mixed effect models. Health Care Manag. Sci. 18(2), 166–172 (2015). https://doi.org/10.1007/s10729-013-9264-9
8. Shahian, D.M., Iezzoni, L.I., Meyer, G.S., Kirle, L., Normand, S.L.T.: hospital-wide mortality as a quality metric: conceptual and methodological challenges. Am. J. Med. Qual. **27**(2), 112–123 (2012). https://doi.org/10.1177/1062860611412358
9. Reeves, M.J., Gargano, J., Maier, K.S. et al.: Patient-level and hospital-level determinants of the quality of acute stroke care: a multilevel modeling approach. Stroke 41(12), 2924–2931 (2010). https://doi.org/10.1161/STROKEAHA.110.598664
10. Kidholm, K., Ølholm, A.M., Birk-Olsen, M. et al.: Hospital managers' need for information in decision-making—an interview study in nine European countries. Health Policy (New York). **119**(11), 1424–1432 (2015). https://doi.org/10.1016/j.healthpol.2015.08.011
11. Baekgaard, M., Serritzlew, S.: Interpreting performance information: motivated reasoning or unbiased comprehension. Public Adm Rev. **76** (2015). https://doi.org/10.1111/puar.12406
12. Dover, D.C., Schopflocher, D.P.: Using funnel plots in public health surveillance. Popul Health Metr. **9**, 58 (2011)
13. Thompson, B.L., O'Connor, P., Boyle, R. et al.: Measuring clinical performance: comparison and validity of telephone survey and administrative data. Health Serv. Res. **36**(4), 813–825 (2001)
14. Mazzali, C., Duca, P.: Use of administrative data in healthcare research. Intern. Emerg. Med. **10**(4), 517–524 (2015). https://doi.org/10.1007/s11739-015-1213-9
15. Chowdhury, T.T., Hemmelgarn, B.: Evidence-based decision-making 6: utilization of administrative databases for health services research. In: Parfrey, P.S., Barrett, B.J. (eds.) Clinical Epidemiology: Practice and Methods. New York, NY. Springer, New York, pp. 469–484 (2015). https://doi.org/10.1007/978-1-4939-2428-8_28
16. Murdoch, T.B., Detsky, A.S.: The inevitable application of big data to health care. JAMA **309** (13), 1351–1352 (2013). https://doi.org/10.1001/jama.2013.393
17. Bottle, A., Middleton, S., Kalkman, C.J., Livingston, E.H., Aylin, P.: Global comparators project: International comparison of hospital outcomes using administrative data. Health Serv. Res. **48**(6 PART1), 2081–2100 (2013). https://doi.org/10.1111/1475-6773.12074
18. Sharabiani, M.T.A., Aylin, P., Bottle, A.: Systematic review of comorbidity indices for administrative data. Med. Care **50**(12), 1109–1118 (2012). https://doi.org/10.1097/MLR.0b013e31825f64d0
19. Mayer, E.K., Bottle, A., Aylin, P., Darzi, A.W., Vale, J.A., Athanasiou, T.: What is the role of risk-adjusted funnel plots in the analysis of radical cystectomy volume-outcome relationships? BJU Int. **108**(6):844–850 (2011). https://doi.org/10.1111/j.1464-410X.2010.09896.x
20. Ene, M., Leighton, E.A., Blue, G.L., Bell, B.A.: Multilevel models for categorical data using SAS® PROC GLIMMIX. Sgf **2015**, 1–12 (2015)

21. McCoy, R. G., Tulledge-Scheitel, S.M., Naessens, J.M. et al.: The Method for performance measurement matters: diabetes care quality as measured by administrative claims and institutional registry. Health Serv. Res. **51**(6), 2206–2220 (2016). https://doi.org/10.1111/1475-6773.12453
22. Tabak, Y.P., Sun, X., Johannes, R.S., Hyde L., Shorr AF., Lindenauer PK.: Development and validation of a mortality risk-adjustment model for patients hospitalized for exacerbations of chronic obstructive pulmonary disease. Med Care. **51**(7), 597–605 (2013). https://doi.org/10.1097/MLR.0b013e3182901982

Part II
Emerging Research in Health Care

Crowding in Paediatric Emergency Department, A Review of the Literature and a Simulation-Based Case Study

Caterina Caprara, Filippo Visintin and Francesco Puggelli

Abstract In this study, we perform a systematic review of the literature concerned with crowding in pediatric emergency departments and assess, via discrete-event simulation, the impact of the implementation of triage-based protocols—which consist in allowing triage nurses to prescribe diagnostic tests before the first physician assessment- on the emergency department performance. The study is based on real data from a leading Italian hospital.

Keywords Healthcare · Paediatric emergency department
Anticipated treatments · Crowding

Emergency Department (ED) overcrowding, is a common problem across the globe. The imbalance between supply and demand for emergency service is becoming bigger and bigger, especially in countries with sluggish economies where healthcare is based on the principle of universal coverage (like Italy). Given its relevance, ED overcrowding has been the object of a sizeable number of studies examining the causes, effects, and solutions to ED overcrowding ([1], [2]). Overcrowding in Paediatric EDs (PEDs) however, is still an under-investigated phenomenon. PEDs differ from general EDs in a number of ways: (i) the overall acuity of patients and (consequently) the number of admissions are lower (8–10% in PEDs versus 15–20% in adult EDs [3]); (ii) arrivals peak in the afternoon, when parents return home from work [3] and in the weekends, when availability of paediatricians is limited; (iii) arrivals present a seasonal variation with peaks in winter months, due to influenza and respiratory illnesses and in the summer months (when kids

C. Caprara · F. Visintin (✉)
IBIS Lab, University of Florence, Florence, Italy
e-mail: filippo.visintin@unifi.it

C. Caprara
e-mail: caterina.caprara@unifi.it

F. Puggelli
Meyer Children's Hospital, Florence, Italy
e-mail: f.puggelli@meyer.it

© Springer International Publishing AG 2017
P. Cappanera et al. (eds.), *Health Care Systems Engineering*, Springer Proceedings in Mathematics & Statistics 210, https://doi.org/10.1007/978-3-319-66146-9_26

play outdoor) due to fractures and lacerations [4]; (iv) nurses and physician need to interact with both patients and their parents.

In this study, we have performed a comprehensive and systematic review of the literature on PED. Starting form a sample of 3607 papers, following a structured classification procedure we consolidated a final sample of 47 papers analysing the causes and proposing organisational solutions to address overcrowding problem in PED. These papers were read and thoroughly coded. The analysis led to the identification of 16 types of causes of overcrowding, 10 different types of organisational solution to reduce or prevent it, as well as to the identification of the most important performance indicators used in the literature to assess PED performances. The literature suggests that a promising way to reduce patient length of stay in the ED is to implement triage-based protocols—referred to as *anticipated treatment*—that allow triage nurse to prescribe diagnostic tests *before* the first physician assessment [5] [6]. The utilization of these protocols is supported by empirical evidences showing that the appropriateness of nurse's prescription doesn't significantly differ from the physician's one [7] [8] [9]. However, the literature concerned with the implementation of triage based protocols in PED is still scant. Indeed, in PED, the adoption of these protocols is discouraged by the anxiety characterizing patient's parents who claim for a physician assessment before child are subject to hurting or harmful tests (e.g. blood or X-ray tests). Undoubtedly, a quantification of the benefits in terms of efficiency that is possible to achieve through the implementation of this protocols can better inform managerial decisions.

For this reason, in this study we have investigated, via simulation [10], the impact on 4 KPIs—waiting time (WT) from arrival to triage, WT from arrival to the first physician assessment, length of stay (LOS) from arrival to discharge and LOS from the first physician assessment to discharge—of the implementation in of anticipated treatments in PEDs for two very common types of paediatric patients: (i) those with suspected fractures, typically requiring X-Ray test; and (ii) those affected by respiratory diseases typically, requiring blood test. The model was developed using Arena and validated referring to the as-is configuration of a PED that currently does not implement any anticipated treatment. Several experiments have been carried out to assess the (main and interaction) effects associated with the implementation of the anticipated treatments on the KPIs.

Preliminary results show that implementing anticipated treatments for trauma patients and for patients affected by respiratory diseases significantly decreases their LOS (up to 40% for white codes and up to 26% for green codes) without significantly affecting the LOS of other patients. The waiting time for physician assessment decreases up to 12% for white codes and doesn't worsen for the other patients; whereas the waiting time for triage increases with the implementation of the respiratory diseases pathway. However, we argue that these benefits may vary seasonally since the number of patients potentially needing these types of treatments varies according with the month of the year, the day of the week and hour of the day. Future research will be thus devoted to develop a methodology to decide when, based on a medium-to-short term forecast of the patient's arrivals (both in terms of mix and volume), certain protocols worth to be activated and when not. In addition an action research project [11] will be carried out to assess the impact of the implemention of the described anticipated treatments in real hospital and to shed light on those actions facilitating or hindering their introduction and use.

References

1. Asplin, B., et al.: Emergency department crowding: high impact solutions. ACEP Task Report on Boarding (2008)
2. Hoot, N.R., Aronsky, D.: Systematic review of emergency department crowding: causes, effects, and solutions. Ann. Emerg. Med. **52**(2), 126–136 (2008)
3. Sinclair, D.: Emergency department overcrowding–implications for paediatric emergency medicine. Paediatr. Child Health **12**(6), 491 (2007)
4. Barata, I., et al.: Best practices for improving flow and care of pediatric patients in the emergency department. Pediatrics **135**(1), 273–283 (2015)
5. Yang, K.K., et al.: Managing emergency department crowding through improved triaging and resource allocation. Oper. Res. Health Care **10**, 13–22 (2016)
6. Wiler, J.L., et al.: Optimizing emergency department front-end operations. Ann. Emerg. Med. **55**(2), 142–160 (2010)
7. Lee, K.M., et al.: Accuracy and efficiency of X-ray requests initiated by triage nurses in an accident and emergency department. Accid. Emerg. Nurs. **4**(4), 179–181 (1996)
8. Fry, M.: Triage nurses order x-rays for patients with isolated distal limb injuries: a 12-month ED study. J. Emerg. Nurs. **27**(1), 17–22 (2001)
9. Seaberg, D.C., MacLeod, B.A.: Correlation between triage nurse and physician ordering of ED tests. Am. J. Emerg. Med. **16**(1), 8–11 (1998)
10. Visintin, F., et al.: Applying discrete event simulation to the design of a service delivery system in the aerospace industry: a case study. J. Intell. Manuf. **25**(5), 1135–1152 (2014)
11. Visintin, F., et al.: Development and implementation of an operating room scheduling tool: an action research study. Prod. Plann. Control **28**(9), 758–775 (2017)

Dealing with Stress and Workload in Emergency Departments

Marta Cildoz, Fermin Mallor, Amaia Ibarra and Cristina Azcarate

Abstract Emergency Departments (EDs) are widely known for their stochastic nature, unpredictable arrivals and—in recent years—overcrowding problems. These could cause stress to physicians and, as a consequence, bad quality of patient's medical care. This problem is investigated in a Spanish ED by developing a dynamic measure of stress, analyzing the patient flow control considering not only waiting time but also physicians' stress, and finally by suggesting a computer-based tool to help triage nurses to manage the workload distribution among physicians.

Keywords Emergency department · Workload · Stress · Patient flow

1 Introduction

ED Physicians are usually exposed to more severe stress than other departments' physicians [1]. Its principal sources are time pressure, critical decisions and amount of work [2]. Patient Flow is managed after triage by nurses, in some cases by using simple rotational rules, which look for workload equity among physicians. Nevertheless, the randomness of patient arrivals results in inequality of stress experienced by physicians. The main purpose of this presentation is to outline the current research conducted to analyze physicians' stress working in an ED. The analysis has been divided in three steps: in the first one the issue of developing a dynamic measure of stress is addressed, secondly policies for the patient flow control are studied by incorporating the dynamic stress of physicians as a criterion (in addition to the usual patient waiting time), and finally a computer-based tool implements the best policy to assist the triage nurses to manage physicians' workload.

M. Cildoz (✉) · F. Mallor · C. Azcarate
Public University of Navarre, Pamplona, Navarre, Spain
e-mail: martacildoz@unavarra.es

A. Ibarra
Hospital Compound of Navarre, Navarre, Spain

© Springer International Publishing AG 2017   297
P. Cappanera et al. (eds.), *Health Care Systems Engineering*, Springer Proceedings
in Mathematics & Statistics 210, https://doi.org/10.1007/978-3-319-66146-9_27

2 Measuring Stress and Workload

Physicians in ED reported that the stress they feel is caused mainly by three components: workload (pending patients), uncertainty (unpredictable arrivals or illness of patients not seen) and time pressure (overcrowding). It is being developed a methodology to measure stress among physicians through their assessment of real scenarios. These scenarios account for the workload assigned to a physician disaggregated by type of patients (severity), their stage of medical care process, waiting time targets and teaching duties. The collected scores are the primary data used to estimate the stress assessment function.

3 Patient Flow Control

Patient Flow is controlled by the triage nurses. As soon as patients are triaged, they are assigned to a physician rotationally. This control could be improved by analyzing different ways of distributing patients among physicians in order not only to optimize patient waiting time but also stress among physicians.

4 Implementation

Our purpose is to implement a new tool to support decision making at triage based on the previous flow control analysis. This system should be fair and easy to use as patient flow management does not fall within the competence of triage nurses.

Acknowledgements This research has been supported by grant MTM2016-77015-R.

References

1. Estryn-Behar, M., et al.: Emergency physicians accumulate more stress factors than other physicians-results from the French SESMAT study. Emerg. Med. J. **28**(5), 397–410 (2011)
2. Phipps, L.: Stress among doctors and nurses in the emergency department of a general hospital. CMAJ **139**(5), 6–375 (1988)

Patient-Bed Allocation in Large Hospitals

Fabian Schäfer, Manuel Walther and Alexander Hübner

Abstract *General Background.* Allocating patients to beds is an everyday task in hospitals, which is first of all driven by total bed capacity, patient compatibility, fluctuations in lengths of stay (LOS), and emergency arrival rates. Historically speaking, hospitals managed their patient-bed allocations on a first-come first-serve basis. In addition, wards were typically dedicated to single departments with head nurses and doctors managing actual patient-bed-assignment. To ensure profitability, occupancy levels have to be held high to ensure high utilization of hospital resources. High occupancy rates however, greatly increase the probability of overflow situations which require disproportionate amounts of additional organizational work. This holds especially true for large maximum-care hospitals, which by definition are obligated to treat all incoming patients and have to deal with a lot of uncertainty due to a high share of emergency patients.

Keywords Bed management · Operational patient allocation
Uncertainty management · Overflow management

Specific Problem Setting and Solution Approach. The solution space for assigning a given patient to a bed usually does not comprise every available bed in the hospital. Instead, wards are assigned to departments or groups of departments. Today, large hospitals are starting to adopt shared ward spaces dedicated to specific groups of departments (e.g., internal medicine) in an attempt to balance patient occupancy levels across multiple wards (see for example [3]). This leads to pooled bed capacities of sometimes more than 100 beds per department group. The challenge in this context, is to devise a system that efficiently manages patient-bed allocations all while anticipating future patient arrivals, optimizing workload for hospital staff, taking into account current occupancies, and last but not least ensuring patient satisfaction. Furthermore, such a system will need to be able to react instantaneously if it is to be used on an operational level considering the high levels of volatility and uncertainty regarding patient-specific parameters.

F. Schäfer (✉) · M. Walther · A. Hübner
Catholic University Eichstätt-Ingolstadt, Operations Management, Auf der Schanz 49,
85049 Ingolstadt, Germany
e-mail: fabian.schaefer@ku.de

© Springer International Publishing AG 2017   299
P. Cappanera et al. (eds.), *Health Care Systems Engineering*, Springer Proceedings
in Mathematics & Statistics 210, https://doi.org/10.1007/978-3-319-66146-9_28

To this end we have developed a decision support model to assign inpatients to beds. The main advantage of our approach compared to approaches found in the literature (see for example [1, 2]) is that it is designed to propose a specific bed to the bed manager for every incoming patient at the exact time of their arrival, based on all the information known at that particular moment. The idea is not to predetermine an occupancy plan for a defined time period (e.g., a week) that is set in stone, but to have a system with maximum flexibility that can update in real-time, while still considering the impact of future inpatient arrivals. Furthermore, our solution approach includes dedicated overflow buffer space, hence making it a reliable decision support model for any real-life situation (e.g., personnel shortages, deliberate deviations from preconceived occupancy plans, short-term cancellations).

Objectives and Constraints. There are typically three different stakeholders when assigning patients to beds, namely patients, medical staff, and hospital management. Patients want their stay to be as pleasant as possible while receiving top-notch medical care. In our modeling approach this translates to having roommates within their age group and with a similar illness or severity of their medical conditions. Patients should also not be forced to change rooms if there is no medical need for it. This is especially true for elective patients, as they are less likely to accept that a room and bed is not "reserved" for their entire stay. One of the main issues for medical staff is having a balanced workload. As nurse rosters and shift scheduling typically cannot be changed on a short-term basis it is important to level out workload per ward and week. Workload in this context is dependent on both the amount of patients as well as the severity and types of illnesses exhibited. Finally, doctors want all of their patients to only be on dedicated wards within their department-ward combination to avoid having to cover large distances during rounds.

Last but not least, three hard constraints are considered. First, it is not allowed to mix genders within a single room on a specific day. Second, medical isolation requirements have to be adhered to. For example, infectious patients have to be separated from immunocompromised patients. Third, infrastructural requirements have to be met for every patient (e.g., room has to be equipped with ECG-monitoring system for certain cardiac diseases). We have built and tested our model using different heuristic approaches as a full enumeration is not feasible in light of problem size and scope. First results seem very promising and allow for high-quality solutions within seconds.

References

1. Ceschia, S., Schaerf, A.: Local search and lower bounds for the patient admission scheduling problem. Comput. Oper. Res. **38**(10), 1452–1463 (2011)
2. Demeester, Peter, et al.: A hybrid tabu search algorithm for automatically assigning patients to beds. Artif. Intell. Med. **48**(1), 61–70 (2010)
3. Hübner, A. et al.: Approach to Clustering Clinical Departments. Health Care Systems Engineering for Scientists and Practitioners. Springer International Publishing, pp. 111–120 (2016)

Printed by Printforce, the Netherlands